Applied Plant Genomics and Biotechnology

Related titles

From plant genomics to plant biotechnology
(ISBN: 978-1-907568-29-9)
An introduction to biotechnology
(ISBN: 978-1-907568-28-2)
MATLAB® in bioscience and biotechnology
(ISBN: 978-1-907568-04-6)

Woodhead Publishing Series in Biomedicine: Number 72

Applied Plant Genomics and Biotechnology

Edited by

Palmiro Poltronieri
Yiguo Hong

ELSEVIER

AMSTERDAM • BOSTON • HEIDELBERG • LONDON
NEW YORK • OXFORD • PARIS • SAN DIEGO
SAN FRANCISCO • SINGAPORE • SYDNEY • TOKYO
Woodhead Publishing is an imprint of Elsevier

WP WOODHEAD PUBLISHING

Woodhead Publishing Limited is an imprint of Elsevier
80 High Street, Sawston, Cambridge, CB22 3HJ, UK
225 Wyman Street, Waltham, MA 02451, USA
Langford Lane, Kidlington, OX5 1GB, UK

ISBN: 978-0-08-100068-7 (print)
ISBN: 978-0-08-100071-7 (online)

British Library Cataloguing-in-Publication Data
A catalogue record for this book is available from the British Library

Library of Congress Cataloging-in-Publication Data
A catalog record for this book is available from the Library of Congress

Library of Congress Control Number: 2014948687

For Information on all Woodhead Publishing publications
visit our website at http://store.elsevier.com/

Printed in the United States of America

Contents

List of figures

miRNAs, microRNAs; SLP, SQUAMOSA PROMOTER BINDING PROTEIN-LIKE; SMZ, SCHLAFMÜTZE; SNZ, SCHNARCHZAPFEN; SOC1, SUPPRESSOR OF CONSTANS1; TOE1-3, TARGET OF EAT1-3.

Figure 8.1 Overview of DNA methylation controls in plants. Summary of enzymes and processes involved in the regulation of DNA methylation of plant genomic DNA as determined in the *Arabidopsis* model. *De novo* DNA methylation is set up by DRM1 and DRM2, which are targeted to specific genomic loci by small interference RNA (24 nt siRNA). The RNA-directed DNA methylation (RdDM) requires several additional proteins involved in the production of these small RNAs, a subset of which is indicated. This process is described in detail in Law and Jacobsen (2010). Briefly, RNA-Dependant RNA Polymerase 2 (RDRP2) is thought to produce dsRNA that are most likely the substrate of Dicer Like 3 (DCL3), which produces 24 nt siRNA. The Argonaute protein AGO4 binding to the 24 nt siRNAs is required for RdDM. MET1 and CMT3 are then necessary for maintenance of methylation respectively in the CG and CNG context, and requires Cedrease in DNA Methylation 1 (DDM1) for their activity. CNN methylation in heterochromatic regions is performed by CMT2, which is targeted to DNA through its interaction with DDM1, whereas CNN methylation of TE in euchromatic regions is done by DRM2, which requires Defective in RNA-Directed DNA methylation (DRD1). Methylation removal is thought to occur by simple dilution in all contexts when maintenance of methylation is not active, a process which therefore on active replication. Active demethylation occurs independently of replication and involves DNA Glycosylase-lyases that can remove methylated cytosine independently of replication. Active DNA methylation can target specific loci or affect all genome. The dynamic of DNA methylation determines gene transcriptional states and plays important functions during plant development and adaptation to environmental constraints. **136**

Figure 8.2 Variations of DNA methylation in tomato fruits. (a) Fruit development is initiated following flower pollination and occurs in three main phases; the cell division phase (black arrow) followed by the cell elongation phase (green arrow) and ripening (red arrow). Black bars indicate the period of maximum endoreduplication activity. dpa, days post anthesis; Br., Breaker; Tur., Turning; Or., Orange and RR., Red Ripe stages. (b) Description of major 5 MeC content and profile variations in tomato fruits. 1 – Total percentage of 5 MeC (upper panel) in pericarp (grey area) is higher in pericarp of developing fruits (30%) than in the locular tissues (20%). After the breaker stage, the 5 MeC content of the pericarp genomic DNA drops down to a level similar to the methylation level of the locular tissue genomic DNA. No decrease is observed in the locular tissue (Teyssier et al., 2008). 2 – CNG methylation was shown to increase sharply at the 5S rDNA and GYPSY transposons in the pericarp during fruit development, but not in the locular tissue (Teyssier et al., 2008). 3 – Transposon methylation was reported to decrease in the CG **139**

List of tables

About the editors

Palmiro Poltronieri is a researcher at CNR-ISPA, department of food, agriculture, fisheries and biotechnology, Italian National Research Council. He graduated in biology in 1987 and with a PhD in cellular and molecular biology and pathology in 1995 at the Institute of Chemical Biology, medical faculty of Verona University. He holds a postdoctoral position in Japanese Society for Promotion of Science and has worked at the Institute of Basic Medical Sciences, Tsukuba University from 1996 to 1997. He was a contract scientist at the Institute of Researches in AgroFoood Biotechnology since 1998, with a permanent position since 2001. In 2002, he co-founded Biotecgen – a service company involved in European projects developing molecular tools such as Ribochip DNA arrays and protein chips. He is an associate editor to *BMC Research Notes* and in the editorial board of *The Scientific World Journal*. In 2013, he got a fellowship for short-term stay at the Nagahama Institute of Bioscience. His current interest is on environmental responses in legume plants, phytohormone cross-talk signaling, including regulation by non-protein coding RNAs. He is co-author of 35 international publications, 10 book chapters and editor of *From Plant Genomics to Plant Biotechnology*, Woodhead, Cambridge, 2013.

Yiguo Hong obtained his PhD in plant molecular virology from Beijing Institute of Microbiology, Chinese Academy of Sciences in 1990. He had been a principal investigator at Horticultural Research International (HRI) and then Warwick HRI, University of Warwick since 1999 after his postdoctoral researches at the Scottish Crop Research Institute, Imperial College and John Innes Centre in the United Kingdom. In 2012, as a teaching professor, he taught a plant virology technology course in University of Bordeaux, France. He is now a professor and director of the Research Centre for Plant RNA Signalling in the College of Life and Environmental Sciences, Hangzhou Normal University, People's Republic of China. He currently holds a visiting professorship at University of Warwick. He is an editor for *Frontiers in Plant Physiology* – research topic 'RNA Signalling in Plants' and an associate editor for *Journal of Horticultural Science and Biotechnology*, and serves as an editorial board member for *ISRN Virology* and *The Scientific World Journal*. His research interests include intra-, intercellular and systemic RNA and protein signalling in flowering induction, fruit development and RNA silencing-based antiviral defence. He has published more than 60 peer-reviewed papers in international journals such as *Nature Genetics*, *Scientific Reports*, *Journal of Virology*, *PLoS Pathogens* and *Plant Cell* and co-edited *Plant Virology Protocols*, *Methods in Molecular Biology* (Humana Press) and was editor of a special issue on RNA signalling in plants for *Frontiers in Plant Science*.

About the contributors

Li Bin obtained his MSc degree from the College of Life and Environmental Sciences, Hangzhou Normal University, Zhejiang, People's Republic of China. He is now a research associate in a biotechnology company. His main research interest focuses on intercellular RNA silencing and siRNA functions in plant defence against virus infection.

Lisa Boureau did her Ph.D. thesis title "Analyse fonctionnelle de la protéine Enhancer of zeste, SlEZ2, chez la tomate Solanum lycopersicum" collaborating with Philippe Gallusci on topics of tomato fruit ripening, metabolomics and flavonoid studies, at Bordeaux University, UMR 1332 de Biologie du Fruit et Pathologie, Villenave d'Ornon, France. Presently she is post-doc fellow at McGill University, Quebec city, Canada.

Matteo Busconi is an assistant professor at the Institute of Agronomy, Genetics and Field Crops of the Università Cattolica del Sacro Cuore of Piacenza, Italy. He achieved a postdoc in molecular biotechnology at the Università Cattolica del Sacro Cuore in 2005. His scientific activity is mainly focused on food genomics, biodiversity with particular focus on plant biodiversity, molecular marker development, horizontal gene flow of herbicide resistance from cultivated rice to weedy relatives, gene expression, epigenetics and transgenesis. His teaching includes courses on agricultural genetics, biotechnology and plant breeding at the Università Cattolica del Sacro Cuore. He has taught biotechnology for molecular farming at the University of Modena and Reggio-Emilia (Modena, Italy). He is a member of the Società Italiana di Genetica Agraria and the Società di Ortoflorofrutticoltura Italiana. He has experience as a reviewer since 2005 for several indexed journals. He is author and co-author of publications in international and national journals.

Weiwei Chen is a lecturer in the College of Life and Environmental Sciences and a junior research leader in the Research Centre for Plant RNA Signaling, Hangzhou Normal University, Zhejiang, People's Republic of China, after she obtained her PhD in plant biology from the College of Life Sciences, Zhejiang University, Zhejiang, People's Republic of China. She is interested in functional dissection of RNA-dependent DNA methylation pathways in the regulation of tomato fruit ripening and plant development. Her research also covers microRNAs and their roles in plants.

Qi Cheng got his bachelor degree in biochemistry at Zhejiang University, Zhejiang, People's Republic of China, in 1989, master degree in biophysics at

Chinese Academy of Agricultural Sciences (CAAS), Beijing, People's Republic of China, in 1991 and a PhD in biochemistry at University of East Anglia, United Kingdom, in 1999. Over the past two decades of research career, he was employed by CAAS, University of Durham and University of Cambridge among others and has published over 60 papers and 1 book. He has developed his research expertise on various protein catalysts in the fields of nitrogen fixation engineering, lipid degradation, DNA amplification and so on. Since 2010, he is back in China as a professor and CEO, focusing his research and R&D on plant genetics and epigenetics.

Charlotte Degraeve-Guibault has obtained her master degree at Bordeaux University in the field of plant biology and biotechnology. Her work focused on the analysis of DNA demethylation during plant vegetative development.

Corrado Fogher, PhD, has been an associate professor of genetics and was responsible for the transgenic plants sector of the Observatory of Transgenic Organisms in Agriculture at the Università Cattolica del Sacro Cuore of Piacenza, Italy. He was NATO Fellow (1982–83) at the department of biochemistry, University of Missouri, Columbia. Researcher (1984–85) at the department of cellular physiology and molecular genetics of the Pasteur Institute, Paris, and visiting scientist at The Scripps Research Institute, La Jolla, California. He has authored more than 70 peer-reviewed papers. He was director of three SMEs, Plantechno, Incura and SunChem. His SMEs have started extremely interesting partnerships with corporate companies, one patented product will be tested as oral vaccine in a clinical trial, and a second research product will be exploited for the production of kerosene fuels. He died at the end of 2013.

Rupert Fray is an associate professor in plant molecular biology at the University of Nottingham, United Kingdom. He specializes in post-transcriptional gene regulation, focusing on the role of adenosine methylation in mRNA. He also has a long-standing interest in tomato fruit development and ripening regulation, with a particular interest in the synthesis of carotenoids and the control genes encoding the carotenoid biosynthetic genes during the ripening process.

Wolfgang Friedt is a professor emeritus at the Institute of Agronomy and Plant Breeding at the Justus Liebig University of Giessen, Germany. He has previously served as a dean of the faculty for agriculture and environment preservation. In addition, he has been serving as a member of the advisory boards and consulting committees of various scientific organizations and professional associations in the field of agronomy, plant breeding and general agriculture. He has published or co-authored many research papers mainly dealing with the genetic analysis of major complex traits of crop plants such as barley, sorghum, wheat and oilseed rape aiming at the genetic crop enhancement of crops to provide better varieties for agriculture.

Philippe Gallusci is professor at The University of Bordeaux. Since 2007, he has developed a research project focusing on the role of epigenetic mechanisms in plants. He obtained his PhD in Plant Molecular Biology in 1991 at the University of Toulouse and was a post doc at the Max Planck Institut for Züchtungforshung (Cologne, Germany). During this time his work focused on the transcriptional control of maize grain storage proteins. He was hired at the University of Bordeaux 1 in 1994 as an associate professor and studied plant isoprenoids before developing his own research project on the role of epigenetic mechanisms in the control of tomato fruit development and quality.

Juan Hao is a lecturer in College of Life and Environmental Science at the School of Hangzhou Normal University, Hangzhou, Zhejiang, People's Republic of China. She was awarded a PhD in crop genetics and breeding from the National Key Laboratory of Crop Genetic Improvement, Huazhong Agriculture University, Wuhan, Hubei, People's Republic of China. She has been engaged in cotton molecular breeding and has mainly focused on the research of the molecular mechanism of cotton fibre development.

Meiling He graduated from College of Life and Environmental Sciences, Hangzhou Normal University, Zhejiang, People's Republic of China. She is currently studying RNA and protein signalling in flowering.

Yoshihisa Ikeda is a junior researcher at the Centre of the Region Haná for Biotechnological and Agricultural Research, department of molecular biology, faculty of science, Palacký University in Olomouc, Czech Republic. He obtained a PhD degree in the field of bioscience at Nara Institute of Science and Technology, Japan. He was postdoc at The Rockefeller University, New York; Iowa State University, Ames, Iowa and the Swedish University of Agricultural Science, Sweden. He became a researcher at Umeå University, Sweden and Tohoku University, Japan. He was a Marie Curie Incoming International Fellowship while at the Swedish University of Agricultural Science.

Stephen Jackson is an associate professor since 2009 at the School of Life Sciences, Warwick University, East Anglia, United Kingdom. From 1990 to 1993, he was postdoc at the Institute für Genbiologische Forschung GmbH, Berlin, Germany; from 1993 to 1996, postdoc at the Centro de Investigacion y Desarrollo, C.S.I.C., Barcelona, Spain and from 1996 to 2009, Research leader at HRI, Wellesbourne, Warwick, United Kingdom. His group is investigating the control of plant development by light and photoperiod, a programme to isolate flowering mutants. This programme has identified several mutants that are specifically altered in their response to photoperiod, characterization of these mutants has shown them to represent novel genes in this pathway and work is underway to establish their cellular localization and interaction with known key regulators of the pathway such as CONSTANS. He is working on the role of the FLOWERING LOCUS T (FT) protein which is a component of the mobile flowering signal (florigen) in plants. He showed that *FT* mRNA is systemically mobile within the plant and may also play a role in the control of flowering time.

Hua Jiang is an associate research fellow at Zhejiang Academy of Agricultural Science, Hangzhou, People's Republic of China. He received his PhD in genetics and crop breeding from Yangzhou University, Yangzhou, People's Republic of China, (Supervisor: Qian Qian) in 2008. He has worked as a postdoctoral fellow at Zhejiang Academy of Agricultural Science (2008–10). Jiang's current research interests are plant pathology and molecular plant pathology, focusing on prevention and control of diseases of rice and functional gene analysis of fungal disease in rice.

Eva Jiskrová graduated in molecular and cell biology from the faculty of science, Palacký University in Olomouc, Czech Republic, in 2011, and then she has started her doctoral study programme in biochemistry as a student research assistant at the Centre of the Region Haná for Biotechnological and Agricultural Research, Hana, Czech Republic. Her research is focused on cytokinin metabolism in cereals.

Junhua Kong is an MSc student in the Research Centre for Plant RNA Signaling, Hangzhou Normal University, Zhejiang, People's Republic of China. She is working on epigenetic control in tomato fruit ripening and plant development and is also an expert in plant virus technology.

Ivona Kubalová graduated in genetics from the faculty of natural science, Comenius University in Bratislava, Slovakia, in 2013, and enrolled in doctoral study programme in biochemistry. Presently, she works as a research assistant at the Centre of the Region Haná for Biotechnological and Agricultural Research. Her research is focused on genetic control of pluripotency in plants.

Tongfei Lai obtained his BSc in bioengineering from the College of Life Science, Dalian Nationalities University, Liaoning, People's Republic of China, in 2002; MSc in agronomy from Institute of Cotton Research of the Chinese Academy of Agricultural Sciences, Beijing, People's Republic of China in 2007 and PhD in developmental biology from the Institute of Botany, the Chinese Academy of Sciences, Beijing, People's Republic of China in 2011. Since then, Lai has been recruited as a junior research leader of Research Centre for Plant RNA Signaling, Hangzhou Normal University. His research works mainly focuses on the mechanism of transcriptional regulatory networks on fruit ripening and the molecular basis of fruit quality of tomato.

Margit Laimer is a professor and head of the Plant Biotechnology Unit (PBU), University of Natural Resources and Life Sciences (BOKU), Vienna, Austria, since 1987. She is an expert in plant biotechnology. She holds a PhD in botany and zoology (University of Vienna, 1985) and two habilitations in two related fields – plant biotechnology (1991, BOKU, Vienna) and plant virology (1993, University of Lisbon, Portugal). She has acquired extensive competence in plant engineering since 1985 when confronted with the challenge for the production of healthy food for the development of methods for rapid detection of plant pathogens, and the reduction of the use of biocidal chemicals via the employment of strategies that improve the natural resistance of plants. Since 1990, strategically important viruses were included into her research portfolio, as well as plant tissue cultivation and vaccine production in plants. She has published a series of highly relevant scientific manuscripts in renowned plant and horticultural journals (more than 160 papers), and actively contributed to international conferences. She has published a book, many book chapters and she serves on the editorial boards of many journals. She has also invested considerable time into communicating issues connected with GMO technology to the general public in Austria and in Europe.

Ruie Lui is studying for her PhD at the department of Biological and Health Sciences University of Bordeaux, France. Her work focuses on the analysis of DNA methylation during tomato fruit development. She has obtained her master degree in the field of Genetic at the College of Life and Environmental Sciences at HangZhou Normal University, in China.

Fatemeh Maghuly is an associate professor and deputy head of the Plant Biotechnology Unit (PBU), University of Natural Resources and Life Sciences (BOKU), Vienna, Austria, is an expert in plant functional genomics, genetic populations and molecular marker development. As principal investigator she coordinated the efforts to analyse several hundred transgenic stone fruits and grapevine plants. She was responsible for the genetic characterization of the largest collection of apricot accessions leading to a molecular conformation of the geographic spread of apricot. She contributed to the allergen research in fruits, with the aim to develop improved detection methods for traces of food allergens in fresh and processed plant-derived products. She is also responsible for a new research area, involving important bioenergy plants, e.g. *Jatropha curcas*. Since this non-edible plant is at a nondomesticated level, it requires the development of new *-omics*' tools for its breeding and selection.

Mariangela Marudelli is an assistant professor at the Institute of Agronomy, Genetics and Field Crops of the Università Cattolica del Sacro Cuore of Piacenza (Italy). Her scientific activity is mainly focused on the recovery of enzymatic activities from soil microbiota and on the application of procedures of marker-assisted selections to improve breeding. She has taught the course of agricultural genetics at the Università Cattolica del Sacro Cuore. Since 1995–2005 she was in charge of Observatory on Transgenic Plants of the Institute of Plant Genetics at the Università Cattolica del Sacro Cuore.

Slavica Matic graduated in agricultural sciences in 2001 (University of East Sarajevo, Republika Srpska, Bosnia and Herzegovina) and completed her MSc degree in 2004 in 'Integrated Pest Management of Mediterranean Fruit Tree Crops' at the Mediterranean Agronomic Institute of Bari, Italy. During her PhD studies she specialized in plant protection (2008), with curriculum in plant virology. From 2008 to 2009 she held a postdoc position in plant virology at the International Centre for Genetic Engineering and Biotechnology in Treviso, Italy, and then, from 2009 to 2011 at the Institute of Plant Virology IVV-CNR in Torino, National Research Council of Italy. She has been under contract from 2012 to 2013 at the Centre of Competence for the Innovation in the Agro-environmental Sector – AGROINNOVA, University of Torino. She has published several peer-reviewed publications in high-impact journals.

Masanao Miwa is a professor and the president of the Nagahama Institute of Bioscience and Technology, Nagahama, Japan. His interests are in PARP proteins and their involvement in DNA repair, signalling, transcription and epigenetics, using several approaches, from cell cultures, *Drosophila melanogaster*, animal models of disease and knockout mice. He was researcher at the National Institute of Cancer in Tokyo and head of the department of biochemistry and oncology at the Institute of Basic Medical Sciences, Tsukuba University. He has published more than 50 peer-reviewed publications in high-impact journals such as *PNAS* and the *Journal of Biological Chemistry*.

Atef Mohamad is a lecturer in plant pathology in Fayoum University, Egypt. He obtained his BSc from Cairo University, Egypt, an MSc from the Mediterranean Agronomic Institute of Chania, Greece and a PhD from the University of Warwick, United Kingdom. His research interests include virus–plant interactions with focus on small RNA-mediated antiviral defence in crops.

Emanuela Noris graduated in agricultural sciences in Torino in 1986, where she completed her PhD in cell biology and immunology (viruses and cellular senescence). She was researcher at Institute Guido Donegani, SpA Montedison/Enichem and currently is a researcher at the Institute of Plant Virology of the National Research Council of Italy (IVV-CNR, now Institute for Sustainable Plant Protection, IPSP) in Torino since 2001. She is author of more than 35 publications in international journals (h-index 2013 = 20, WoS) and co-author of 3 international patents. Her scientific interests are addressed towards plant viruses, mainly geminiviruses, focusing on their epidemiology, diagnosis, virion formation, induction of resistance and plant/virus interaction. She is also interested in the use of plants for producing proteins of pharmaceutical or industrial interest, such as vaccines.

Christian Obermeier is a senior scientist at the department of plant breeding at Justus Liebig University in Giessen, Germany. He received his PhD in biology from the Technical University Braunschweig, Germany. He has been working in the fields of plant virology and resistance breeding of agricultural and horticultural crops at the Federal Research Centre for Agriculture and Forestry, now Julius Kühn-Institut, in Braunschweig, Germany; at the Agricultural Research Service of the US Department of Agriculture in Salinas, California; at the Institute for Agricultural and Horticultural Sciences of the Humboldt University Berlin, Germany and at Horticulture Research International, now Warwick Crop Centre, of the University of Warwick, United Kingdom. Since 2005, his research within the department of plant breeding at Julius Liebig University Giessen focuses in close collaboration with the German plant breeding industry on genomics-based breeding of oilseed rape.

Toba Osman is a professor of plant molecular virology at Fayoum University, Egypt. He obtained his BSc and MSc degrees from Cairo University and a PhD in plant molecular virology from Imperial College London, United Kingdom. Over 30 years, he has published significant papers in high-impact factor scientific journals in the field of virus research such as *Journal of Virology*, *Virology* and *Journal of General Virology*. His main research interests focus on plant virus genome replication and virus–plant interactions.

Qian Qian is a professor and deputy president of the China National Rice Research Institute, CAAS, People's Republic of China. He received his PhD in genetics from Graduate School of CAAS in 1995. Qian is currently the head of the State Key Lab of Rice Biology, People's Republic of China. His areas of specialization are rice genetics and breeding with focus on rice genetic materials creation, molecular mechanism dissection of important agronomy traits and molecular application. He has long-time international cooperation experiences such as China–Japan or China–IRRI collaboration projects. His books include *Rice Genetics and Functional Genomics* (Science Press, 2006), *Rice Breeding by Gene Design* (Science Press, 2007) and *Technology Review of Molecular Breeding in Rice* (Science Press, 2011), and his writings include numerous articles on rice molecular genetics and breeding.

Cheng Qin received his PhD degree from the College of Life Sciences, Zhejiang University in 2008. He is now a research leader in the Research Centre for Plant RNA Signaling, College of Life and Environmental Sciences, Hangzhou Normal University. His research interest includes functional analysis of florigen signaling and plant development. His works were published in international journals such as *PNAS* and *Scientific Reports*.

Ida Barbara Reca is a postdoctoral fellow at CNR-ISPA in Lecce (Italy). She graduated in biology and biotechnology from 'La Sapienza' University in Rome in 2004, earned two PhD titles in 2008, one in plant biotechnology from 'La Tuscia' University, Viterbo (Italy) and one in nutrition, molecular and cellular aspects, from 'Paul Cezanne' University, Aix-Marseille (France). She worked from 2009 to 2013 with Markus Paulyteam at the Great Lake Bioenergy Research Center, Plant Research Laboratory, Michigan State University, on the mechanism of hemicellulose *O*-acetylation in plant cell-wall biosynthesis, and with Dr. Kenneth Keegstra Laboratory, on the identification of three transcription factors responsible for hemicellulose biosynthesis. Presently, she studies amino acid substitutions and pathogen recognition by LRR receptors in model plants.

Giampaolo Ricci graduated in medicine and surgery in 1977 and the specialization in hygiene and preventive medicine (1980) and in paediatrics (1984) at University of Bologna, Italy. His main research activity has focused on the field of paediatric allergy and immunology. Presently, he is an assistant professor at the department of medical and surgical sciences of University of Bologna and is responsible of the paediatric allergy section (Paediatric Unit – University of Bologna). He is author, in collaboration with other groups, of several papers published in national and international journals.

Dominique Rolin has been a professor at Bordeaux University, France, since 1994 and has recognized expertise in the fields of NMR spectroscopy, plant biology, metabolism, metabolomics and fluxomics. He is widely involved in the structuring of research efforts in the emerging field of functional genomics especially in metabolomics. He is heading the 'Bordeaux Functional Genomics Centre', a federation of seven technological platforms devoted to the study of living organisms at tissular, cellular and molecular scales. In 2002, he set up and led the Metabolome Facility of Bordeaux (BMP) until 2008. He is deeply involved in the promotion of metabolomics and fluxomics in France as co-founder, treasurer (2007–2010) and now president of the French Metabolomics and Fluxomics Network (RFMF), https://www.bordeaux.inra.fr/ifr103/reseau_metabolome/accueil.htm, (a French Scientific Society created in 2005 that aims at the promotion of metabolomics and fluxomics in France). He was co-organizer of eight national metabolomics congresses and two scientific schools for metabolomics. Since 2013, he is the coordinator of the MetaboHUB, the French infrastructure for metabolomics and fluxomics.

Federica Savazzini graduated in agricultural sciences in 1990 and during her PhD she specialized in molecular biotechnology and plant genetics (1997), with particular interests in plant cell culture, plant physiology and plant defence/pathology. She worked for several years at Tsukuba University, Japan, and in several plant genetics laboratories, in Piacenza, San Michele all'Adige and at Plantechno. Her work experiences were focused towards the food technology area, in particular in GMO traceability in wine and grape products, and biochemistry of wheat seeds. In the last years, she has been studying apple allergens with the aim to find out the determinants of this fruit allergy, in collaboration with allergologists at the Bologna University.

Tiziana Sgamma is currently a lecturer and research fellow at De Montfort University, United Kingdom. After obtaining her BSc and MSc in agricultural and industrial biotechnology in 2005 and 2008 from the Università degli Studi della Tuscia in Italy, she obtained her PhD in plant and environmental sciences from the University of Warwick, United Kingdom, in 2012, where she investigated the role of the gene Tempranillo in juvenility phase in different plants. Her primary research interests include the genes' expressions that occur when plants interact with the environment during their development, in particular during their juvenility, and the regulation of flowering time in horticultural species. She is currently applying novel DNA-based techniques to the identification and authentication of medicinal plants and phytomedicines.

Nongnong Shi is a professor at the School of Life and Environmental Sciences, Hangzhou Normal University, People's Republic of China. She received her BSc in horticulture in 1985 from Zhejiang Agricultural University and a PhD in plant virology in 1999 from Rothamsted Research/The Open University, United Kingdom. She teaches genetics (bilingual), genetic experiments and molecular virology courses to undergraduate and postgraduate students. She has been leading scientific programs funded by Natural Science Foundation of China and Natural Science Foundation of Zhejiang Province. She has published more than 30 scientific research papers. Her current research interests include plant molecular genetics, virus–plant interactions and disease resistance mechanisms.

Linda Stammitti has obtained her PhD in plant biology at the University of Nancy (France) where she studied plant cuticular permeability. Since 1998 she works as an engineer at Bordeaux University (France). She first studied terpenoid biosynthesis and carotenoid accumulation in tomato fruit, and since 2004 her work focused on the analysis of the epigenetic control of tomato fruit development.

Yuqiang Sun is a professor at the College of Life and Environmental Science at Hangzhou Normal University. He obtained his BSc (2000) and PhD degrees (2005) in plant genetics and breeding from the National Key Laboratory of Crop Genetic Improvement of Huazhong Agricultural University in Wuhan, People's Republic of China. He worked as a lecturer and an associate professor (July 2005–August 2008) in the College of Agriculture and Biotechnology of Zhejiang University in Hangzhou, Zhejiang Province, People's Republic of China. He did a postdoc (August 2008–August 2010) at Umeå University in Sweden. He is principal investigator for cotton molecular genetics and biotechnology in Hangzhou Normal University, has a passion for understanding the genomic expression profile in cotton somatic hybrids (from protoplast fusion between two cotton species/wild species), cotton biotechnology as transgenic cotton, protoplast culture and fusion and genetic improvement in cotton. Many works were focused on creating new germplasm sources in cotton, he successfully developed an efficient method for cotton somatic embryogenesis and plant regeneration in upland cotton and especially in wild species; experimented with protoplast culture and plant regeneration in cotton and tobacco; carried on somatic hybridization via protoplast fusion and created several somatic hybrids between wild cotton species and cultivars and also performed functional studies on genes and their modification into cotton plants. Currently, his focus is on functional genomics and transcriptomics of fibre development and gene expression pattern after polyploidization in somatic hybrids. Over 40 papers about cotton work were published by him.

Stefano Tartarini graduated in agricultural sciences in 1990 and did his PhD research on the development of molecular markers in apple (1994). He is currently a research scientist at the department of agricultural sciences (DipSA) of Bologna University (since 1999). He worked mainly on molecular genetics and applied biotechnology in fruit trees starting from the development of molecular markers linked to key traits (resistance to pathogens and fruit quality traits), to candidate genes identification and to the proof of gene function by the development of genetically modified plants. In the last years, he worked on apple allergens in order to identify not only the main determinants of allergenicity but also the role of some allergens/pathogenesis-related proteins in resistance response.

Emeline Teyssier is an assitant professor at the University of Bordeaux, performing her research work at the Laboratory of Fruit Biology and Pathology, at the INRA, focusing on fruit biology and epigenetics. She obtained her PhD in 1997 in the field of plant science in the University of Grenoble. She was a post-doc in the University of Pennsylvania, Philadelphia, PA in USA. She became first an assistant professor at the University of Angers in 1999, and then moved to the University of Bordeaux in 2008.

Md. Sorof Uddin was awarded a PhD in biochemistry and molecular biology from the Biotechnology Research Institute of Chinese Academy of Agricultural Sciences, Beijing, People's Republic of China and an MSc from Bangladesh Agricultural University. He is working as senior scientific officer of the Bangladesh Agricultural Research Institute (BARI). He has authored 12 research papers, has co-authored 20, 70 articles in agricultural journals and other contributions for national daily newspapers and agricultural magazines. He has published three booklets on mango and Ber cultivation. His areas of specialization are mango breeding, mango molecular breeding and post-harvest of mango. He isolated 23 mangoes 18SrRNA genes and registered them in GenBank at NCBI. He patented three mango varieties in Bangladesh including one variety with export potential.

Johann Vollmann is an associate professor at the plant breeding division, department of crop sciences, University of Natural Resources and Life Sciences (BOKU) at Tulln an der Donau, Austria. He has a strong background in plant breeding and is mainly dealing with breeding research in oilseed crops including soybean, camelina and oil pumpkin. His research interest is in the genetics and improvement of quality features such as oil and protein content, fatty acid composition or food safety characters (reduction of trypsin inhibitors, allergens, heavy metals, etc.). Consequently, his major activity is the evaluation and utilization of genetic resources and introgression of traits of interest into breeding material using marker-assisted selection and other breeding procedures.

Dawei Xue is an associate professor at College of Biology and Environmental Science, Hangzhou Normal University, People's Republic of China. He received his PhD in biochemistry and molecular biology from the China National Rice Research Institute, CAAS, (Supervisor: Qian Qian) in 2006. He has worked as a postdoctoral fellow at Zhejiang University (2006–2008). His scientific activity is mainly focused on plant genetics, molecular markers and molecular mechanism dissection of important agronomic traits in rice. His teaching includes courses on molecular biology and plant molecular breeding. Since 2005, he has written more than 30 papers on this topic.

Zhiming Yu obtained his PhD in plant science from Zhejiang University in 2010. His PhD thesis is about roles of *EXPA7* subclade members in regulation of rice root hair elongation. He did a postdoc in medicine college of Zhejiang University and he is now a junior research leader at the Research Centre For Plant RNA Signaling, Hangzhou Normal University. His research interest focuses on long distance movement of *Flowering Locus T* mRNA and development of novel technology to directly visualize RNA trafficking in plants.

Pengcheng Zhang is a research assistant in the Research Centre for Plant RNA Signaling, Hangzhou Normal University. He received an MSc degree in plant virology. He works on virus–plant interactions and siRNA-mediated antiviral defence in plants.

Qi Zhang is a postgraduate student in the College of Life and Environmental Sciences, Hangzhou Normal University. Her current research project is about the epigenetic role in plant antiviral defence.

Xian Zhang is a lecturer of microbiology in Hangzhou Normal University. He obtained his BSc in biotechnology, MSc in chemical biology and molecular biology from Zhejiang University and his PhD in applied microbiology from Institute of Plant Sciences and Resources, Okayama University. He is interested in the interactions between geminiviruses and host plants.

Silin Zhong graduated in biochemistry in 2002 and did her PhD research on tomato ethylene hormone receptor and kinase in 2008. He is currently an assistant professor in the Chinese University of Hong Kong. His work mainly focuses on genome sequencing, functional genomics and epigenetics. In the last year, he sequenced the tomato fruit methylome at different developmental stages and discovered a unique epigenome reprogramming event during fruit ripening.

List of abbreviations

5 AzaC	5 azacytidine
5MeC	cytosine methylation in C5
ABA	abscisic acid
ACC	1-aminocyclopropane-1-carboxylate
ACMV	African cassava mosaic virus
ACO1	ACC oxidase 1
AFLP	amplified fragment length polymorphism
AGL24	agamous-like 24
AGO1	argonaute1
AHP	pseudo-phosphotransfer protein
AMV	alfalfa mosaic virus
AP1	apetala1
ApoA-IMilano	apolipoprotein $A\text{-}I_{Milano}$
BC	backcross populations
bHLH	basic helix-loop-helix
BPV	bovine papilloma virus
BR	brassinosteroids
BRCT	BRAC1 carboxy terminal domain
CaLCV	cabbage leaf-curl geminivirus
CaMV	cauliflower mosaic virus
CBSV	cassava brown streak virus
CCA1	circadian clock associated 1
CDF1	cycling DoF factor 1
CESA	cellulose synthase
ChIP	chromatin immuno-precipitation
CKX	cytokinin oxidase
CLF	curly leaf
CsCMV	cassava common mosaic virus
CSL	cellulose-synthase-like
CMD	cassava mosaic disease
CMG	cassava mosaic geminiviruses
CMT	chromomethylases
CMV	cucumber mosaic virus
CNR	colorless non ripening
CO	constans
COP1	constitutive photomorphogenic 1
CP	coat protein
CPMV	cowpea mosaic virus
CRISPR	clustered, regularly interspaced, short palindromic repeats
CRY	cryptochrome

Cry1Ab	*Bacillus thuringensis* crystal 1Ab toxin
CYP735A	cytochrome P450 family 735, subfamily A
DArT	diversity arrays technology
DDB1	UV-damaged DNA-binding protein 1
DDM1	decrease in DNA methylation 1
DET1	de-etiolated 1
DH	doubled haploid polulation
DLC3	dicer-like 3
DMR	differentially methylated regions
DNA-GL	DNA glycosylase-lyase
DRD1	decrease in RNBA-dependent DNA methylation 1
DRM	domain rearranged methyltransferases
DSB	DNA double-strand breaks
DST	drought and salt tolerance
DUS	distinctness, uniformity and stability
EACMV	East African cassava mosaic virus
EACMZV	East African cassava mosaic Zanzibar virus
EMF2	embryonic flower 2
ER	endoplasmic reticulum
Esc	extra sex combs
ESR2	enhancer of shoot regeneration 2
EST	expressed sequence tag
ET	ethylene
E(Z)	enhancer of Zeste
FDL	flowering locus D-like2
FEC	friable embryogenic callus
FIE	fertilization-independent endosperm
FIS2	fertilization-independent seed 2
FKF1	flavin-binding, kelch repeat F-box 1
FLC	flowering locus C
FLK	flowering locus K
FLM	flowering locus M
FRI	frigida
FT	flowering locus T
FUL	fruitful
GAA	acid alpha glucosidase (acid maltase)
gbM	gene body methylation
GCase	acid β-glucosidase
GC-MS	gas chromatography mass detection
GEBV	genomic estimated breeding values
GI	gigantea
GID1	gibberellin insensitive dwarf1
GMP	genetically modified plants
GSDII	glycogen storage disease type II
GT	gene targeting
GWAS	genome-wide association analysis
GWS	genome-wide selection
HAT	histone acetyltransferase
HB-1	homeobox protein 1

HBcAg	HB core antigen
HBsAg	HB surface antigen
HBV	hepatitis B virus
HD1	heading-date1
HDAC	histone deacetylases
HDL	high density lipoprotein
HIV	human immunodeficiency virus
HK	histidine kinase
HMT	histone methyl transferases
HPt	histidine-containing phosphotransporter
HPV	human papilloma virus
HR	homologous recombination
HR	hypersensitive response
HRM	high-resolution melting curve analysis
HST	hasty
HSV	homeologous sequence variants
HYL1	hyponastic leaves1
IAA	indol-acetic acid
IAV	influenza A virus
IBM1	increase in Bonzai methylation
IL	introgressive lines
IPA1	ideal plant architecture 1
IPT	isopentenyl transferase
ISSR	inter simple sequence repeats
JA	jasmonic acid
KYP/SUVH4	kryptonite
LD	luminidependens
LDP	long-day plants
LFY	leafy
LHY	late elongated hypocotyl
LINE	long interspersed nuclear elements
LTP	lipid transfer protein
LTR	long teminal repeats
MAF2	MADS affecting flowering
MAS	marker-assisted selection
McrBC	methylation-dependent endonuclease
MEA	MEDEA
MET1	DNA methyltransferase 1
MITEs	miniature inverted repeat transposable elements
Mr VIGS	microRNA-mediated virus-induced gene silencing
MSI 1	multisuppressor of IRA 1
MtRR1	*Medicago truncatula* Mt RR1 response regulator
MYB	myeloblastosis DNA-binding domain proteins
NAD	nicotinamide adenine dinucleotide
NAM	nested association mapping
NDV	Newcastle disease virus
NHEJ	non-homologous end joining
NOR	non-ripening
NPP	novel plant products

NUDIX	nucleoside diphosphate linked to some moiety X hydrolases
OSR	oilseed rape
PAP1	production of anthocyanin pigment 1
PAR	poly ADP-ribose
PARP	poly (ADP-ribose) polymerase
PBZ	PAR-binding zinc finger
PCW	primary cell wall
PEP1	perpetual flowering 1
PHB	phaboluta
PHY	phytochrome
PIF4	phytochrome interacting factor4
PMP	plant-made pharmaceuticals
PPV	plum pox virus
PR	pathogenesis-related (protein)
PRC2	polycomb repressor complex 2
PRR5	pseudo response regulators 5
PSY1	phytoene synthase
PTM	post-translational modifications
PVT	plant virus technology
PVX	potato virus X
QTL	quantitative trait loci
R2R3MYB	R-repeat containing MYB
RAA	recombinase-aid amplification method
RAD	restriction-site associated DNA marker
RAPD	random amplification polymorphic DNA
RCD-1	radical induced cell death 1
RdDM	RNA-dependent DNA methylation
RDRP2	RNA-dependent RNA polymerase 2
REF6	relative of early flowering6
RFLP	restriction fragment length polymorphism
RIL	recombinant inbred line
RIN	ripening inhibitor
RISC	RNA-induced silencing complex
RNAi	RNA interference
ROS	reactive oxygen species
SA	salicylic acid
SCAR	sequence characterized amplified region
SCF	Skf, Cullin, F-box
SCoT	start codon targeted primers
SDG	set-domain group
SDN	site-directed nucleases
SE	serrate
SEP3	sepallata3
SHR	short-root
SIMR	stress-induced morphogenetic response
SINE	short interspersed nuclear elements
Sir VIGS	siRNA-mediated virus-induced gene silencing
SLCMV	Sri Lankan cassava mosaic virus
SLP	squamosa promoter-binding protein-like

SMZ	schlafmütze
SNP	single nucleotide polymorphism
SNZ	schnarchzapfen
SOC1	suppressor of constans1
SQN	squint
sRNA	small RNA
SRO	similar to RCD one
SSR	simple sequence repeat
STM	shoot meristemless
Su(z)12	suppressor of Zeste12
SusA	sucrose synthase
SVP	short vegetative phase
SWN	swinger
TAG	triacylglycerols
TALEN	transcription activator-like effector nucleases
TCP	teosynthe branched, cycloidea and PCF transcription factors
TE	transposable element
Tet	ten-eleven translocation
TFL1	terminal flower 1
TGM	targeted genome modification
TILLING	targeting-induced local lesions in genomes
TLP	thaumatin-like protein
TMV	tobacco mosaic virus
TOC1	timing of CAB expression 1
TOE1	target of eat1
TRV	tobacco rattle virus
TSF	twin sister of FT
TSP	total soluble proteins
UCBSV	Ugandan cassava brown streak virus
VbMS	virus-based microRNA silencing
VIGE	virus-induced genome editing
VIGS	virus-induced gene silencing
VIN3	vernalization insensitive 3
ViTGS	virus-induced transcriptional gene silencing
VLCFA	very long chain fatty acid
VLP	virus-like particles
VRMA	virus-based RNA mobility assay
WD40	tryptophan glutammic acid domain transcription factors
WRKY	tryptophan, arginine, lysine, tyrosine-based 60 amino acid domain transcription factor family
XTH	xyloglucanase
ZFN	zinc finger nucleases
ZTL	zeitlupe

Introduction

Applied Plant Genomics and Biotechnology reviews the recent advancements in the postgenomic era, discussing how different varieties respond to abiotic and biotic stresses, investigating epigenetic modifications and epigenetic memory through analysis of DNA methylation states, applicative uses of RNA silencing and RNA interference in plant physiology and in experimental transgenics and plants modified to specific aims (production of high-value pharmaceutical proteins and metabolites in plants). This book provides an overview of research advances in plant biology, functional genomics, application of RNA silencing and RNA interference, RNA-based vaccines, RNA signaling at distance, plant epigenetics and exploring plants as factories for useful products and pharmaceuticals. This book reviews and discusses plant functional genomic studies and the developments in the field, exploring the new technologies supporting the genetic improvement of plants and the production of plant varieties more resistant to biotic and abiotic stresses. The specific crops are analyzed in detail to provide a glimpse on the most up-to-date methods and topics of investigation.

Chapter 1 presents a review on current state of GMO application in the world, and the regulation on GMO with the difference in rules between national authorities, and the efforts carried on to harmonise them in order to pave the way for the introduction of safe technologies such as cisgenic plants and novel plant products.

In Chapter 2, cytokinin metabolism and hormone signaling pathways are discussed in various tissues, with a presentation of the latest findings and improvements using barley modified plants.

Chapter 3 discusses the problems associated with apple consumption by allergic individuals, with an overview on the most allergic isoforms and the potential approaches to reduce the allergen content in fruits.

Chapter 4 introduces the various protocols used to produce transgenic plants modified for specific aims such as production of high-value pharmaceutical proteins, with several examples of enzymes and oral vaccines today under testing in new clinical trials through agreements by pharma industries.

In Chapter 5, the expression of virus proteins and virus-like particles in plants is presented, with potential as therapeutic virus vaccines.

Chapter 6 introduces three plants belonging to Euphorbiaceae, castor bean, cassava, and *Jatropa*. It discusses the genomic structure, the disease and viruses, and the potential methods of transformation.

Chapter 7 offers an overview on the knowledge of the regulatory networks involved in flowering, such as control by vernalization in *Arabidopsis* and cereals, and microRNAs controlling transcription factors expression.

Chapters 8 and 9 discuss the development and ripening processes in tomato, with a focus on epigenetics, and the exploitation of genomics and fruit modeling to establish fleshy fruit traits of interest.

Chapter 10 discusses rice genomics, important agricultural traits, and potential of biotechnology and engineering to improve rice varieties.

Chapter 11 presents an overview on the methods applied to study the tomato genome-wide DNA methylation, its regulation by small RNAs, the reprogramming during fruit ripening and the ethylene-dependent and -independent DNA methylation changes.

Chapter 12 discusses an important fruit crop of tropical countries, mango, with an overview of molecular methods for variety differentiation, and problems in fruit improvement by traditional and biotechnology methods.

Chapter 13 is dedicated to the genomics of cotton, and the problems faced by cotton biotechnology.

Chapter 14 presents the most up-to-date information on the exploitation of virus technology in plant functional genomics studies. With presentation of several cases and applications, the chapter discusses virus-based transient gene expression systems, virus-induced gene complementation (VIGC), virus-induced gene silencing (Sir VIGS, Mr VIGS), virus-based microRNA silencing (VbMS) and virus-based RNA mobility assays (VRMA).

Chapter 15 presents an overview on PARP-domain containing enzymes involved in stress-induced morphogenetic response as well as in regulation of NAD signaling and ROS-dependent synthesis of anthocyanins.

Chapter 16 presents a synthesis of the technologies involved in oilseed rape studies, from genetic diversity to quality traits, genetic maps, genomic selection, and comparative genomics for improvement of varieties.

Transgenic, cisgenic and novel plant products: Challenges in regulation and safety assessment

1

Palmiro Poltronieri and Ida Barbara Reca
CNR-ISPA, Department of Food, Agriculture, Fisheries and Biotechnology, National Research Council, Italy

1.1 Genetically modified plant products in the United States

The growing area of genetically modified (GM) crops has substantially expanded since they were first commercialized in 1996. Correspondingly, the adoption of GM crops has brought huge economic and environmental benefits (processed proteins and carbohydrates, soy sauce, proteins for feeds). All these achievements have been primarily supported by two simple traits of herbicide tolerance and insect resistance in the past years (Chen and Lin, 2013). The populations of at least nine pest species have evolved resistance to *Bacillus thuringiensis* (Bt) toxins in the field. It has been reported that widespread control failures of Bt cotton associated with pink bollworm (*Pectinophora gossypiella*) resistance to Cry1Ac have happened in the state of Gujarat in western India. Moreover, the wide adoption of HR crops in the United States also appeared to accelerate the evolution of resistance weeds to glyphosate compared with areas not growing GM crops (Tabashnik et al., 2012; Carriere et al., 2010; Dhurua and Gujar, 2011). Twenty-four glyphosate-resistant weed species have been identified since Roundup-tolerant crops were introduced in 1996. However, studies on Palmer amaranth (*Amaranthus palmeri*) showed that the plant can easily grow in transgenic cotton fields since 2008. Palmer amaranth is a weed especially in the south-eastern United States. It outcompetes cotton for moisture, light and soil nutrients and can quickly take over fields. Farmers had historically used multiple herbicides, which slowed the development of resistance, and controlled weeds through ploughing and tilling, practices that deplete topsoil and release carbon dioxide but do not encourage resistance. Monsanto has changed its recommendations on glyphosate use, so that farmers change and use a mix of chemical products and ploughing.

For GM plants (GMPs), it takes almost 6 years and US$ 35 million to generate the data for a regulatory dossier, limiting the use of this technology to the major agrobiotechnology companies and to high-value crops and traits (Podevin et al., 2013).

Applied Plant Genomics and Biotechnology.

1.2 GMP products in Europe

The purpose of European Union (EU) legislation on GM food and feed products is to protect not only the environment but also the public by ensuring food safety and to uphold consumers' rights to choose between GM and non-GM through food labelling (EC No 1829/2003, 1830/2003). The enforcement of EU legislation requires reliable qualitative and quantitative analytical methods. There are several different aspects in risk assessment of GM plants and derived products that must be discussed with EU authorities in the accomplishment of a regulatory dossier: a thorough analysis of the environmental risk assessment (persistence and invasiveness assessment including agronomic and phenotypic performance, interactions with target and non-target organisms, horizontal gene transfer, impacts on biogeochemical processes, impacts on agricultural management practices); food/feed safety assessment (composition, toxicology, allergenicity, nutritional assessment, dietary exposure) and molecular characterization of genetic modifications (transformation process, molecular analysis and expression of inserted DNA, inheritance and stability of inserted DNA, bioinformatic analysis).

Public attention is mainly focussed on the introduction and marketing of GM crops, food and seeds. Since 1980, the regulation of health safety and environmental risks are generally much stricter in EU than in the United States. The EU introduced the Cartagena protocol on biosafety (http://bch.cbd.int/protocol), which extends the United Nations Convention on Biological Diversity (CBD) and implemented it in its legislation. While the US Food and Drug Administration (FDA) delegates the determination of safety of production of modified plants to the company filing the submission, it regulates the environmental release of GM organisms (GMOs) that could deliver plant pests or genes from plant pests under the Plant Protection Act. Depending on the nature and intended use of the plant it may still be subject to other regulatory authorities such as the Environmental Protection Agency (USEPA) and FDA.

The precautionary principle and recently the social and economic aspects that were included in the decision process of GM risk assessment are the main reasons that delay GM plants approval in Europe, while in other advanced countries these constraints are less tight.

At a subsequent step, most regulatory authorities require that GMO be devoid of unnecessary DNA, especially vector backbone sequences (VBSs), containing bacterial resistance genes and origins of replication (OR).

In 2013, only one transgenic crop (Bt maize) is commercially planted in EU. Bt maize was planted in six countries (Spain, Czech Republic, Slovakia, Portugal, Romania and Poland) in about 300 ha.

Considering field trials in EU countries, both for intended commercialization and research purposes, many projects were performed recently (e.g. potatoes with modified sugar content, changed composition of starch polysaccharides or resistance against infection, virus-resistant plum trees, flax with changed linseed oil properties, trees for bioremediation, resistance against illnesses, growth acceleration or changed technological properties of wood).

New field trials have been stopped since the year 2008, according to EU directive 2001/18/EC (http://www.efsa.eu.int/science/gmo/gmoopinions/384en.html) (EPEC, 2011).

Further studies with different purpose are realized in the category of the 'contained use' (laboratory scale). The most promising are the experiments testing the application of transgenic plants for the production of various recombinant proteins (molecular farming, e.g. antibodies).

Under the EU legislation for authorization of release of new GMOs, detection methods are also required to be submitted to the EU reference laboratory on GMOs for assessment and validation. The European Network of GMO Laboratories (ENGL) serves the analytical laboratory community with up-to-date information on the latest developments in GMO testing.

Several previous EU projects have sought to develop such methods to satisfy a wide range of requirements, e.g. high quantification accuracy, on-site detection, high throughput and wide target range. These projects included QPCRGMOFOOD (FP5), SIGMEA (FP6), COEXTRA (FP6) and GMOSeek (SAFEFOOD ERA-NET).

Today, therefore, there is a very wide choice of GMO analytical methods and advice for most situations (van de Bulcke et al., 2010). However, as GMO technology has advanced in both volume and application, and as new knowledge on the nature of GMOs has become available, there are several problematic areas of GMO testing that now require new research and the development of new methods in order to ensure continued consumer safety and choice. The problematic areas are as follows:

- Increased GMO volume. As the number of GMOs released increases, so does the need to screen them simultaneously and identify them in a cost-effective and accurate way.
- Unauthorized GMO proliferation. As the number of authorized GMOs increases, so too does the number in development in non-EU countries, and the risk of unauthorized GMOs entering the EU. Superficially, many of these GMOs many appear very similar to authorized ones (same marker genes) and may therefore evade current screening methods.

1.2.1 Novel GMP producing methods and risk mitigation

Plant products produced by conventional breeding are more familiar to consumers. Reverse breeding is a technology that transiently produces parental lines that are hybridized with non-transformed plants to obtain segregants devoid of the bacterial vector, and of the gene variants inducing the desired traits. Reverse breeding is applied to reconstitute parental lines starting with an elite F1 hybrid whose genetic material is unknown. Reverse breeding combines several other techniques such as RNA interference (RNAi) to suppress meiotic recombination, tissue culture to regenerate plants from cells and the double haploidization technique to create double haploid plants, which are used as the respective parental lines to produce new elite F1 hybrids.

These novel plant products (NPPs) include GMPs that do not use traditional methods of genetic transformation and may not contain stable transgenes, so that in most cases standard molecular methods cannot detect any non-self gene or mutation. Transient or secondary genetic alterations caused by the novel transformation process may still be detected (through transcriptomic analyses and next-generation sequencing [NGS]).

Another concern about transgenic crops relates to the mixing of genetic materials between species that cannot hybridize by natural means.

1.2.2 Targeted genome modification using site-specific nucleases

In the case of higher plants, an efficient RNA delivery system has not been established. *Agrobacterium*-mediated transformation of DNA or direct injection of DNA using a particle gun are commonly used to express proteins of interest. In *Agrobacterium*-mediated transformation, a single-stranded transfer DNA (T-DNA) protected by coat proteins is delivered from *Agrobacterium* to the plant cell nucleus, where naked single-stranded molecules are converted to double strands. These double-stranded T-DNAs are used as templates for homologous recombination (HR) (Chen and Gao, 2014; Figure 1.1). The critical step is the introduction of DNA double-strand breaks (DSBs) at given genomic sites. Engineered nucleases generate DSBs and consequently activate DNA repair to seal the breaks along with any modifications such as mutations, insertions, replacements and chromosomal rearrangements.

Because transcription and translation of engineered nucleases needs some time, it is not easy to harmonize the timing of expression of engineered nuclease tightly with the existence of the HR template. In addition, *Agrobacterium*-mediated gene targeting (GT) is only applicable to plant species in which an efficient transformation system is established. A novel GT method called *in planta* GT (Fauser et al., 2012; Ayar et al., 2013) promises to solve both problems. In this system, a GT donor vector flanked by two engineered nuclease recognition sites is first stably integrated into the plant genome. Site-specific nuclease (SSN) or site-directed

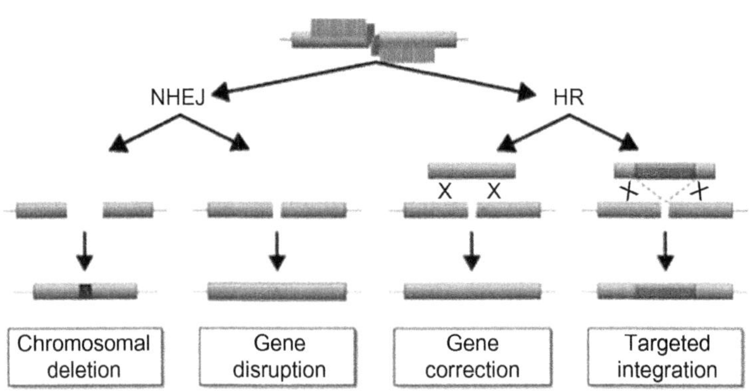

Figure 1.1 Schematic representation of targeted genome modifications using site-specific nucleases.

nuclease (SDN) exploit the presence of an oligonucleotide sequence complementary to the gene region to be targeted for modification (i.e. meganuclease, endonuclease). Once the site-specific endonuclease is expressed or introduced inside the cell, it cuts within the target and also excises the chromosomal transgenic donor if the recognition sequence of an engineered nuclease present at both ends of the GT donor is the same as that in the target gene. HR between the excised GT donor vector and target locus results in *in planta* GT. Conventional GT approaches rely on the generation of a very large number of transformation events, while the *in planta* GT system requires only a minimum transformation efficiency because, once plant materials in which the HR template and an inducible expression cassette for the engineered nuclease are integrated into the plant genome are obtained, these transgenic cells can be clonally propagated for use in inducing GT. Thus, this *in planta* GT system is suitable for plant species with low transformation efficiency (Endo and Toki, 2014).

1.2.3 New technologies applied to plant genome editing

Conventional plant breeding exploits existing genetic variability and introduces new variability by mutagenesis. This has been up to now the method of choice to sustain the food supply.

The recent evolution of targeted mutagenesis and DNA insertion techniques based on tailor-made SDN provides opportunities to rapidly engineer new plant genomes.

Plant breeding companies are exploiting SDNs to develop a new generation of crops with new and improved traits (Podevin et al., 2013). New technologies – robust, affordable and easy to engineer in terms of precision and controlled activity – have been recently applied to plant engineering. Genome editing technologies have been performed exploiting the zinc finger nucleases (ZFNs) as well as transcription activator-like effector nucleases (TALENs) (Deng et al., 2012; Mak et al. 2012; Mak et al., 2013) which possess greater cutting efficiency and less off-target activity (Chen et al., 2013; Tsuji et al., 2013) than ZFN technology (Gaj et al., 2013). However, ZFNs could occasionally cut areas of the genome in addition to the intended site, an activity that could result in undesirable effects.

A new method has been established for delivering ZFN proteins directly to cells (Chen et al., 2013) moving from the endosome into the cytosol; due to the ZFN's high positive charge. From there, the ZFN targeted the nucleus and cleaved a specific sequence of DNA. The method of delivery was highly efficient, providing better temporal control and control of the ZFN quantity delivered while maintaining the same cutting efficiency as a ZFN expressed from complementary DNA (cDNA) (Chen et al., 2013). Vectors such as pTAL10 have been engineered to coexpress the two recombinant TALEs (left and right) that form the active dimmer, in order to avoid transformation of two plasmids in the same cell, and to increase transformation efficiency (Mariano et al., 2014).

More recently, the clustered, regularly interspaced, short palindromic repeats (CRISPR) and the CRISPR-associated protein Cas9 system has been applied to plants to generate genome modifications at defined positions in the genomes, based on DNA sequences complementary to the desired DNA regions (Lusser et al., 2011; Xie and Yang, 2013; Belhaj et al., 2013; Chen and Gao, 2014). Cas9 forms a complex with a synthetic single-guide RNA (sgRNA), consisting of a fusion of CRISPR RNA (crRNA) and trans-activating crRNA. The sgRNA guides Cas9 to recognize and cleave target DNA. Cas9 has an HNH (histidine-asparagine-histidine) nuclease domain and an RuvC-like domain; each cleaves one strand of a double-stranded DNA. It can be used as an RNA-guided endonuclease to perform sequence-specific genome editing in bacteria and eukaryotic organisms. Customizable sgRNAs directing Cas9 have been shown to induce sequence-specific genome modifications in rice (*Oryza sativa*) and common wheat (*Triticum aestivum*) (Shan et al., 2013; Upadhyay et al., 2013), and in *Arabidopsis*, tobacco (Nekrasov et al., 2013), sorghum and rice (Jiang et al., 2013).

Engineered nucleases and technologies such as nuclease-based GTg (NBGT) are posing specific and novel safety and regulatory considerations, shifting the concerns from the organisms – which are not containing external genetic material – to 'novel foods' that have not a tradition of consumption (EC Directive 258/97), thus not being supported by a long and sustained use as food (Lusser and Davies, 2013; Pauwels et al., 2014).

The critical step in NBGT is the introduction of DNA DSB at given genomic sites. It was soon discovered that engineered nucleases could generate DSBs and consequently activate DNA repair to seal the breaks along with any modifications such as mutations, insertions, replacements and chromosomal rearrangements.

The production of single and double breaks in the genome induce the DNA repair machinery through homology-directed repair (HDR). Nuclease-mediated DNA strand breaks can increase the HR favouring it in respect to non-homologous end joining (NHEJ), which is a predominant pathway in most higher eukaryotes. A possible approach to obtaining targeted integrations is to provide a donor with 5′ overhangs complementary to those created by nuclease cleavage (Pauwels et al., 2014; Figure 1.1).

SSNs allow precise modification of genomes, and therefore offer the prospect of application in crop improvement. The SSN-1 and SSN-2 techniques can introduce subtle modifications, such as small deletions and single-base substitutions of target genes (indels). The final plants derived by SSN-1 and SSN-2 techniques are similar to natural variants, or those produced by physical or chemical mutagenesis in conventional breeding. SSN-3 techniques allow scientists to insert foreign genes at predefined sites, and this site-specific gene addition should prevent the 'position effects' associated with random insertion of genes into plant genomes. The delivery of SSN DNAs using *Agrobacterium* or other delivery methods requires that the SSN DNAs integrate at different loci from the target loci; hence these foreign SSN DNAs can be easily eliminated from the genome during the segregation and recombination accompanying sexual reproduction. The final plants generated by SSN-1 or SSN-2 should fall outside the existing definitions and regulation affecting GM crops (Lusser et al., 2012; Podevin et al., 2013).

The results of a written survey of a number of plant biotech companies revealed that ZFN technology had been used in breeding of maize, oilseed rape and tomatoes

(Lusser et al., 2012). Also, Dow AgroSciences has received assurance from the US Department of Agriculture (USDA) that their GM corn developed by the ZFN technique will not require regulatory oversight. Evaluation of risk assessment will still be required before commercialization to ensure food and environment safety (Chen and Gao, 2014).

Potential uses of SDN for plant genome editing are very large. DNA shuffling can generate sequence variation producing proteins with desirable properties (kinetics, substrate specificity, temperature or pH optimum and ligand binding) including improved disease resistance (e.g., R gene shuffling). SDNs can target the most effective changes required.

Furthermore, different techniques such as targeting induced local lesions in genomes (TILLING), EcoTILLING and NGS/whole genome sequence analysis have been applied to identify candidate genes and sequences polymorphisms that can be used in SDN applications for targeted phenotype development.

There are many examples that show the potential applications of SDN (Podevin et al., 2013), which are as follows:

- Virus resistance (targeting translation initiation factors)
- Herbicide tolerance (target – acetolactate synthase gene)
- Lowering antinutritional compounds (erucic acid in Brassicas – targeting fatty acid elongases) and allergens (target – conglutin genes in peanut; *Mal d 1*, an apple Pathogenesis Response protein allergen) (Gambino and Gribaudo, 2012)
- Improved nutritional value via elevated carotenoids, modified carotenoid balance (target – zeaxanthin epoxidase)
- Modified starches and fats for food and non-food uses (targeting starch synthases, branching enzymes and fatty acid desaturases)
- Longer shelf life/reduced wastage (targets include aminocyclopropane [ACC] oxidase and polygalacturonase)
- Improved quality by reducing enzymic (target polyphenol oxidases) and nonenzymic browning (high quality and low acrylamide potato – target invertase genes)
- Yield benefits via modified RuBisCO genes, increasing catalytic activity and/or decreasing oxygenation activity and improved seed set (e.g. producing novel barley seed properties by targeting homeodomain leucine zipper genes)
- Improved biomass conversion for biofuels (lower lignin-target caffeic acid *O*-methyltransferase gene)

Recently, the European Food Safety Authority (EFSA) published a scientific opinion to address the safety assessment of plants developed through cisgenesis and intragenesis (EFSA, 2012a), and on NPPs produced by SDN (EFSA, 2012b).

1.2.3.1 Transient introduction of recombinant DNA

Techniques that introduce recombinant DNA molecules transiently to plants are ZFNs introduced into the cell with or without a repair template (ZFN1 and ZFN2), oligonucleotide-directed mutagenesis (ODM) (Breyer et al., 2007) and agro-infiltration that makes use of *Agrobacterium* to inject several foreign DNA molecules into the plant cells. These processes resemble transgenesis – *in vitro* synthesized nucleic acids and DNA delivery methods – but the end products are similar to, and indistinguishable from, plants obtained through conventional plant breeding. Therefore, NPPs are in most cases undetectable (Lusser et al., 2011). Definitions according to the EU working group on new techniques are as follows: ODM uses oligonucleotides for targeted (site-specific) induction of point mutations (Breyer et al., 2007); ZFN1 generates site-specific random mutations by NHEJ; ZFN2 generates site-specific desired point mutations by DNA repair processes through HR.

1.2.3.2 Stable introduction of recombinant DNA during an intermediate step in the development of NPPs

Techniques that use stable GM intermediates include ZFN1 and ZFN2, RNA-dependent DNA methylation (RdDM) and reverse breeding. Intermediate plants are GMPs, but the end products are similar to and indistinguishable from plants obtained through conventional plant breeding. Therefore, the NPP is in most cases undetectable (Lusser et al., 2011).

Definitions, according to the EU working group on new techniques, are as follows: ZFN1, ZFN2, TALEN and CRISPR/Cas9 complexes (Gaj et al., 2013), including the oligonucleotides complementary to the DNA target, delivered inside cells by agro-infiltration or receptor-mediated uptake; RNA-dependent DNA methylation (RdDM) is a technique that uses the effect of small RNA sequences to alter gene expression through methylation of specific DNA sequences without changing the nucleotide sequence itself (epigenetic change).

1.2.3.3 Stable integration of recombinant DNA

Integration-based plant breeding techniques include cisgenesis, intragenesis, grafting and ZFNs (ZFN3). The process of generating cisgenic plants resembles transgenesis (random DNA insertion), but the product is similar to plants obtained through conventional breeding. Detection might be challenging. ZFN3 technique targets delivery of transgenes (insertions) by HR. The transformation protocol then can be followed by crossing with a different variety; by pollen fertilization; by grafting. Grafting a non-GM scion onto a GM rootstock results in a fruit that does not contain the insert.

It is important to consider whether any unintended changes arising from these techniques are specific to NPPs and differ from those caused by conventional breeding. The *in vitro* procedures (e.g. cell and tissue culture) used to obtain NPPs are also used in conventional plant breeding, so unintended changes owing to somaclonal variation will be similar in both cases. Only unintended changes attributed to the stable or transient presence of recombinant DNA in NPPs are new compared with conventional breeding, and therefore merit further investigation. In the case of NPPs obtained after transient expression of recombinant DNA or after the use of a GM intermediate, it will be essential to verify that the recombinant DNA is no longer present in the genome of the selected plant.

1.2.4 *RNA, RNA-mediated control of gene expression and post-translational modification: From transgenic plants to NPP and cisgenic plants*

In several applications for the marketing of GM plants, the intended genetic alteration is obtained through RNAi-mediated down-regulation of specific target genes *in planta* or in insect pests.

Since RNAi mechanisms could potentially lead to unintended *in planta* or cross-species effects, there is a need to evaluate whether existing risk assessment strategies currently applied to GM plants remain appropriate, or whether new or complementary risk assessment strategies should be developed for RNAi-based GM plants. In June 2014, the EFSA organized a scientific workshop on RNAi-based GM plants in Brussels to discuss the potential risks associated with the use of RNAi mechanisms in GM plants along with the approaches and challenges in assessing those risks. The objectives of this workshop were to discuss RNAi mechanisms in plants, current and future RNAi applications and risk assessment approaches for such plants.

Gene overexpression, gene disruption, gene silencing, RNAi, targeting specific mRNAs by inducible microRNAs and 'miRNA sponges' that sequester specific microRNA species are further potential tools to control gene expression, and will be developed further more in the next years. The exploitation of RNA silencing and antisense technologies for controlling gene expression have been already translated in new plant phenotypes and tree populations with novel traits.

One such approach, aimed at up-regulating a microRNA, OsmiR397, in rice, was shown to improve rice grain yield and size (Zhang et al., 2013).

Gene silencing or gene overexpression technologies deploy at first a plasmid vector, to deliver the gene and integrate it, but during the final breeding process the plants carrying the inserted transgene are segregated out. Double haploid plants screened for the absence of RNAi constructs produce by breeding plants devoid of genetical modification-related DNA sequences (Lusser and Davies, 2013).

Epigenetics is currently recognized as one of the most exciting fields of modern science. There is an urgent need for considering a role for the EU research community with respect to this new and emerging discipline of biology. It is expected that epigenetics will have a positive impact in human health and well-being through the advances of agriculture and food biotechnology.

Carol Auer summarized the state of the art of plant biotechnologies with a special focus on new approaches based on small RNAs, RNAi and production of RNA-mediated traits in plants (Auer, 2011). The potential of RNA-regulated traits in non-food plants and biofuel producing plants is well acknowledged. Accordingly, new methods for risk analysis are required to perform analyses of off-target effects and persistence of RNAs in the environment.

NPPs attenuate the sharp distinction between GMP and non-GMP, and introduce a new continuum between genetic engineering and conventional breeding. The exploitation of gene modification, and use of antisense technologies have already translated to new plant varieties adapted to abiotic stresses.

1.2.5 Cisgenesis

To overcome the inter-species differences in gene content, two new transformation approaches, cisgenesis and intragenesis, were developed as alternatives to transgenesis (Holme et al., 2013). These approaches imply that plants must only be transformed with genetic material derived from the species itself or from closely related species capable of sexual hybridization, including P-borders (right and left) with similarity to T-borders from *Agrobacterium*, remaining in plant genomes at the integration sites.

Definitions, according to the EU working group on new techniques, are as follows:

> *Cisgenesis* is genetic modification of a recipient organism with a gene (cisgene) from a crossable, sexually compatible organism;
> *Intragenesis* is genetic modification of a recipient organism that involves the insertion of a reorganized, full or partial coding region of a gene combined frequently with a promoter and/or terminator from another gene of the same species or a crossable species.

Cisgenic plants and NPPs seek to maintain this familiarity by relying on the existing genetic variation in the breeders' gene pool. A survey about the perception of biotechnology in the EU highlighted that 55% of EU citizens support cisgenic products compared with only 22% support for transgenic plants. Overall, cisgenic products are perceived to be more natural, less problematic for the environment and generally safer and more promising. Another major premise for the development of NPPs is the regulation of GM plants. Cisgenic NPPs could fall outside the definitions of GMPs in the United States (according to USDA regulations) and in some case in Australia. Thus, in these countries they are not subjected to regulatory oversight beyond that applied to other conventionally bred plants (Lusser and Davies, 2013).

The 'breeders' gene pool' is the total of all genes, or genetic information, in any population that can be used by conventional breeders to improve their crops. Breeders might need to overcome barriers to gene transfer depending on the source of the genes used. The primary gene pool comprises species that interbreed freely with the plant of interest. The secondary gene pool includes species that cross-breed with the plant of interest, but only with difficulty, and which produce at least some fertile hybrids. The tertiary gene pool comprises species that are more distantly related with the plant of interest, and which cross-breed solely by using advanced techniques such as embryo rescue, induced polyploidy and bridge crosses.

In a 2010 report for the EU Commission, Directorate L-Science, Economy and Society, the public acceptance of cisgenic crops was analysed. In contrast to perceptions of the indices of transgenic and cisgenic apples, it was shown that, across EU 27, 55% supported cisgenesis, some 22% more than those who supported transgenics. As it was found, cisgenic apples with resistance to scab, canker and mildew diseases were more positively perceived on all the indices, while people felt less uneasy than with transgenic apples (Gaskell et al., 2010).

Pastoral Genomics, New Zealand, has registered the trademark Cisgenics® and uses this trademark for their future GM ryegrass (Bajaj et al., 2008, 2010). The

DNA introduced in the modified plants is not completely in agreement with the cisgenesis definition, since the plant-derived P-DNA borders have been introduced using a vector backbone of bacterial (*Agrobacterium*) origin, so that in some countries a definition of intragenesis may be applied.

In EU, the field trials on the cisgenic potato line 'Modena', (AV-43-6-G7) producing higher amounts of amylopectin, developed by Avebe, a Dutch company, and transferred to BASF, with a pending approval process, has ceased in Germany and Czech Republic, and the field trial was moved to the United States. The notification for the field release of the 'Modena' potato in 2007 (B/NL/07/04) included a potato with the *GBSS* terminator instead of the *nos*-terminator, being thus a complete cisgenic line. This cisgenic potato line will remain in field trials in the Netherlands until 2015.

Furthermore, advances have been made in the direction of transformation of plants difficult to be transformed or producing very few transformants. *Agrobacterium*-mediated transformation is the most widely used technique for generating transgenic plants. However, many crops remain recalcitrant. An *Arabidopsis* myb family transcription factor (MTF1) inhibited plant transformation susceptibility. The Gelvin team showed that cytokinins from *Agrobacterium tumefaciens* decreased the expression of MTF1 through activation of the cytokinin response regulator ARR3 (Sardesai et al., 2013). AT14A encodes a transmembrane receptor for cell adhesion molecules. Plants overexpressing AT14A offer a strong anchoring mechanism to Agrobacteria. Thus, promoting bacterial attachment and transformation of recalcitrant plant species or treating roots with cytokinins may increase transformation efficiency with a higher number of positive plantlets.

1.2.6 Modification of trees, stone fruit trees and plants with long vegetative periods

Senesco proprietary technology (http://www.senesco.com) has been extensively applied to produce transgenic plants and trees in the United States, Israel and Australia, countries with less strict GM regulation. Senesco patented a method able to regulate polyamine modification of eIF5A, an Eukaryotic Translation Initiation Factor that acts at a critical point in the process of cell death; eIF5A is a partner of proteins involved in endoplasmic reticulum (ER)-to-Golgi vesicle transport of the ribosome-bound nascent proteins (co-translational translocation). A block in eIF5A activation up-regulates stress-induced chaperones. By modulating the expression of eIF5A they obtained an increase in plant size, in seed yield, in shelf life of perishable products and in growth rates, with a reduction of harmful effects of environmental stresses. Senesco's present portfolio includes companies dealing with production of banana, corn and soybean, forestry species (cold-adapted eucalyptus trees), alfalfa, turfgrass, ornamental bedding plants and plants for ethanol production.

FuturaGene, a Brazilian forestry company, signed recently an agreement with the Donald Danforth Plant Science Center, in Saint Louis, to exploit their agbiotech technology, already tested in GM eucalyptus and poplar, to boost plant biomass levels, improve crop adaptation to climate changes and facilitate processing for animal feed in strategic crops. The technology is based on the endo-β-1,4-glucanase *CEL1* gene from *Arabidopsis*, with activity in cell wall metabolism. Expressing the gene produce a relaxation of the crystalline matrix of cells facilitating cell expansion. The regulatory trails are envisaged for eucalyptus (ongoing), the model grass such as setaria, and subsequently millet, sorghum and cassava. The FuturaGene scientific advisory board is chaired by Marc Van Montagu, recent winner of the 2013 World Food Prize.

In wood tree, fruit tree and stone fruit tree research, the major difficulty is to shorten the long juvenile phase to accelerate the selection of recombinant clones, where phenotypes can be observed only at flowering or fruiting time (Gambino and Gribaudo, 2012). The incorporation of flowering genes may result in reduction of generation time and in juvenile phenotypes. For example, *BpMADS4* has been overexpressed in birch (*Betula pendula* Roth.), *APETALA1* (*AP1*) and *LEAFY* genes have been expressed in citrus. *TERMINAL FLOWER* (*MdTFL1*) gene from apple, an inhibitor of flowering that maintains the identity of inflorescence shoot meristems, was introduced as an RNAi cassette targeting the native pear genes *PcTFL1-1* and *PcTFL1-2* (Freiman et al., 2012), producing a transgenic line of Spadona variety, Early Flowering Spadona (EF-Spa). Pollination of EF-Spa trees generated normal-shaped fruits with viable seeds. F1 seedlings formed shoots and produced flowers 1–33 months after germinations, and were found devoid of T-DNA, being thus non-transgenic. The *FLOWERING LOCUS T 1* (*MdFT1*) gene from apple was positively used to produce an early flowering phenotype in apple (Kotoda et al., 2010), in pear or in poplar, without adverse pleiotropic effects such as morphological changes in flowers (Gambino and Gribaudo, 2012).

Exploiting the properties of modification of tree life cycle, genetically engineered resistance to *Plum pox virus* (causative agent of Sharka disease) infection was induced in stone fruit trees (peach, plum and apricot) (Ilardi and Di Nicola-Negri, 2011). In the frame of COST (European Cooperation in Science and Technology) action FA0806: 'Plant virus control employing RNA-based vaccines: a novel non-transgenic strategy', research is being carried on through plant virus control using gene silencing approaches. This will produce efficient and cost-effective methods for reactive and proactive response to viral diseases of plants. Specific primary objectives are to develop novel non-transgenic control strategies for managing plant viral diseases, and to optimize protocols for high-throughput production and delivery of suitable resistance inducer molecules.

1.2.7 Product-based regulatory frameworks

Canada and the United States opted to regulate all plants or products with new traits developed either through genetic engineering or any other plant breeding techniques under the same, yet existing, regulatory system. The transformation techniques were not considered inherently risky. Therefore, the focus of product-based regulatory systems is on the risks of products and new traits or attributes introduced into a plant, rather than the method of production.

New biotechnology-based plant breeding techniques and their derived products raise several regulatory challenges, as they do not necessarily fit into known product definitions, regulatory frameworks and risk assessment approaches for GMPs. Regulators and policy makers will have to decide whether NPPs are actually GMPs as categorized by standard definitions. If they were to be classified as GMPs, it raises the question of whether product definitions should be modified to take into account these new techniques and any future advances in plant breeding methods. It also raises the question of whether the regulatory frameworks and risk assessment approaches implemented for GMPs provide a sustainable and proportionate approach for the regulation and safety assessment of NPPs. In addition, regulators and policy makers have to consider whether the frameworks put in place provide an optimal balance between policy objectives, international harmonization and equal regulatory oversight for different products that raise similar safety concerns.

The first challenge is a regulatory frameworks fit for purpose. Frameworks that use process-based definitions as a trigger for regulatory oversight might not be functional over time. New biotechnology-based plant breeding techniques might rapidly outgrow the definitions for GMPs.

The second challenge is to ensure that regulatory frameworks and risk assessment approaches for NPPs remain proportional to the level of risk that these plants might pose to human and animal health and the environment. The genomic changes are often similar to those obtained by conventional breeding, such that the end products are indistinguishable from conventionally bred plants. From a product-based perspective, the risk profile of certain NPPs resembles conventionally bred plants. In 2012, the EFSA published a scientific opinion to address the safety assessment of plants developed through cisgenesis and intragenesis. The EFSA Panel on GMO concluded that cisgenic and conventionally bred plants represent similar hazards, whereas intragenic and transgenic plants could raise new hazards.

The EFSA GMO Panel considered that its existing risk assessment guidelines for plants and products developed through transgenesis would generally apply to cisgenic and intragenic plants, but that it would require less event-specific data, depending on the specific case (EFSA, 2012a).

The third challenge is to develop regulatory frameworks and risk assessment practices able to stimulate the innovation required to meet other policy objectives such as food security, economic development and building consumer trust. The new biotechnological plant breeding techniques offer the opportunity of economic growth and company development, while support and improve consumer attitudes. In addition, the European Commission (EC) has commissioned several reports on specific aspects of new biotechnological plant breeding techniques. In the United States, the USEPA has opened up a debate on a draft rule that would exempt certain cisgenic organisms from registration as GMOs.

The fourth challenge is the international harmonization of regulatory frameworks for NPPs. So far, there is a lack of consensus on detailed product definitions for NPPs. At an international workshop organized by the EC's Joint Research Centre, it became clear that the definition of a GMO differs between jurisdictions and that this determines whether or not NPPs are classified as GMO. Products generated through targeted mutation, e.g. will probably be considered as non-GMO in many countries, although they might be defined as a GMO in other jurisdictions. The US DA does not have the authority to oversee cisgenic plants created without the help of a plant pathogen, whereas the Australian Office of the Gene Technology Regulator considered that certain cisgenic plants might not be regulated. EU member states and the EC are also considering developments in plant breeding and discussing whether new biotechnology-based plant breeding techniques would be included or excluded from the existing definition of GMOs (Lusser and Davies, 2013).

The fifth challenge is the need to avoid disparities in risk assessment between products with equal potential to cause harm. Conventionally bred products can raise safety concerns similar to those for their transgenic counterparts (Podevin et al., 2012). The Canadian legislative approach enables consistent evaluation of plants with similar new traits, irrespective of the techniques used. Proportionate risk assessment practices requires regulatory requirements for NPPs based on risk assessments associated with the plant species, traits, receiving environments and intended uses, and the combination of these characteristics, rather than the production method itself.

References

Auer, C., 2011. Small RNAs for crop improvement. In: Erdmann, V.A., Barciszewski, J. (Eds.), Non Coding RNAs in Plants. Springer-Verlag, Berlin, pp. 461–484.

Ayar, A., Wehrkamp-Richter, S., Laffaire, J.-B., Le Goff, S., Levy, J., Chaignon, S., et al., 2013. Gene targeting in maize by somatic ectopic recombination. Plant Biotechnol. J. 11, 305–314.

Bajaj, S., Puthigae, S., Templeton, K., Bryant, C., Gill, G., Lomba, P., et al., 2008. Towards engineering drought resistance in perennial ryegrass using its own genome. Sixth Canadian Plant Genomics Workshop, Toronto, 26-27/06/2008, Abstract, p. 62.

Bajaj, S., Puthigae, S., Bryant, C., Whittaker, D., Elborough, K., Hanley, Z., 2010. Ryegrass genes- a potential source for improving rice and grass species. AsPac. J. Mol. Biol. Biotechnol. 8, 71–73.

Belhaj, K., Chaparro-Garcia, A., Kamoun, S., Nekrasov, V., 2013. Plant genome editing made easy: targeted mutagenesis in model and crop plants using the CRISPR/Cas system. Plant Methods. 9, 39.

Breyer, D., Herman, P., Brandenburger, A., Gheysen, G., Remult, E., Soumillion, P., et al., 2007. Genetic modification through oligonucleotide-mediated mutagenesis. A GMO regulatory challenge? Environ. Biosafety. Res. 8, 57–64.

Carriere, Y., Crowder, D.W., Tabashnik, B.E., 2010. Evolutionary ecology of insect adaptation to Bt crops. Evol. Appl. 3, 561–573.

Chen, H., Lin, Y.J., 2013. Promise and issues of genetically modified crops. Curr. Opin. Plant. Biol. 16, 255–260.

Chen, K., Gao, C.X., 2014. Targeted genome modification technologies and their applications in crop improvements. Plant Cell Rep. 33 (4), 575–583 (advanced online). Available from: http://dx.doi.org/10.1007/s00299-013-1539-6.

Chen, Z., Jaafar, L., Agyekum, D.G., Xiao, H., Wade, M.F., Kumaran, R.I., et al., 2013. Receptor-mediated delivery of engineered nucleases for genome modification. Nucleic Acids Res. 41 (19), e182.

Deng, D., Yan, C., Pan, X., Mahfouz, M., Wang, J., Zhu, J.K., et al., 2012. Structural basis for sequence-specific recognition of DNA by TAL effectors. Science. 335, 720–723.

Dhurua, S., Gujar, G.T., 2011. Field-evolved resistance to Bt toxin Cry1Ac in the pink bollworm, *Pectinophora gossypiella* (Saunders) (Lepidoptera: *Gelechiidae*), from India. Pest Manag. Sci. 67, 898–903.

EFSA Panel on Genetically Modified Organisms (GMO), 2012a. Scientific opinion addressing the safety assessment of plants developed through cisgenesis and intragenesis. EFSA J. 10, 2561.

EFSA Panel on Genetically Modified Organisms (GMO), 2012b. Scientific opinion addressing the safety assessment of plants developed using zinc finger nucleases and other site-directed nucleases with similar function. EFSA J. 10, 2943. Available from: http://dx.doi.org/doi:10.2903/j.efsa.2012.2943.

Endo, M., Toki, S., 2014. Toward establishing an efficient and versatile gene targeting system in higher plants. Biocatal. Agric. Biotechnol. (advanced online). Available from: http://dx.doi.org/10.1016/j.bcab.2013.10.002.

EPEC, 2011. European Policy Evaluation Consortium Evaluation of the EU Legislative Framework in the Field of Cultivation of GMOs under Directive 2001/18/EC and Regulation (EC) No 1829/2003, and the Placing on the Market of GMOs as or in Products under Directive 2001/18/EC. Available from: http://ec.europa.eu/food/food/biotechnology/evaluation/index_en.htm.

Fauser, F., Roth, N., Pacher, M., Ilg, G., Sánchez-Fernández, R., Biesgen, C., et al., 2012. Proc. Natl. Acad. Sci. USA. 109, 7535–7540.

Freiman, A., Shlizerman, L., Golobovitch, S., Yablovitz, Z., Korchinsky, R., Cohen, Y., et al., 2012. Development of a transgenic early flowering pear (*Pyrus communis* L.) genotype by RNAi silencing of *PcTFL1-1* and *PcTFL1-2*. Planta. 235, 1239–1251.

Gaj, T., Gersbach, C.A., Barbas, C.F., 2013. ZFN, TALEN, and CRISPR/Cas-based methods for genome engineering. Trends Biotechnol. 31, 397–405.

Gambino, G., Gribaudo, I., 2012. Genetic transformation of fruit trees: current status and remaining challenges. Transgenic Res. 21, 1163–1181.

Gaskell, G., et al., 2010. Europeans and Biotechnology in 2010. Winds of Change? EU Commission Brussels. A report to the European Commission's Directorate-General for Research. Available from: http://ec.europa.eu/public_opinion/archives/ebs/ebs_341_winds_en.pdf.

Holme, I.B., Wendt, T., Holm, P.B., 2013. Intragenesis and cisgenesis as alternatives to transgenic crop development. Plant Biotechnol. J. 11, 395–407.

Ilardi, V., Di Nicola-Negri, E., 2011. Genetically engineered resistance to *Plum pox virus* infection in herbaceous and stone fruit hosts. GM Crops. 2, 24–33.

Jiang, W., Zhou, H., Bi, H., Fromm, M., Yang, B., Weeks, D.P., 2013. Demonstration of CRISPR/Cas9/sgRNA-mediated targeted gene modification in Arabidopsis, tobacco, sorghum and rice. Nucleic Acids Res. 41 (20), e188.

Kotoda, N., Hayashi, H., Suzuki, M., Igarashi, M., Hatsuyama, Y., Kidou, S., et al., 2010. Molecular characterisation of flowering locus T-like genes of apple (*Malus x domestica* Borkh.). Plant Cell Physiol. 51, 561–575.

Lusser, M., Davies, H.V., 2013. Comparative regulatory approaches for groups of new plant breeding techniques. N. Biotechnol. 30, 437–446.

Lusser, M., Parisi, C., Plan, D., Rodríguez-Cerezo, E., 2011. New Plant Breeding Techniques: State-of-the-Art and Prospects for Commercial Development. Publications Office of the European Union, Luxembourg. Available from: http://ftp.jrc.es/EURdoc/JRC63971.pdf.EUR.24760.

Lusser, M., Parisi, C., Plan, D., Rodríguez-Cerezo, E., 2012. Deployment of new biotechnologies in plant breeding. Nat. Biotechnol. 30, 231–239.

Mak, A.N., Bradley, P., Cernadas, R.A., Bogdanove, A.J., Stoddard, B.L., 2012. The crystal structure of TAL effector PthXo1 bound to its DNA target. Science. 335, 716–719.

Mak, A.N., Bradley, P., Bogdanove, A.J., Stoddard, B.L., 2013. TAL effectors: function, structure, engineering and applications. Curr. Opin. Struct. Biol. 23, 93–99.

Mariano, A., Xu, L., Han, R., 2014. Highly efficient genome editing via 2A-coupled co-expression of two TALEN monomers. BMC Res. Notes, 7, 628. Available from: http://dx.doi.org/10.1186/1756-0500-7-628.

Nekrasov, V., Staskawicz, B., Weigel, D., Jones, J.D.G., Kamoun, S., 2013. Targeted mutagenesis in the model plant *Nicotiana benthamiana* using Cas9 RNA-guided endonuclease. Nat. Biotechnol. 31, 691–693.

Pauwels, K., Podevin, N., Breyer, D., Carroll, D., Herman, P., 2014. Engineering nucleases for gene targeting: safety and regulatory considerations. Nat. Biotechnol. 31, 18–27.

Podevin, N., Davies, H.V., Hartung, F., Nogué, F., Casacuberta, J.M., 2013. Site-directed nucleases: a paradigm shift in predictable, knowledge-based plant breeding. Trends Biotechnol. 31, 375–383.

Podevin, N., Devos, Y., Davies, H.V., Nielsen, K.M., 2012. New biotechnology-based plant breeding techniques and the regulatory landscape. EMBO Rep. 13, 1057–1061.

Sardesai, N., Lee, L.-Y., Chen, H., Yi, H.C., Olbricht, G.R., Stirnberg, A., et al., 2013. Cytokinins secreted by agrobacterium promote transformation by repressing a plant myb transcription factor. Sci. Signal. 6 (302), ra 100.

Shan, Q., Wang, Y., Li, J., Zhang, Y., Chen, K., Liang, Z., et al., 2013. Targeted genome modification of crop plants using a CRISPR-Cas system. Nat. Biotechnol. 31, 686–688.

Tabashnik, B.E., Wu, K., Wu, Y., 2012. Early detection of field-evolved resistance to Bt cotton in China: cotton bollworm and pink bollworm. J. Invertebr. Pathol. 110, 301–306.

Tait, J., Barker, G., 2011. Global food security and the governance of modern biotechnologies. EMBO Rep. 12, 763–768.

Tsuji, S., Futaki, S., Imanishi, M., 2013. Creating a TALE protein with unbiased 5′-T binding. Biochem. Biophys. Res. Commun. 441, 262–265.

Upadhyay, S., Kumar, J., Alok, A., Tuli, R., 2013. RNA-guided genome editing for target gene mutations in wheat. G3 (Genes, Genomes, Genetics). 3, 2233–2238.

Van de Bulcke, M., Lievens, A., Barbau-Piednoir, E., Mbongolo Mbella, G., Roosens, N., Sneyers, M., et al., 2010. A theoretical introduction to "combinatory SYBRGreen qPCR screening", a matrix-based approach for the detection of materials derived from genetically modified plants. Anal. Bioanal. Chem. 396, 2113–2123.

Xie, K., Yang, Y., 2013. RNA-guided genome editing in plants using a CRISPR-Cas system. Mol. Plant. 6, 1975–1983.

Zhang, Y.-C., Yu, Y., Wang, C.-Y., Li, Z.-Y., Liu, Q., Liao, J.-Y., et al., 2013. Overexpression of microRNA OsmiR397 improves rice yield by increasing grain size and promoting panicle branching. Nat. Biotechnol. 31, 848–852.

What turns on and off the cytokinin metabolisms and beyond

2

Eva Jiskrová, Ivona Kubalová and Yoshihisa Ikeda
Department of Molecular Biology and Centre of Region Hana for Biotechnological and Agricultural Research, Faculty of Science, Palacky University, Olomouc, Czech Republic

Acronyms

AHP *ARABIDOPSIS* HISTIDINE PHOSPHOTRANSFER PROTEIN
CKX CYTOKININ OXIDASE
CYP735A CYTOCHROME P450 FAMILY 735 SUBFAMILY A
DST DROUGHT AND SALT TOLERANCE
ESR2 ENHANCER OF SHOOT REGENERATION 2
HK HISTIDINE KINASE
IPT ISOPENTENYL TRANSFERASE
MtRR1 *MEDICAGO TRUNCATULA* RESPONSE REGULATOR 1
PHB PHABULOSA
SHR SHORT-ROOT
STM SHOOT MERISTEMLESS
VLCFA very long chain fatty acid

2.1 Introduction

For all living organisms, regardless of their kingdom, the ability to sense and to adapt themselves to the changing environment is fundamental to their survival and propagation. Over time organisms have developed mechanisms to perceive and to respond to fluctuating surroundings. It is particularly crucial for sessile organisms such as plants to develop tissue- and cell-plasticity to cope with fluctuating environmental conditions. Their above-ground parts are structured to effectively capture light and CO_2 as a whole and in the cellular level several photoreceptors sensing the changing light conditions have been evolved. Their below-ground parts are specialized for the absorption and adaptation to the fluctuating availabilities of water and inorganic minerals. The local response to minerals in below-ground parts is then systemically transmitted to the other parts of the body and phytohormones are likely to act as candidate transmitters in some cases (Vieten et al., 2007; Sakakibara et al., 2006). For instance, the nitrogen response is mediated by cytokinins (Sakakibara et al., 2006).

Applied Plant Genomics and Biotechnology.

It is interesting to note that prokaryotes use two-component module for sensing changes inside and outside of their cell (Wuichet et al., 2010), whereas plants utilize the same system for sensing internal signals, cytokinins (Inoue et al., 2001). Cytokinin actions in plant development and differentiation (Perilli et al., 2010, Bianco et al., 2013), mechanisms of signal transduction pathway (Argueso et al., 2010; Schaller et al., 2011), enzymatic properties of biosynthesis and breakdown of cytokinins (Galuszka et al., 2000; Frebort et al., 2011) and crosstalks with the other hormones (El-Showk et al., 2013) are reviewed elsewhere. The objective of this chapter is to summarize the factors having an impact on cytokinin biosynthesis and degradation at the transcriptional level. The role of all the biosynthesis and the breakdown of active cytokinins is converged to either turn on or off the activity of differentially expressed cytokinin receptors. Understanding the spatial and temporal transcription regulation of genes involved in cytokinin metabolism will reveal the intrinsic plasticity of plant cells to cope with surrounding environments – inside and outside of the cell, for local and systemic communication. At last, we discuss the potential application by engineering the expression of cytokinin metabolism genes for coping with a challenging environment.

2.2 Regulation of cytokinin biosynthesis

A major cytokinin biosynthesis is comprised of two major pathways – tRNA and adenosine monophosphate (AMP). The rate-limiting initial step for isoprenoid cytokinin biosynthesis is catalyzed by adenosine phosphate-isopentenyltransferase (IPT), producing N^6-(Δ^2-isopentenyl) adenine (iP) riboside 5′-tri, di- or monophosphate (iPRTP, iPRDP or iPRMP, respectively). The next step is the hydroxylation of these molecules to *trans*-zeatin (tZ) riboside 5′-tri-, di- or monophosphate (tZRTP, tZRDP or tZRMP, respectively). These reactions are catalyzed by a cytochrome P450 monooxygenase, CYP735A. Additional two-step reactions, catalyzed by nucleotidase and nucleosidase, are required to produce active cytokinins that are free-base form, such as iP and tZ (Chen and Kristopeit, 1981a,b). The rice gene LONELY GUY (LOG) has been demonstrated to have phosphoribohydrolase activity, being able to directly convert iPRMP or tZRMP to iP or tZ, respectively (Kurakawa et al., 2007).

Due to its instability or low abundance in plant cells, *IPT* genes were originally identified in gall-forming phytopathogenic bacteria and extensively studied in last century. In 2001, two Japanese groups have independently reported *IPT* genes in *Arabidopsis* through *in silico* analysis. Biochemical activities and *in vivo* gain-of-function analysis were also carried out (Takei et al., 2001; Kakimoto, 2001). Their expressions differ spatially and temporally (Takei et al., 2001; Kakimoto, 2001).

2.2.1 *The role of cytokinins in plant development*

Back in 1957, cytokinin has been implicated to play an important role in controlling shoot apical meristem (SAM) activity (Skoog and Miller, 1957). Class I *KNOTTED* genes were shown to induce *AtIPT* expression. In rice, *OsIPT2* and *OsIPT3* were induced by OSH1 (Sakamoto et al., 2006). In *Arabidopsis*, *AtIPT7* transcript was up-regulated by SHOOT MERISTEMLESS (STM) function (Yanai et al., 2005). Transcript of *AtIPT7* was also directly up-regulated by HD ZIPIII transcription factor, PHABULOSA (PHB). Expression of *AtIPT7* is increased in *phb-1d*, gain-of-function mutant, in which the expression of *PHB* mRNA was increased and broadened due to its insensitivity to microRNA-mediated post-transcriptional repression, whereas its expression was reduced in the *phb-13;phv-11* loss-of-function mutant. Direct activation of *AtIPT7* transcription was confirmed by chromatin immunoprecipitation (ChIP) analysis by using miRNA-insensitive version of *PHB* fused to *GFP*, driven under the control of its own *PHB* promoter (Dello Ioio et al., 2012).

The gain-of-function activation-tagging mutant, *sob5-D* (*SUPPRESSOR OF PHYTOCHROME B DOMINANT*) – caused by over-expressing plant specific proteins in previously uncharacterized family – displayed phenotypes similar to those obtained by *Agrobacterium ipt* over-expression. The elevated expression of *AtIPT3* and *AtIPT7* was observed in a tissue-specific manner (Zhang et al., 2006).

In epidermis the very long chain fatty acids (VLCFAs) are synthesized and used for cuticular wax formation. Defects in VLCFA synthesis exhibited a variety of phenotypes, ranging from the deformation of cuticles resulting in the alternation of pathogen–plant interaction (Reina-Pinto and Yephremov, 2009) and post-embryonic organ fusion (Sieber et al., 2000, Chen et al., 2003) to retarded plant growth with abnormal morphology (Panikashvili et al., 2007). A series of recent works showed that the decrease of VLCFA either by inhibitor treatments or by the mutation resulting in defects in VLCFA content could increase expression of *AtIPT3* in *Arabidopsis* (Nobusawa et al., 2013a,b). The expression of *AtIPT3* in wild-type vascular bundles was extended to the surrounding tissues by treatment with a VLCFA synthesis inhibitor, cafenstrole. Active cytokinin contents were increased by cafenstrole treatments and in the *pasticcino2* (*pas2*) mutant, in which a gene encoding *3-hydroxy acyl-CoA dehydratase* (*HCD*) involved in VLCFA synthesis was disrupted (Bach et al., 2008; Nobusawa et al., 2013a,b).

The relationship between auxin and cytokinin has been extensively studied. Auxin treatment induced *IPT5* and *IPT7* expression in roots within 4 h (Miyawaki et al., 2004). *De novo* organogenesis in an *in vitro* tissue culture is a well-established system to study these hormones. The *AUXIN RESPONSE FACTOR 3* (*ARF3*) that was considered to be a transcriptional repressor (Guilfoyle and Hagen, 2007) was shown to negatively regulate *AtIPT5* expression during shoot induction by directly binding to the *AtIPT5* promoter (Cheng et al., 2013). The expression of *AtIPT5*, as well as *AtIPT3* and *AtIPT7*, was mis-expressed in *arf3* (Cheng et al., 2013).

2.2.2 Nitrogen and cytokinin metabolism

Environmental cues are known to have an impact on the regulation of cytokinin metabolisms. Drew (1975) showed that a nutrient-rich 'patch' could elicit a localized increase in lateral root initiation and elongation in barley. Later on the same response was observed in *Arabidopsis* (Zhang and Forde, 1998). The authors also showed that nitrate itself, rather than its downstream metabolites, appeared to be the stimulus signal for the lateral root elongation (Zhang and Forde, 1998). Nitrate ions and N-assimilation products (i.e. amino acids) were also known to act as long-range signalling molecules so that plants can respond and adjust themselves to the fluctuating nutrient availability in soil (Sakakibara et al., 2006; Ruffel et al., 2011). Nitrogen sensing, in particular, draw the attention of hormone researchers since auxin, abscisic acid and cytokinin signalling pathways lie downstream of nitrate response pathway and nitrogen availability could control cytokinin biosynthesis (Malamy, 2005).

Under nitrogen-limited condition, iP-type and tZ-type CKs in the xylem of roots and shoots were found to accumulate in response to NO_3^- supply. This is due to the rapid up-regulation of *AtIPT3* by NO_3^- supply and the NO_3^--dependent accumulation of the above-mentioned molecules was substantially diminished in *Atipt3-1* mutant shoots and roots (Takei et al., 2004). *De novo* protein synthesis was not required for the rapid up-regulation of *AtIPT3* (Miyawaki et al., 2004). It is also noted that in the *Arabidopsis chl3-5/nia1-2* nitrate reductase (NR)-null mutant, in which *NIA1* and *NIA2* genes are disrupted, the up-regulation of *AtIPT3* by NO_3^- supply is still responsive. The result suggests that nitrate itself serves as a cue to induce *AtIPT3* expression (Wang et al., 2004). The expression of *AtIPT5* remained unchanged in the short-term NO_3^- supply, however, the authors found a positive correlation between *AtIPT5* transcript accumulation and long-term nitrogen availability of both NO_3^- and NH_4^+ (Takei et al., 2004). An alternative pathway controlling the temporal expression of *IPT* genes is suggested. Nitrate treatment induced *CYP735A2* expression only in roots (Wang et al., 2004). The same response was observed in rice, as well. Exogenous application of nitrate and ammonium stimulated cytokinin accumulation both in roots and shoots. The expression of *OsIPT4*, *OsIPT5*, *OsIPT7* and *OsIPT8* were responsive to exogenously applied nitrate and ammonium (Kamada-Nobusada et al., 2013). Interestingly their expression by nitrate and ammonium application was impaired by L-methionine sulfoximine, a potent glutamine synthase inhibitor, suggesting that glutamine or its related metabolite, rather than nitrate and ammonium, is the inducer (Kamada-Nobusada et al., 2013). Long-term application of Nitrogen fertilizer (urea) into soil induced *OsIPT* expression in rice panicles, which was found to be responsible for increased cytokinin content (Ding et al., 2014). Ruffel et al. (2011) extended the research towards the better understanding of the relationship between cytokinin and nitrogen signalling by employing genetic and genomic approaches in a 'split-root' framework. They corroborated the specific role of cytokinin as a part of systemic nitrogen signalling in plants as a whole and claimed that the cytokinin is a crucial component of a root-to-shoot signalling relay mechanism involved in conveying the NO_3^- status.

2.2.3 Defence response, development and cytokinin

The relationship between cytokinin and plant defence was reported in the study of *Arabidopsis* semidominant mutant, *uni-1D*, in which coiled-coil nucleotide-binding leucine-rich-repeat protein was mutated (Igari et al., 2008). The mutation caused the constitutive activation of the protein. It belongs to the disease-resistant protein family involved in pathogen recognition. However, the *uni-1D* plants exhibited stunted growth, bushy architecture and delayed senescence. A part of morphological phenotypes such as the formation of ectopic meristems in *uni-1D/+* was attributed to elevated cytokinin contents. The up-regulation of *CYP735A2* transcripts was also observed in *uni-1D/+*. Changes in gene expression of cytokinin biosynthesis are summarized in Table 2.1.

2.2.4 Regulatory mechanism of cytokinin degradation

An alternative way of fine-tuning cytokinin contents is an irreversible breakdown of active cytokinins, catalyzed by cytokinin oxidase/dehydrogenases (CKX; EC 1.5.99.12). The enzyme was considered to be copper-containing amine oxidase (Hare and Vanstaden, 1994). Later on, a genuine nature of the enzyme was revealed to be flavin adenine dinucleotide (FAD) containing dehydrogenase (Bilyeu et al., 2001; Galuszka et al., 2001). Several differential expression pattern and distinct subcellular localization have been reported so far (Miyawaki et al., 2004; Werner et al., 2003; Werner et al., 2006).

Cytokinin is known to regulate root growth and development (Bianco et al., 2013; Miyashima et al., 2013). Cytokinin perception and primary signalling are known to regulate vascular tissue differentiation and its patterning (Mahonen et al., 2006). The *wooden leg* (*wol*) allele of cytokinin receptor, *CYTOKININ RESPONSE 1* (*CRE1*) and the triple-knockout mutant for all three CRE-family receptor genes (*CRE1/AHK4*, *AHK2* and *AHK3*) perturbed periclinal procambial cell division that is required for the proliferation of vascular cell files.

Genetic studies in the root patterning have revealed several important players, *SHORT-ROOT* (*SHR*), involved in controlling both endodermis specification and asymmetric cell division (Helariutta et al., 2000) and *SCARECROW* (*SCR*), required for asymmetric cell division (Di Laurenzio et al., 1996). Both genes belong to GRAS family transcription factors and act in a non cell-autonomous manner. SHR proteins are generated in the stele, move to the endodermis and quiescent centre (QC), where *SCR* is expressed. In the endodermis where both SHR and SCR are present they activate the expression of two microRNAs, *miR165a* and *miR166b*. When *miR165/miR166* are transcribed in the endodermis, they travel to the vascular cylinder and degrade mRNAs of *HD-ZIP III* transcription factor members in the endodermis and periphery of vascular cylinder (Carlsbecker et al., 2010).

Genome-wide direct target analysis of SHR identified *AtCKX3* (Cui et al., 2011). Cytokinin content was elevated both in *shr-2* root and shoot and it was due to the decreased expression of *AtCKX3*. SHR proteins occupy the promoter region of *AtCKX3*, directly repress its transcription and control vascular patterning.

2.2.5 Transcription factors regulating cytokinin metabolisms

Several transcription factors were reported as regulators of the expression of cytokinin metabolism genes. *AtCKX6* was also identified as direct target of *ESR2*

Table 2.1 **Summary of effector and its effect on the regulation of genes related to cytokinin metabolism**

Organism	Gene	Effector	Effect	Reference
Arabidopsis	*IPT1*	in *atmyb2* background	Up-regulation	Guo and Gan, 2011
	IPT3	in *pas2* background	Up-regulation	Nobusawa et al., 2013a
		VLCFA synthesis inhibitor cafenstrole	Up-regulation	Nobusawa et al., 2013b
		NO_3^-	Up-regulation	Takei et al., 2004; Wang et al., 2004
		in *arf3* background	Up-regulation	Cheng et al., 2013
		in *sob5-D* background	Up-regulation	Zhang et al., 2006
	IPT4	in *atmyb2* background	Up-regulation	Guo and Gan, 2011
	IPT5	long-term nitrogen availability of NO_3^- and NH_4^+	Up-regulation	Takei et al., 2004
		in *arf3* background	Up-regulation	Cheng et al., 2013
		in *atmyb2* background	Up-regulation	Guo and Gan, 2011
	IPT6	in *atmyb2* background	Up-regulation	Guo and Gan, 2011
	IPT7	STM	Up-regulation	Yanai et al., 2005
		PHB	Up-regulation	Dello loio et al., 2012
		in *arf3* background	Up-regulation	Cheng et al., 2013
		in *sob5-D* background	Up-regulation	Zhang et al., 2006
	IPT8	in *atmyb2* background	Up-regulation	Guo and Gan, 2011
	CYP735A2	in *uni-1D/+* background	Up-regulation	Uchida et al., 2011
		cytokinin	Up-regulation	Bhargava et al., 2013
	UGT76C2	cytokinin	Up-regulation	Bhargava et al., 2013
	CKX2	in *iku1* and *iku2* background	Down-regulation	Li et al., 2013a
		crossing in high paternal to maternal genome ratio	Down-regulation	Tiwari et al., 2010
	CKX3	cytokinin	Up-regulation	Bhargava et al., 2013
		SHR	Down-regulation	Cui et al., 2011
	CKX4	cytokinin	Up-regulation	Bhargava et al., 2013
		Auxin	Down-regulation	Goda et al., 2004

	CKX5	cytokinin	Up-regulation	Bhargava et al., 2013
	CKX6	Auxin	Up-regulation	Yoshida et al., 2009
		ESR2	Up-regulation	Ikeda et al., 2006
	CKX7	cytokinin	Up-regulation	Bhargava et al., 2013
Rice	*IPTs*	long-term effect of urea	Up-regulation	Ding et al., 2014
	IPT2	OSH1	Up-regulation	Sakamoto et al., 2006
	IPT3	OSH1	Up-regulation	Sakamoto et al., 2006
	IPT4	auxin in shoot	Up-regulation	Tsai et al., 2012
		glutamine or its related metabolite	Up-regulation	Kamada-Nobusada et al., 2013
	IPT5	glutamine or its related metabolite	Up-regulation	Kamada-Nobusada et al., 2013
	IPT7	auxin in root	Down-regulation	Tsai et al., 2012
		glutamine or its related metabolite	Up-regulation	Kamada-Nobusada et al., 2013
	IPT8	glutamine or its related metabolite	Up-regulation	Kamada-Nobusada et al., 2013
		auxin in root	Up-regulation	Tsai et al., 2012
	CYP735A4	auxin in shoot	Down-regulation	Tsai et al., 2012
	CYP735A4	auxin in shoot	Down-regulation	Tsai et al., 2012
	CKX1	DST	Down-regulation	Huang et al., 2009
		cytokinin in root	Up-regulation	Tsai et al., 2012
	CKX2	cytokinin in shoot and root	Up-regulation	Tsai et al., 2012
		auxin in root	Up-regulation	Tsai et al., 2012
		DST	Down-regulation	Huang et al., 2009
	CKX3	cytokinin in shoot	Up-regulation	Tsai et al., 2012
		DST	Down-regulation	Huang et al., 2009
	CKX4	cytokinin in shoot and root	Up-regulation	Tsai et al., 2012
		DST	Down-regulation	Huang et al., 2009
	CKX5	cytokinin in root	Down-regulation	Tsai et al., 2012
	CKX6	cytokinin in shoot and root	Down-regulation	Tsai et al., 2012
	CKX8	cytokinin in shoot and root	Up-regulation	Tsai et al., 2012
		DST	Down-regulation	Huang et al., 2009
	CKX9	DST	Down-regulation	Huang et al., 2009
	CKX10	DST	Down-regulation	Huang et al., 2009

(*Continued*)

Table 2.1 (Continued)

Organism	Gene	Effector	Effect	Reference
	CKX11	cytokinin in root	Up-regulation	Tsai et al., 2012
	LOGL8	auxin in root	Up-regulation	Tsai et al., 2012
Zea Mays	*IPT1*	salinity stress in shoot and root (72 h)	Up-regulation	Vyroubalova et al., 2009
	IPT4	cytokinin in root (0.5 h)	Up-regulation	Vyroubalova et al., 2009
	IPT5	osmotic stress in shoot and root (72 h)	Up-regulation	Vyroubalova et al., 2009
		salinity stress in shoot (72 h)	Up-regulation	Vyroubalova et al., 2009
	IPT6	osmotic stress in root (0.5, 72 h)	Up-regulation	Vyroubalova et al., 2009
	IPT7	osmotic stress in root (0.5 h)	Up-regulation	Vyroubalova et al., 2009
		osmotic stress in root (72 h)	Down-regulation	Vyroubalova et al., 2009
	IPT8	osmotic stress in shoot (72 h)	Up-regulation	Vyroubalova et al., 2009
		salinity stress in shoot (72 h)	Up-regulation	Vyroubalova et al., 2009
	IPT9	osmotic stress in root (72 h)	Up-regulation	Vyroubalova et al., 2009
	IPT10	salinity stress in shoot (72 h)	Up-regulation	Vyroubalova et al., 2009
	CKX1	cytokinin in shoot (0.5, 72 h) and root (0.5, 72 h)	Up-regulation	Vyroubalova et al., 2009
		osmotic stress in shoot (72 h)	Up-regulation	Vyroubalova et al., 2009
		salinity stress in shoot (72 h)	Up-regulation	Vyroubalova et al., 2009
	CKX2	cytokinin in shoot (0.5, 72 h) and root (0.5, 72 h)	Up-regulation	Vyroubalova et al., 2009
		osmotic stress in root (0.5, 72 h)	Up-regulation	Vyroubalova et al., 2009
		salinity stress in shoot (72 h) and root (0.5, 72 h)	Up-regulation	Vyroubalova et al., 2009
	CKX3	cytokinin in shoot and root (72 h)	Down-regulation	Vyroubalova et al., 2009
		osmotic stress in root (72 h)	Up-regulation	Vyroubalova et al., 2009
		salinity stress in shoot and root (72 h)	Up-regulation	Vyroubalova et al., 2009
	CKX4	cytokinin in shoot (0.5, 72 h) and root (0.5, 72 h)	Up-regulation	Vyroubalova et al., 2009

		osmotic stress in shoot (72 h) and root (0.5 h)	Up-regulation	Vyroubalova et al., 2009
	CKX4b	cytokinin in shoot (0.5, 72 h) and root (0.5, 72 h)	Up-regulation	Vyroubalova et al., 2009
		osmotic stress in shoot (72 h) and root (0.5 h)	Up-regulation	Vyroubalova et al., 2009
		salinity stress in root (0.5 h)	Up-regulation	Vyroubalova et al., 2009
	CKX6	cytokinin in root (0.5 h)	Up-regulation	Vyroubalova et al., 2009
		salinity stress in shoot and root (72 h)	Up-regulation	Vyroubalova et al., 2009
	CKX8	cytokinin in shoot (0.5 h) and root (72 h)	Up-regulation	Vyroubalova et al., 2009
		osmotic stress in root (0.5 h)	Up-regulation	Vyroubalova et al., 2009
		salinity stress in shoot and root (72 h)	Up-regulation	Vyroubalova et al., 2009
	CKX9	cytokinin in shoot (72 h)	Up-regulation	Vyroubalova et al., 2009
		osmotic stress in shoot (72 h)	Up-regulation	Vyroubalova et al., 2009
		salinity stress in shoot (72 h)	Up-regulation	Vyroubalova et al., 2009
	CKX10	cytokinin in root (0.5, 72 h)	Up-regulation	Vyroubalova et al., 2009
		osmotic stress in shoot (72 h)	Up-regulation	Vyroubalova et al., 2009
		salinity stress in shoot (72 h)	Up-regulation	Vyroubalova et al., 2009
	CKX11	cytokinin in root (72 h)	Up-regulation	Vyroubalova et al., 2009
		osmotic stress in root (72 h)	Up-regulation	Vyroubalova et al., 2009
		salinity stress in shoot and root (72 h)	Up-regulation	Vyroubalova et al., 2009
	CKX12	cytokinin in root (0.5, 72 h)	Up-regulation	Vyroubalova et al., 2009
		salinity stress in shoot and root (72 h)	Up-regulation	Vyroubalova et al., 2009
Lupinus albus	*CKX*	in phosphorous deficient cluster roots	Up-regulation	O'Rourke et al., 2013
Medicago truncatula	*CKX1*	MtRR1	Up-regulation	Ariel et al., 2012
Triticum aestivum L.	*CKXs*	brassinosteroid	Down-regulation	Yuldashev et al., 2012

AHP, ARABIDOPSIS HISTIDINE PHOSPHOTRANSFER PROTEIN; ARF, AUXIN RESPONSE FACTOR; CKX, CYTOKININ OXIDASE; CYP735A, CYTOCHROME P450 FAMILY 735 SUBFAMILY A; DST, DROUGHT AND SALT TOLERANCE; ESR2, ENHANCER OF SHOOT REGENERATION 2; iku, haiku; IPT, ISOPENTENYL TRANSFERASE; MtRR1, MEDICAGO TRUNCATULA MT RR1 RESPONSE REGULATOR 1; pas2, pasticcino2; PHB, PHABULOSA; SHR, SHOOT-ROOT; sob5-D, suppressor of phytochrome b dominant; STM, SHOOTMERISTEMLESS; VLCFA, very long chain fatty acid.

(*ENHANCER OF SHOOT REGENERATION 2*) (Ikeda et al., 2006). Ectopic expression of *ESR2* in cultured *Arabidopsis* root explants conferred cytokinin-independent shoot regeneration (Ikeda et al., 2006), whereas loss-of-function *esr2* exhibited reduced shoot regeneration (Matsuo et al., 2011). *ESR2* acts downstream of the cytokinin receptor *CRE1* and directly up-regulates *CUP-SHAPED COTYLEDON1* (*CUC1*). Estradiol-mediated translocation of ESR2 combined with transcriptome also identified direct up-regulation of *AHP6* (Mahonen et al., 2006), encoding a pseudo-transmitter and acting as an inhibitor of two-component cytokinin primary pathway. It is likely that induced expression of *CKX6* and *AHP6* constitutes a negative feedback control.

The *atmyb2* exhibited branchy and bushy phenotypes with elevated levels of cytokinins. This is due to the increased and local expression of *AtIPT1, 4, 5, 6* and *8* in the stem. The expression of *AtCKX1* under the control of *AtMYB2* promoter in *atmyb2* restored *atmyb2* to wild-type phenotypes, suggesting that the observed phenotypes were caused by the elevated cytokinin level (Guo and Gan, 2011).

2.2.6 Regulation of grain yield and seed size by cytokinin

In rice, *Gn1a/OsCKX2* (*Grain number 1a/Cytokinin oxidase 2*) was known as a major quantitative trait locus associated with increased grain productivity (Ashikari et al., 2005). Reduced expression of *OsCKX2* was responsible for the increased cytokinin accumulation in inflorescence meristems. Genetic screening for elucidating molecular component controlling *OsCKX2* identified a unique mutant, *regulator of Gn1a 1* (*reg1*), with elevated meristem activity and increased grain number. The *reg1* phenotypes were caused by a semidominant mutation of *DST* (*DROUGHT AND SALT TOLERANCE*), resulting in reduced expression of several *OsCKX* genes and subsequent elevated cytokinin contents. *DST* encodes zinc finger protein and regulates drought and salt tolerance via stomatal aperture control (Huang et al., 2009). The authors showed that the expression of *OsCKX2* is directly up-regulated by DST and DST^{reg1} is a dominant negative allele; $\mathrm{DST}^{\mathrm{reg1}}$ proteins occupy the *OsCKX2* promoter region, outcompete DST and repress *OsCKX2* expression in a semidominant fashion (Li et al., 2013b). Another mutant with increased grain yield in rice, *large panicle* (*lp*), in which the Kelch repeat-containing F-box protein coding gene is disrupted, showed reduced expression of *OsCKX2* (Li et al., 2011).

The seed size is known to be under the control of cytokinin signalling (Riefler et al., 2006; Argyros et al., 2008). In *Arabidopsis* endosperm size depends on the coordination of the genetic pathway HAIKU (IKU) with epigenetic controls comprising genome dosage, DNA methylation and trimethylated lysine 27 on histone H3 (H3K27me3) deposition (Li et al., 2013a). The high level of cytokinin activity was found in endosperm, especially in syncytial endosperm in *Arabidopsis*, maize and rice. Li et al. showed that the gradient of cytokinin activity was established by the different expression of genes responsible for synthesis and degradation across endosperm. Based on the expression profile during seed development (Belmonte et al., 2013), transcripts from *CKX1*, *CKX2*, *IPT4* and *IPT8* were present in syncytial endosperm. The transcripts of *IPT4* and *IPT8* were predominantly detected in chalazal endosperm; especially, the confined expression of *IPT4* to chalazal endosperm was confirmed by *β-GLUCURONIDASE* and *GFP* reporter constructs (Li et al., 2013a). The decreased expression of *CKX2* in *iku1* and *iku2* suggested the role of the IKU pathway in controlling cytokinin homoeostasis. The direct binding of IKU transcription factor, WRKY10 encoded by *MINI3*, to the *CKX2* promoter was shown (Li et al., 2013a). The facts that the increased maternal genome dosage, which resulted in smaller seeds with reduced endosperm growth similar to that of *iku* mutant seeds, reduced *CKX2* expression, along with enriched H3K27me3, suggested epigenetic control over the *CKX2* expression. The results were consistent with the previous work of transcriptional profiling of patent-of-origin effects in *Arabidopsis* seeds, showing that the expression of *CKX2* was increased in high maternal to paternal genomes (maternal excess) and decreased in high paternal excess (Tiwari et al., 2010).

2.2.7 Establishment of symbiosis, cluster root formation and CKX

The availability of phosphorus, which is the most limiting micronutrient in soils due to its slow diffusion and high fixation in soils, affects plant growth and development (Shen et al., 2011). Plants have evolved a variety of adaptive strategies to cope with phosphorus deficiency (Hermans et al., 2006; Lynch, 2011). Around 80% of plants have established symbiotic association with mycorrhizal fungi to aid phosphorus acquisition (Burleigh et al., 2002), whereas the others have evolved a mechanism to form cluster roots that are closely spaced tertiary lateral rootlets (Neumann, 2010). An RNAseq analysis of phosphorus deficiency response in white lupin (*Lupinus albus*) revealed that two *CKX* transcripts were up-regulated 3-fold in phosphorus-deficient cluster roots. Treatment with the cytokinin benzyladenine (BA) inhibited cluster root formation induced by phosphorus deficiency. Knockdown of *CKX* transcripts by RNA interference (RNAi) exhibited an interruption of the cluster rootlet developmental pattern along the primary root (O'Rourke et al., 2013). These results suggest the role of cytokinin degradation in the cluster root formation in response to phosphorus deficiency.

In legume crops cytokinins play an essential role in symbiotic nodule organogenesis to undergo atmospheric nitrogen fixing (Frugier et al., 2008). Among the primary cytokinin responsive genes, rapid induction of *CKX1* expression depended on the activity of a cytokinin receptor, CRE1. The nodule-enhanced *Medicago truncatula* response regulator (MtRR1), involved in primary cytokinin signalling, can occupy promoter region of *CKX1* (Ariel et al., 2012).

2.2.8 Crosstalk with other hormones

Expression profiling analysis revealed that *CKX* genes in *Arabidopsis* (Bhargava et al., 2013), rice (Tsai et al., 2012) and maize (Vyroubalova et al., 2009) can respond to cytokinin. Besides, other hormone treatments are shown to affect *CKX* expression. Auxin has an impact on the regulation of cytokinin metabolism gene expressions in *Arabidopsis* (Goda et al., 2004; Yoshida et al., 2009) and in rice (Tsai et al., 2012). The activity and expression of *CKX* genes from wheat were shown to be down-regulated by brassinosteroid treatment, which were responsible for two-fold accumulation of cytokinins (Yuldashev et al., 2012). Responses of *CKX* genes to phytohormones or other stimuli are summarized in Table 2.1.

2.3 Application of altered cytokinin metabolisms to improve agricultural traits

Changing the cytokinin contents has the great potential for improving grain yields (Ashikari et al., 2005; Li et al., 2011; Li et al., 2013b; Zhang et al., 2011), susceptible resistance to pathogens (Naseem and Dandekar, 2012) and resistance to abiotic stresses (Nishiyama et al., 2011). The feasibility of technologies applied to genetically engineering genes related to cytokinin metabolisms, perception and signalling was discussed elsewhere (Zalabak et al., 2013). Phenotypes caused by increased cytokinins are obtained either through over-expressing *IPT* or decreased expression of *CKX*, whereas opposite phenotypes related to cytokinin deficiency are created through decreased expression of *IPT* or over-expression of *CKX*. Either way, the obtained phenotypes are the results of expected and undesired trait combinations caused by gene manipulation. Undesired phenotypes can be overcome by employing tissue-specific or inducible promoters. The *CKX* expression under the control of root-specific promoter conferred profitable traits of enlarged root system in *Arabidopsis* (Werner et al., 2010; Vercruyssen et al., 2011), barley (Mrizova et al., 2013), as well as resistance to drought and heat stress in tobacco (Mackova et al., 2013). In addition, the negative effects on shoot growth and development, observed in constitutively over-expressing plants, were diminished by using tissue-specific or inducible promoters (Werner et al., 2010; Mackova et al., 2013).

To reduce pathogen infection in roots a unique and alternative way of employing CKX as a scaffold protein was reported (Fang et al., 2006). Selected defence peptides that function as inhibitory ligands were fused to maize *CKX*, expressed in tomato roots and the resulting scaffold proteins were secreted to the apoplast and the rhizosphere. Pathogen infection was significantly inhibited in the transgenic tomato roots (Fang et al., 2006).

In contrast to other hormones, cytokinins are seldom used in horticulture as active components of commercial growth regulator preparations. Inhibitors of ethylene perception and gibberellin biosynthesis as well as compounds with gibberellin activity are commercially used for controlling longevity of ornamental crops and improving fruit yield and quality in pomiculture. Better understanding of the fundamental regulatory mechanisms of cytokinin metabolism genes will be the stepping stone to the design of compounds that can manipulate cytokinin contents for improving crop yields, resistance to biotic and abiotic stresses and delayed senescence.

Acknowledgement

Due to the limitation of space, we apologize for not being able to include an additional bibliography and other articles by world leading scientists. No conflict of interest is declared. We thank Dr. Vaclav Mik for critical reading of our manuscript. This work was supported by the grant No. LO1204 from the National Program of Sustainability I,MEYS, Czech Republic. Y.I. was supported by the Operational Program Education for Competitiveness - European Social Fund (project CZ.1.07/2.3.00/20.0165).

References

Argueso, C.T., Raines, T., Kieber, J.J., 2010. Cytokinin signaling and transcriptional networks. Curr. Opin. Plant. Biol. 13, 533–539.

Argyros, R.D., Mathews, D.E., Chiang, Y.H., Palmer, C.M., Thibault, D.M., Etheridge, N., et al., 2008. Type B response regulators of Arabidopsis play key roles in cytokinin signaling and plant development. Plant Cell. 20, 2102–2116.

Ariel, F., Brault-Hernandez, M., Laffont, C., Huault, E., Brault, M., Plet, J., et al., 2012. Two direct targets of cytokinin signaling regulate symbiotic nodulation in *Medicago truncatula*. Plant Cell. 24, 3838–3852.

Ashikari, M., Sakakibara, H., Lin, S., Yamamoto, T., Takashi, T., Nishimura, A., et al., 2005. Cytokinin oxidase regulates rice grain production. Science. 309, 741–745.

Bach, L., Michaelson, L.V., Haslam, R., Bellec, Y., Gissot, L., Marion, J., et al., 2008. The very-long-chain hydroxy fatty acyl-CoA dehydratase PASTICCINO2 is essential and limiting for plant development. Proc. Natl. Acad. Sci. USA. 105, 14727–14731.

Belmonte, M.F., Kirkbride, R.C., Stone, S.L., Pelletier, J.M., Bui, A.Q., Yeung, E.C., et al., 2013. Comprehensive developmental profiles of gene activity in regions and subregions of the Arabidopsis seed. Proc. Natl. Acad. Sci. USA. 110, E435–E444.

Bhargava, A., Clabaugh, I., To, J.P., Maxwell, B.B., Chiang, Y.H., Schaller, G.E., et al., 2013. Identification of cytokinin-responsive genes using microarray meta-analysis and RNA-Seq in Arabidopsis. Plant Physiol. 162, 272–294.

Bianco, M.D., Giustini, L., Sabatini, S., 2013. Spatiotemporal changes in the role of cytokinin during root development. New Phytol. 199, 324–338.

Bilyeu, K.D., Cole, J.L., Laskey, J.G., Riekhof, W.R., Esparza, T.J., Kramer, M.D., et al., 2001. Molecular and biochemical characterization of a cytokinin oxidase from maize. Plant Physiol. 125, 378–386.

Burleigh, S.H., Cavagnaro, T., Jakobsen, I., 2002. Functional diversity of arbuscular mycorrhizas extends to the expression of plant genes involved in P nutrition. J. Exp. Bot. 53, 1593–1601.

Carlsbecker, A., Lee, J.Y., Roberts, C.J., Dettmer, J., Lehesranta, S., Zhou, J., et al., 2010. Cell signalling by microRNA 165/6 directs gene dose-dependent root cell fate. Nature. 465, 316–321.

Chen, C.M., Kristopeit, S.M., 1981a. Metabolism of cytokinin: dephosphorylation of cytokinin ribonucleotide by 5′-nucleotidases from wheat germ cytosol. Plant Physiol. 67, 494–498.

Chen, C.M., Kristopeit, S.M., 1981b. Metabolism of cytokinin: deribosylation of cytokinin ribonucleoside by adenosine nucleosidase from wheat germ cells. Plant Physiol. 68, 1020–1023.

Chen, X., Goodwin, S.M., Boroff, V.L., Liu, X., Jenks, M.A., 2003. Cloning and characterization of the WAX2 gene of Arabidopsis involved in cuticle membrane and wax production. Plant Cell. 15, 1170–1185.

Cheng, Z.J., Wang, L., Sun, W., Zhang, Y., Zhou, C., Su, Y.H., et al., 2013. Pattern of auxin and cytokinin responses for shoot meristem induction results from the regulation of cytokinin biosynthesis by Auxin Response Factor3. Plant Physiol. 161, 240–251.

Cui, H., Hao, Y., Kovtun, M., Stolc, V., Deng, X.W., Sakakibara, H., et al., 2011. Genome-wide direct target analysis reveals a role for Short-Root in root vascular patterning through cytokinin homeostasis. Plant Physiol. 157, 1221–1231.

Dello Ioio, R., Galinha, C., Fletcher, A.G., Grigg, S.P., Molnar, A., Willemsen, V., et al., 2012. A PHABULOSA/cytokinin feedback loop controls root growth in Arabidopsis. Curr. Biol. 22, 1699–1704.

Ding, C., You, J., Chen, L., Wang, S., Ding, Y., 2014. Nitrogen fertilizer increases spikelet number per panicle by enhancing cytokinin synthesis in rice. Plant Cell Rep. 33, 363–371.

Drew, M.C., 1975. Comparison of effects of a localized supply of phosphate, nitrate, ammonium and potassium on growth of seminal root system, and shoot, in barley. New Phytol. 75, 479–490.

El-Showk, S., Ruonala, R., Helariutta, Y., 2013. Crossing paths: cytokinin signalling and crosstalk. Development. 140, 1373–1383.

Fang, Z.D., Laskey, J.G., Huang, S., Bilyeu, K.D., Morris, R.O., Schmidt, F.J., et al., 2006. Combinatorially selected defense peptides protect plant roots from pathogen infection. Proc. Natl. Acad. Sci. USA. 103, 18444–18449.

Frebort, I., Kowalska, M., Hluska, T., Frebortova, J., Galuszka, P., 2011. Evolution of cytokinin biosynthesis and degradation. J. Exp. Bot. 62, 2431–2452.

Frugier, F., Kosuta, S., Murray, J.D., Crespi, M., Szczyglowski, K., 2008. Cytokinin: secret agent of symbiosis. Trends Plant Sci. 13, 115–120.

Galuszka, P., Frebort, I., Sebela, M., Pec, P., 2000. Degradation of cytokinins by cytokinin oxidases in plants. Plant Growth Regul. 32, 315–327.

Galuszka, P., Frebort, I., Sebela, M., Sauer, P., Jacobsen, S., Pec, P., 2001. Cytokinin oxidase or dehydrogenase? Mechanism of cytokinin degradation in cereals. Eur. J. Biochem. 268, 450–461.

Goda, H., Sawa, S., Asami, T., Fujioka, S., Shimada, Y., Yoshida, S., 2004. Comprehensive comparison of auxin-regulated and brassinosteroid-regulated genes in Arabidopsis. Plant Physiol. 134, 1555–1573.

Guilfoyle, T.J., Hagen, G., 2007. Auxin response factors. Curr. Opin. Plant. Biol. 10, 453–460.

Guo, Y., Gan, S., 2011. AtMYB2 regulates whole plant senescence by inhibiting cytokinin-mediated branching at late stages of development in Arabidopsis. Plant Physiol. 156, 1612–1619.

Hare, P.D., Vanstaden, J., 1994. Cytokinin oxidase – biochemical features and physiological significance. Physiol. Plant. 91, 128–136.

Helariutta, Y., Fukaki, H., Wysocka-Diller, J., Nakajima, K., Jung, J., Sena, G., et al., 2000. The SHORT-ROOT gene controls radial patterning of the Arabidopsis root through radial signaling. Cell. 101, 555–567.

Hermans, C., Hammond, J.P., White, P.J., Verbruggen, N., 2006. How do plants respond to nutrient shortage by biomass allocation? Trends Plant Sci. 11, 610–617.

Huang, X.Y., Chao, D.Y., Gao, J.P., Zhu, M.Z., Shi, M., Lin, H.X., 2009. A previously unknown zinc finger protein, DST, regulates drought and salt tolerance in rice via stomatal aperture control. Genes Dev. 23, 1805–1817.

Igari, K., Endo, S., Hibara, K., Aida, M., Sakakibara, H., Kawasaki, T., et al., 2008. Constitutive activation of a CC-NB-LRR protein alters morphogenesis through the cytokinin pathway in Arabidopsis. Plant J. 55, 14–27.

Ikeda, Y., Banno, H., Niu, Q.W., Howell, S.H., Chua, N.H., 2006. The Enhancer of Shoot Regeneration 2 gene in Arabidopsis regulates CUP-Shaped Cotyledon 1 at the transcriptional level and controls cotyledon development. Plant Cell Physiol. 47, 1443–1456.

Inoue, T., Higuchi, M., Hashimoto, Y., Seki, M., Kobayashi, M., Kato, T., et al., 2001. Identification of CRE1 as a cytokinin receptor from Arabidopsis. Nature. 409, 1060–1063.

Kakimoto, T., 2001. Identification of plant cytokinin biosynthetic enzymes as dimethylallyl diphosphate: ATP/ADP isopentenyltransferases. Plant Cell Physiol. 42, 677–685.

Kamada-Nobusada, T., Makita, N., Kojima, M., Sakakibara, H., 2013. Nitrogen-dependent regulation of de novo cytokinin biosynthesis in rice: the role of glutamine metabolism as an additional signal. Plant Cell Physiol. 54, 1881–1893.

Kurakawa, T., Ueda, N., Maekawa, M., Kobayashi, K., Kojima, M., Nagato, Y., et al., 2007. Direct control of shoot meristem activity by a cytokinin-activating enzyme. Nature. 445, 652–655.

Laurenzio, D.I., Wysocka-Diller, L., Malamy, J., Pysh, J.E., Helariutta, L., Freshour, Y., et al., 1996. The Scarecrow gene regulates an asymmetric cell division that is essential for generating the radial organization of the Arabidopsis root. Cell. 86, 423–433.

Li, J., Nie, X., Tan, J.L., Berger, F., 2013a. Integration of epigenetic and genetic controls of seed size by cytokinin in Arabidopsis. Proc. Natl. Acad. Sci. USA. 110, 15479–15484.

Li, M., Tang, D., Wang, K., Wu, X., Lu, L., Yu, H., et al., 2011. Mutations in the F-box gene LARGER PANICLE improve the panicle architecture and enhance the grain yield in rice. Plant Biotechnol. J. 9, 1002–1013.

Li, S., Zhao, B., Yuan, D., Duan, M., Qian, Q., Tang, L., et al., 2013b. Rice zinc finger protein DST enhances grain production through controlling Gn1a/OsCKX2 expression. Proc. Natl. Acad. Sci. USA. 110, 3167–3172.

Lynch, J.P., 2011. Root phenes for enhanced soil exploration and phosphorus acquisition: tools for future crops. Plant Physiol. 156, 1041–1049.

Mackova, H., Hronkova, M., Dobra, J., Tureckova, V., Novak, O., Lubovska, Z., et al., 2013. Enhanced drought and heat stress tolerance of tobacco plants with ectopically enhanced cytokinin oxidase/dehydrogenase gene expression. J. Exp. Bot. 64, 2805–2815.

Mahonen, A.P., Bishopp, A., Higuchi, M., Nieminen, K.M., Kinoshita, K., Tormakangas, K., et al., 2006. Cytokinin signaling and its inhibitor AHP6 regulate cell fate during vascular development. Science. 311, 94–98.

Malamy, J.E., 2005. Intrinsic and environmental response pathways that regulate root system architecture. Plant Cell Environ. 28, 67–77.

Matsuo, N., Makino, M., Banno, H., 2011. Arabidopsis Enhancer of Shoot Regeneration (ESR)1 and ESR2 regulate in vitro shoot regeneration and their expressions are differentially regulated. Plant Sci. 181, 39–46.

Miyashima, S., Sebastian, J., Lee, J.Y., Helariutta, Y., 2013. Stem cell function during plant vascular development. EMBO J. 32, 178–193.

Miyawaki, K., Matsumoto-Kitano, M., Kakimoto, T., 2004. Expression of cytokinin biosynthetic isopentenyltransferase genes in Arabidopsis: tissue specificity and regulation by auxin, cytokinin, and nitrate. Plant J. 37, 128–138.

Mrizova, K., Jiskrova, E., Vyroubalova, S., Novak, O., Ohnoutkova, L., Pospisilova, H., et al., 2013. Overexpression of Cytokinin Dehydrogenase genes in barley (*Hordeum vulgare* cv. Golden Promise) fundamentally affects morphology and fertility. PLoS One. 8, e79029.

Naseem, M., Dandekar, T., 2012. The Role of Auxin-Cytokinin antagonism in Plant-Pathogen interactions. PLoS Pathog. 8, e1003026.

Neumann, G., 2010. Mining for nutrients – regulatory aspects of cluster root function and development. New Phytol. 187, 879–882.

Nishiyama, R., Watanabe, Y., Fujita, Y., Le, D.T., Kojima, M., Werner, T., et al., 2011. Analysis of cytokinin mutants and regulation of cytokinin metabolic genes reveals important regulatory roles of cytokinins in drought, salt and abscisic acid responses, and abscisic acid biosynthesis. Plant Cell. 23, 2169–2183.

Nitz, I., Berkefeld, H., Puzio, P.S., Grundler, F.M., 2001. Pyk10, a seedling and root specific gene and promoter from *Arabidopsis thaliana*. Plant Sci. 161, 337–346.

Nobusawa, T., Okushima, Y., Nagata, N., Kojima, M., Sakakibara, H., Umeda, M., 2013a. Restriction of cell proliferation in internal tissues via the synthesis of very-long-chain fatty acids in the epidermis. Plant Signal Behav. 8, e25232.

Nobusawa, T., Okushima, Y., Nagata, N., Kojima, M., Sakakibara, H., Umeda, M., 2013b. Synthesis of very-long-chain fatty acids in the epidermis controls plant organ growth by restricting cell proliferation. PLoS Biol. 11, e1001531.

O'rourke, J.A., Yang, S.S., Miller, S.S., Bucciarelli, B., Liu, J., Rydeen, A., et al., 2013. An RNA-Seq transcriptome analysis of orthophosphate-deficient white lupin reveals novel insights into phosphorus acclimation in plants. Plant Physiol. 161, 705–724.

Panikashvili, D., Savaldi-Goldstein, S., Mandel, T., Yifhar, T., Franke, R.B., Hofer, R., et al., 2007. The Arabidopsis Desperado/AtWBC11 transporter is required for cutin and wax secretion. Plant Physiol. 145, 1345–1360.

Perilli, S., Moubayidin, L., Sabatini, S., 2010. The molecular basis of cytokinin function. Curr. Opin. Plant. Biol. 13, 21–26.

Reina-Pinto, J.J., Yephremov, A., 2009. Lipid determinants of cell death. Plant Signal Behav. 4, 625–628.

Riefler, M., Novak, O., Strnad, M., Schmulling, T., 2006. Arabidopsis cytokinin receptor mutants reveal functions in shoot growth, leaf senescence, seed size, germination, root development, and cytokinin metabolism. Plant Cell. 18, 40–54.

Robatzek, S., Somssich, I.E., 2001. A new member of the Arabidopsis WRKY transcription factor family, AtWRKY6, is associated with both senescence- and defence-related processes. Plant J. 28, 123–133.

Ruffel, S., Krouk, G., Ristova, D., Shasha, D., Birnbaum, K.D., Coruzzi, G.M., 2011. Nitrogen economics of root foraging: transitive closure of the nitrate-cytokinin relay and distinct systemic signaling for N supply vs. demand. Proc. Natl. Acad. Sci. USA. 108, 18524–18529.

Sakakibara, H., Takei, K., Hirose, N., 2006. Interactions between nitrogen and cytokinin in the regulation of metabolism and development. Trends Plant Sci. 11, 440–448.

Sakamoto, T., Sakakibara, H., Kojima, M., Yamamoto, Y., Nagasaki, H., Inukai, Y., et al., 2006. Ectopic expression of KNOTTED1-like homeobox protein induces expression of cytokinin biosynthesis genes in rice. Plant Physiol. 142, 54–62.

Schaller, G.E., Shiu, S.H., Armitage, J.P., 2011. Two-component systems and their co-option for eukaryotic signal transduction. Curr. Biol. 21, R320–R330.

Shen, J., Yuan, L., Zhang, J., Li, H., Bai, Z., Chen, X., et al., 2011. Phosphorus dynamics: from soil to plant. Plant Physiol. 156, 997–1005.

Sieber, P., Schorderet, M., Ryser, U., Buchala, A., Kolattukudy, P., Metraux, J.P., et al., 2000. Transgenic Arabidopsis plants expressing a fungal cutinase show alterations in the structure and properties of the cuticle and postgenital organ fusions. Plant Cell. 12, 721–737.

Skoog, F., Miller, C.O., 1957. Chemical regulation of growth and organ formation in plant tissues cultured in vitro. Symp. Soc. Exp. Biol. 11, 118–130.

Takei, K., Sakakibara, H., Sugiyama, T., 2001. Identification of genes encoding adenylate isopentenyltransferase, a cytokinin biosynthesis enzyme, in *Arabidopsis thaliana*. J. Biol. Chem. 276, 26405–26410.

Takei, K., Ueda, N., Aoki, K., Kuromori, T., Hirayama, T., Shinozaki, K., et al., 2004. AtIPT3 is a key determinant of nitrate-dependent cytokinin biosynthesis in Arabidopsis. Plant Cell Physiol. 45, 1053–1062.

Tiwari, S., Spielman, M., Schulz, R., Oakey, R.J., Kelsey, G., Salazar, A., et al., 2010. Transcriptional profiles underlying parent-of-origin effects in seeds of *Arabidopsis thaliana*. BMC Plant. Biol. 10, 72.

Tsai, Y.C., Weir, N.R., Hill, K., Zhang, W., Kim, H.J., Shiu, S.H., et al., 2012. Characterization of genes involved in cytokinin signaling and metabolism from rice. Plant Physiol. 158, 1666–1684.

Uchida, N., Igari, K., Bogenschutz, N.L., Torii, K.U., Tasaka, M., 2011. Arabidopsis ERECTA-family receptor kinases mediate morphological alterations stimulated by activation of NB-LRR-type UNI proteins. Plant Cell Physiol. 52, 804–814.

Vercruyssen, L., Gonzalez, N., Werner, T., Schmulling, T., Inze, D., 2011. Combining enhanced root and shoot growth reveals cross talk between pathways that control plant organ size in Arabidopsis. Plant Physiol. 155, 1339–1352.

Vieten, A., Sauer, M., Brewer, P.B., Friml, J., 2007. Molecular and cellular aspects of auxin-transport-mediated development. Trends Plant Sci. 12, 160–168.

Vyroubalova, S., Vaclavikova, K., Tureckova, V., Novak, O., Smehilova, M., Hluska, T., et al., 2009. Characterization of new maize genes putatively involved in cytokinin metabolism and their expression during osmotic stress in relation to cytokinin levels. Plant Physiol. 151, 433–447.

Wang, R., Tischner, R., Gutierrez, R.A., Hoffman, M., Xing, X., Chen, M., et al., 2004. Genomic analysis of the nitrate response using a nitrate reductase-null mutant of Arabidopsis. Plant Physiol. 136, 2512–2522.

Werner, T., Motyka, V., Laucou, V., Smets, R., Van Onckelen, H., Schmulling, T., 2003. Cytokinin-deficient transgenic Arabidopsis plants show multiple developmental alterations indicating opposite functions of cytokinins in the regulation of shoot and root meristem activity. Plant Cell. 15, 2532–2550.

Werner, T., Kollmer, I., Bartrina, I., Holst, K., Schmulling, T., 2006. New insights into the biology of cytokinin degradation. Plant Biol. (Stuttg.). 8, 371–381.

Werner, T., Nehnevajova, E., Kollmer, I., Novak, O., Strnad, M., Kramer, U., et al., 2010. Root-specific reduction of cytokinin causes enhanced root growth, drought tolerance, and leaf mineral enrichment in Arabidopsis and tobacco. Plant Cell. 22, 3905–3920.

Wuichet, K., Cantwell, B.J., Zhulin, I.B., 2010. Evolution and phyletic distribution of two-component signal transduction systems. Curr. Opin. Microbiol. 13, 219–225.

Yanai, O., Shani, E., Dolezal, K., Tarkowski, P., Sablowski, R., Sandberg, G., et al., 2005. Arabidopsis KNOXI proteins activate cytokinin biosynthesis. Curr. Biol. 15, 1566–1571.

Yoshida, S., Iwamoto, K., Demura, T., Fukuda, H., 2009. Comprehensive analysis of the regulatory roles of auxin in early transdifferentiation into xylem cells. Plant Mol. Biol. 70, 457–469.

Yuldashev, R., Avalbaev, A., Bezrukova, M., Vysotskaya, L., Khripach, V., Shakirova, F., 2012. Cytokinin oxidase is involved in the regulation of cytokinin content by 24-epibrassinolide in wheat seedlings. Plant Physiol. Biochem. 55, 1–6.

Zalabak, D., Pospisilova, H., Smehilova, M., Mrizova, K., Frebort, I., Galuszka, P., 2013. Genetic engineering of cytokinin metabolism: prospective way to improve agricultural traits of crop plants. Biotechnol. Adv. 31, 97–117.

Zhang, H., Forde, B.G., 1998. An Arabidopsis MADS box gene that controls nutrient-induced changes in root architecture. Science. 279, 407–409.

Zhang, J., Wrage, E.L., Vankova, R., Malbeck, J., Neff, M.M., 2006. Overexpression of SOB5 suggests the involvement of a novel plant protein in cytokinin-mediated development. Plant J. 46, 834–848.

Zhang, J., Liu, W., Yang, X., Gao, A., Li, X., Wu, X., et al., 2011. Isolation and characterization of two putative cytokinin oxidase genes related to grain number per spike phenotype in wheat. Mol. Biol. Rep. 38, 2337–2347.

Apple allergens genomics and biotechnology: Unravelling the determinants of apple allergenicity

3

Federica Savazzini[1], *Giampaolo Ricci*[2] *and Stefano Tartarini*[1]
[1]Department of Agricultural Science (DipSA), University of Bologna, Bologna, Italy;
[2]Department of Pediatrics, University of Bologna, Bologna, Italy

3.1 Introduction: Fruit and apple allergies

Among food allergies, cow's milk and eggs still predominate throughout the world. However, fruit allergy importance is increasing: it has been estimated that the perceived adverse reaction towards any fruit varies around 0.4%–6.6% in adult population and up to 11% in the children aged 0–3 years (Zuidmeer et al., 2008). The percentage among children is highly concerning, due to the importance of a regular consumption of fruits and vegetables in the diet. The apple (*Malus* x *domestica* Borkh.) crop is one of the main cultivated fruit species in the world, but in the last years several clinical studies draw attention to apple allergy as an increasing problem. Therefore, allergenicity is becoming an important issue in fruit quality determination and advanced multidisciplinary research was performed to highlight its molecular and clinical bases. Moreover, it is also crucial to understand the biotic/abiotic environmental conditions that can increase the content of these allergen proteins in the fruit in order to evaluate a different strategy during cultivation, storage and processing (see Figure 3.1). Finally, a large genetic variation was found and this opened the way to breed for hypoallergenicity.

Plant allergens belong to few protein families/superfamilies, related mainly to seed-storage (cupins and prolamins) and pathogenesis-related (PR) proteins (Hoffmann-Sommergruber, 2002); additional groups are represented by inhibitors of proteases, peroxidases and profilins (Breiteneder and Ebner, 2000; Mills et al., 2003). To date, according to the rules of the WHO/IUIS Allergen Nomenclature Subcommittee (www.allergen.org), four main classes of allergens have been identified in apple: *Mal d 1*, *2*, *3* and *4*. The first three families belong to PR proteins (PR-10 or *Bet v 1*-homologues, PR-5 or thaumatin-like proteins or TLP and PR-14 or non-specific lipid transfer proteins (nsLTPs)) while *Mal d 4* belong to profilins. The complete apple genome draft sequence confirmed the genetic complexity of these families, which include 31 *Mal d 1*, 36 *Mal d 2*, 32 *Mal d 3* and 17 *Mal d 4*

Applied Plant Genomics and Biotechnology.

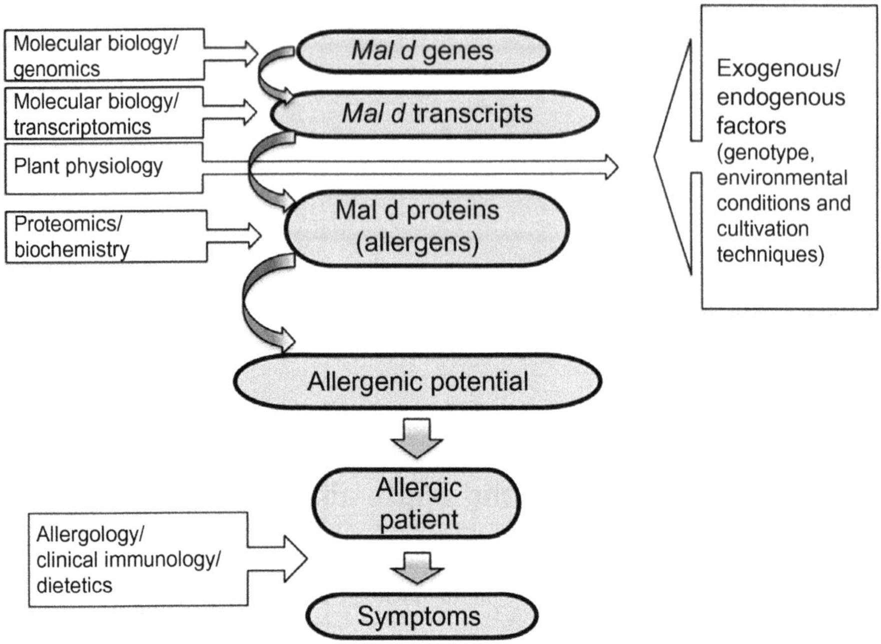

Figure 3.1 Flow chart of the critical steps, from genes to patient's symptoms, involved in apple allergy. On the right side are shown some of the environmental/endogenous factors and on the left side the several disciplines involved in the study of the determinants of apple fruit allergy.

isoforms, dispersed in different chromosomes/linkage groups (LG) or organized in clusters (Gao et al., 2005a,b,c; Pagliarani et al., 2012; Velasco et al., 2010).

3.1.1 Apple allergy symptoms and clinical studies

Apple allergens are responsible of a fast hypersensitive reaction mediated by immunoglobulin E (IgE). According to the allergen surface reacting with the immune system, apple allergies can be distinguished into the following:

- Class-I, identified as gastro-intestinal and systemic symptoms, because the allergen is not degraded by gastric digestion (Pastorello and Rivolta, 2004)
- Class-II, identified as oral allergy syndrome (OAS), in which the pollen proteins are acting as primary sensitizing agents and the fruit ingestion immediately provokes mild symptoms in the mouth (Eriksson et al., 1982) and in the skin in contact with apple during consumption (Fernandez-Rivas et al., 2006)

Cross-sensitization is largely demonstrated between birch pollen allergen *Bet v 1* and apple *Mal d 1* (Fritsch et al., 1998), while it is still to be confirmed for cypress pollen and *Mal d 2* (Cortegano et al., 2004). The wide distribution of *Mal d 3* in plant foods suggests a potential role of these proteins as plant pan-allergen, with cross-reactivity with other non-Rosaceae extracts, such as peanut, walnut, pistachio, broccoli, carrot, tomato, melon and kiwifruit (Asero et al., 2000). Similar cross-reactions are described for fruits and pollen from Mediterranean plant species, such as lettuce (San Miguel-Moncin et al., 2003), olive, mugwort, plane tree, peach and apple (Diaz-Perales et al., 2000). The ingestion of other Rosaceae fruits, in particular peach, can increase the sensitization to *Mal d 3* (Fernandez-Rivas et al., 2006). Inhibition experiments with serum pools from patients with fruit allergies revealed cross-reactivity between *Mal d 4* and other profilins form pear, cherry and birch (Scheurer et al., 2001; Ebner et al., 1995). Finally, a cross-sensitization between Rosaceae fruits profilins and other profilins from diverse pollen spp. (i.e. birch, olive and ryegrass) was demonstrated (van Ree et al., 1995).

Population surveys based only on self-reported symptoms should be suspected with clinical tests such as skin prick tests (SPTs) and confirmed with Double blind, placebo-controlled oral food challenges (DBPCFC), which guarantee more reliability and less false-positives (Zuidmeer et al., 2008). Otherwise, Schmidt-Andersen et al. (2011) found good accordance between confirmed clinically relevant cases and studies where patients were included based solely on patient histories and positive IgE and/or SPT tests. Apple allergens are classified as minor or major allergens according to clinical relevance and prevalence in a population. Northern and Central European populations suffer more frequently of the *Bet v 1-Mal d 1* and *Bet v 2-Mal d 4* OAS syndrome, due to the widespread presence of birch, while the western Mediterranean area, where peaches are more represented in the daily diet, is more subjected to the gastro-intestinal syndrome associated to *Mal d 3* (Fernandez-Rivas et al., 2006; Schmidt-Andersen et al., 2011). The allergy to profilin *Mal d 4* is widespread in both North and South Europe. In conclusion, the cause for the geographic differences in fruit allergies is still unknown, but differences in pollen exposures have the most significant impact on allergy sensitization; additionally, population genetics of population and eating habits have a not secondary role (Schmidt-Andersen et al., 2011).

Apple cultivars have different allergenicity levels, as reported in various clinical studies. Unluckily, the majority of commercial varieties shows high allergenic potential but a few displays a very promising hypoallergenicity, such as Santana, Topaz, Elise, Braeburn (Bolhaar et al., 2005a; Kootstra et al., 2007), Pink Lady (Vlieg-Boerstra et al., 2011), Gloster, McIntosh, Modi (Vlieg-Boerstra et al., 2011; Vlieg-Boerstra et al., 2013) and Durello di Forlì (Ricci et al., 2010). Furthermore, patients diversely react to apple fruit tissues, with skin generally more allergenic than flesh (Figure 3.2). The age of the patient plays always an important role, with young adults and children more allergic than adults (Ricci et al., 2010).

Specific immunotherapy (SIT) is a medical treatment of disease by inducing, enhancing or suppressing an immune response, in which small doses of antigen are given to patients diagnosed with food allergy, with the aim to build immune-tolerance. Natural extracts are commonly used in IT, but they could easily degrade with reduced efficacy (Rudeschko et al., 1995); on the other hand, a single recombinant protein stably expressed in any heterologous system can be easily managed and has a constant concentration. In the last few decades, numerous allergen protein

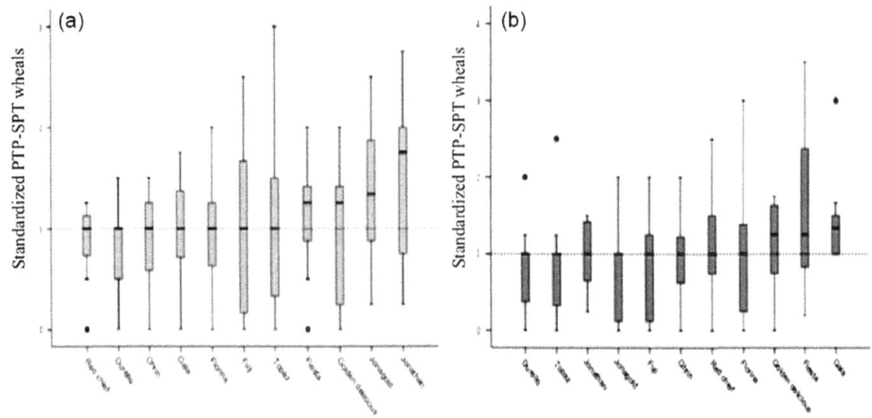

Figure 3.2 Ranking of different apple cultivars after standardization and according to pulp (a) and peel (b) allergenicity. The allergenicity was measured by SPT and prick-to-prick-test (PTP), employing both commercially available apple extracts and separated peel and pulp directly. The dark dots represent the outliers. (a) In pulp, Jonathan, Jonagold, Golden Delicious and Fiesta appeared to be the most allergenic cultivars. (b) In peel, Gala, Fiesta and Golden Delicious apparently caused the larger wheals.

sequences were published and employed to understand the IgE-binding sites on the surface of the proteins (Ebner et al., 1991; Borges et al., 2008); to date, 65 apple allergenic linear epitopes were predicted using consensus sequences and are available on public database (www.iedb.org). The next step is to design and create epitope variants by site-directed mutagenesis; thus, mutated-hypoallergenic variants from *Bet v 1* (Ferreira et al., 1998) and *Mal d 1* (Ma et al., 2006a) or recombinant *Mal d 1* (Vanek-Krebitz et al., 1995; Son et al., 1999), *Mal d 2* (Krebitz et al., 2003), *Mal d 3* (Diaz-Perales et al., 2002) and *Mal d 4* (Ma et al., 2006b) were expressed in heterologous systems and employed as 'disarmed' allergens in several clinical tests. Rapid and significant decrease of allergenic response was experimented in SIT trial, using birch pollen/apple extracts and the corresponding *Bet v 1/Mal d 1* recombinant proteins (Bolhaar et al., 2004, 2005b). These studies demonstrated the feasibility of using the recombinant and/or mutated proteins as an immunotherapy vaccine with reduced anaphylactic risks, or for the production of multiple-allergen preparations for personalized SIT (Cromwell et al., 2011)

3.1.2 Genomics and biochemistry

3.1.2.1 Mal d 1 – PR-10 ribonuclease-like

Mal d 1 is considered a major allergen of apple fruit (Vieths et al., 1995; Vanek-Krebitz et al., 1995) with a putative role in plant defence and development (Atkinson et al., 1996; Puehringer et al., 2000). The translated protein has a final length of 159 aminoacids (aa), as the homologous birch Bet v 1, with exceptions of 158 and 160 AA (Hoffmann-Sommergruber et al., 1997; Gao et al., 2005a).

The PR-10 cytoplasmic protein of 17 kDa has a still putative function of binding-transport of apolar ligands, such as fatty acids, flavonoids and brassinosteroids which are located in a large hydrophobic cavity determined by a highly conserved P-loop rich in glycine residues (Spangfort et al., 1997). Up to now, only one protein was isolated and characterized as a putative binding ligand of *Mal d 1*, MdAP (Puehringer et al., 2003).

Regardless of several nucleotide substitutions and the presence/absence of introns in the full coding *Mal d 1* gene sequences, the protein exhibits a few aminoacidic differences. Substitutions in the sequence would produce some modification in the protein P-loop cavity, modifying the ligand-binding specificity. A reduced or increased affinity with human IgE was also found among *Mal d 1* isoforms from different apple varieties (Son et al., 1999) or by analysing the effects of amino-acidic substitutions on epitope surfaces of the protein (Ma et al., 2006a; Holm et al., 2011), similarly to *Bet v 1* (Ferreira et al., 1998; Bolhaar et al., 2005b; Ebner et al., 1991).

Genetic and physical maps agreed in clustering the majority of *Mal d 1* genes on the two homologous LG 13 and 16, similarly to the *Pru p 1* gene family clustering on LG1 in peaches (Gao et al., 2005a; Chen et al., 2008). Some divergences in gene order were found between the results of Bacterial artificial chromosome (BAC) clones sequencing on LG 16 and the corresponding region on the draft apple genome (Pagliarani et al., 2012; Velasco et al., 2010).

The phylogenetic tree derived from the comparison of nucleotidic sequences is branched into five sub-families (Figure 3.3). A single intron of different size is present in all sub-families except the sub-family IV (Gao et al., 2005a). By testing the allelic diversity of ten apple cultivars of known allergenicity, the *Mald 1.04* and *1.06A* genes were associated to the allergenic potential and a gene dosage effect was also suggested for *1.06A* (Gao et al., 2008).

Some conserved sequences were also identified in promoter regions, like GAGAATC in isoforms of the intron-less sub-family IV (*Mal d 1.03A*, *1.03D*, *1.03E* and *1.03F*) and TCATC upstream the translation initiation codon ATG in the intron-containing genes (*Mal d 1.01*, *1.02*, *1.04* and *1.06A*) (Gao et al., 2005a). Other conserved transcription factor-binding sites (the Box-W motif, ERE-element, TCA-element, AuxRR-core and ERRE-motif) are contained in the regulatory regions of *Mal d 1* and involved in fungal elicitor, ethylene, salicylic acid (SA) production, auxin and biotic elicitor responsive elements (Ziadi et al., 2001). Differences in regulatory regions could be crucial for developmental stage and tissue specificity, as well as for environmental or biotic stress conditions. Tissue specificity is well documented for several *Mal d* genes and their detection in fruit tissues is the first step to understand the allergenic potential of each isoform (Beuning et al., 2004). Scoring expressed sequence tags (ESTs) databases (www.ncbi.nlm.nih.gov), *Mal d 1.01* and *1.02* are highly represented in fruit tissues, followed by the *1.03* (C, E and F isoforms), *1.06* (isoforms A and B), *1.07*, *1.10* and *1.11* groups. Interestingly, *Mal d 1* ESTs are also found in specialized organs where bud/flower (*Mald 1.01*, *1.02* and *1.03E*), roots (*1.03F*, *1.06B*, *1.08* and *1.09*) and leaves are challenged with biotic infection (*1.04* and *1.06*). ESTs data have been confirmed by other studies (Marzban et al., 2005; Pagliarani et al., 2013).

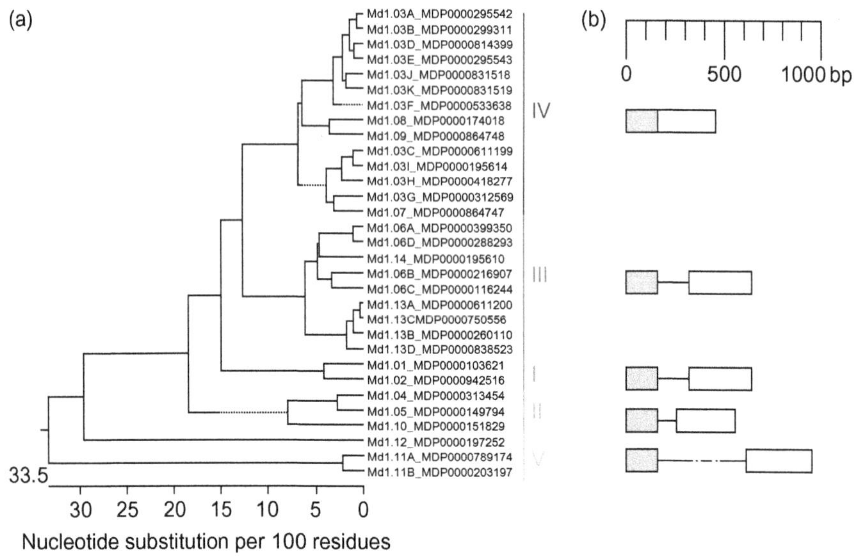

Figure 3.3 ***Mal d 1*** **gene family alignment (a) and structure (b).** (a) Neighbour-joining tree based on the coding sequence of the 31 *Mal d 1* genes retrieved from the 'Golden Delicious' *Malus* x *domestica* v1.0 genome sequence (Velasco et al., 2010). Each gene is described with the genome reference accession number (MDP); vertical lines and roman numerals (I–V) delineate the sub-families. (b) The corresponding gene structure of each sub-family is shown in scale; the line represents the intron, while the grey and white rectangles represent the first and second exons, respectively.

The specific expression of each *Mal d 1* isoforms using qRT-PCR in fruit tissues of Gala and Florina was investigated (Pagliarani et al., 2013). In detail, among the 31 reported isoforms, 11 were not expressed while some isoforms were detected for the first time in fruits; the highest represented mRNA are *Mal d1.01* and *1.02*, preceding *1.03*, *1.06*, *1.07*, *1.08* and *1.11* with enormous differences among isoforms of the same group. *Mal d 1* transcripts localize preferentially in the apple skin, with the exception of *Mal d 1.11* and *1.12* (sub-family V), showing a more equilibrate expression in both tissues (Pagliarani et al., 2013). Because *Mal d 1.01* and *1.02* are highly expressed in both hypo- and high-allergenic varieties, with different intensity, Son et al. (1999) proposed that the total amount of *Mal d 1* proteins is causing the allergenic reactions. Other authors (Gao et al., 2008; Pagliarani et al., 2013) excluded *Mal d 1.01* and *1.02* genes as allergy determinants, because of their widespread expression, even in hypoallergenic cultivars, and focused on other isoforms, such as *Mal d 1.03,1.04, 1.06 and 1.07*, differentially transcribed among cultivars.

3.1.2.2 Mal d 2 – PR-5 thaumatins

Thaumatin-like proteins are considered minor apple allergen, even if they are abundant in mature apples and produce a strong reaction with allergic patent's sera

(Hsieh et al., 1995; Oh et al., 2000). PR-5 are proteins of 23 kDa (Krebitz et al., 2003) with a crystal conformational structure highly conserved among species; eight disulphide bonds are crucial for the protein folding and activity (Krebitz et al., 2003) and confer highly thermal and protease resistance (Smole et al., 2008). A 24–27 aa N-terminal signal sequence drives the protein towards specific compartments: the basic sub-group of thaumatins is localized in vacuolar compartment, and the slightly acidic sub-group in the apoplast (Breiteneder, 2004). Glycosylation sites can be present in several isoforms; nevertheless, the involvement of the carbohydrate moiety in the increasing of allergenic potential has still to be confirmed (Breiteneder, 2004).

Mal d 2 genes are scattered in different LGs: 2, 5, 9 (*Mal d 2.01* and *2.03*), 10, 15 and 17 (*Mal d 2.02*), as deduced from mapping studies and apple genome sequence (Gao et al., 2005b; Velasco et al., 2010). Genes are generally structured in two exons of different length and one intron, the length of which increases in *Mal d 2.04* and other thaumatins; a few isoforms are organized in three exons.

The numerous isoforms have different functions and they are differentially expressed in response to osmotic stress ('osmotins') and fungal infections (Breiteneder, 2004). Numerous EST sequences from *Mal d 2.01*, *2.02* and other isoforms are found in mature fruit tissues and flower. Actually, thaumatin genes are among the mostly expressed in apple fruits and, among varieties, the highest content was found in Braeburn variety (Botton et al., 2008). The presence of *Mal d 2* in pollen was confirmed in cypress (Togawa et al., 2006; Cortegano et al., 2004).

3.1.2.3 Mal d 3 – PR-14 nsLTPs

nsLTPs constitute a major apple allergen (Pastorello et al., 1999; Diaz-Perales et al., 2002) and a highly cross-reactive pan-allergen (Asero et al., 2000). The *Mal d 3* family is divided into two sub-families and the identified allergens belong to nsLTP1, except a pollen *B. napus* nsLTP2. The sub-group LTP1 are characterized by small 9 kDa apoplastic proteins (90–95 aminoacids) and sub-group LTP2 by 7 kDa proteins (70 aminoacids); both present a signal peptide at the N-terminus. Their main function is related to phospholipid transport from the endoplasmic reticulum to organelle membranes or during secretion and deposition of extracellular lipophilic materials as cutin layer (Sterk et al., 1991; de Oliveira Carvalho and Moreira Gomes, 2007). The many isoforms are supposed to be involved in diverse developmental stages, during photosynthesis and in vesicle trafficking (Kader, 1996). Despite the large variability of aminoacidic sequences inside nsLTP family, the lipid binding motifs (Asp-Arg-Gln and Cys-Gly-Val) are shared among the different isoforms. Moreover, the three-dimensional protein structure is highly conserved due to the presence of four disulphide bonds located at fixed position (Kader, 1996); this compact structure confers stability to thermal denaturation as well as resistance towards human proteases (Pastorello and Rivolta, 2004). In analogy to peach *Pru p 3* (Garcia-Casado et al., 2003), four conserved linear IgE-binding epitopes are predicted on the surface of nsLTPs *Mal d 3* (Borges et al., 2008). These four epitopes cover about 50% of the total accessible surface, explaining the strong allergenicity and cross-sensitization towards Rosaceae LTPs, especially in Mediterranean countries (Borges et al., 2008). The ubiquitous proteins nsLTPs are differentially activated during biotic and abiotic stress (Jung et al., 2003); some LTPs display an antibiotic property (Garcia-Olmedo et al., 1995).

Mal d 3.01 and *3.02* are localized on LG 12 and 4, respectively (Gao et al., 2005c); other isoforms are scattered in several LGs on apple genome. *Mal d 3* genes contain generally one single exon, *Mal d 3.01* has two exons, and other genes have more complexes structures (Velasco et al., 2010).

In apple fruit tissues, *Mal d 3* is detected both as cDNA (Diaz-Perales et al., 2002) and protein (Pastorello et al., 1999). *Mal d 3* content increases as fruits reach the physiological maturity, with *Mal d 3.01* transcripts more expressed than *3.02* (Pagliarani et al., 2009) and decreases during storage at 4°C, with a putative involvement of ethylene as signal activator of gene expression (Sancho et al., 2006b). Furthermore, apple LTPs seem to be absent in fruit flesh (Marzban et al., 2005); however, they are localized in fruit skin in the outer cell surface, explaining the contact dermatitis effect on sensitized people (Borges et al., 2006). On the other hand, nsLTP are a wide family and isoforms may have an independent expression: *Mal d 3* EST sequences with 100% similarity are found in flower and fruit tissues at different stages. Rajica cultivar seems to have the highest production of LTP transcripts in fruits, while other commercial cultivars Fuji, Golden Delicious and Granny Smith have a lower content (Botton et al., 2008).

3.1.2.4 *Mal d 4* – Profilins

Profilins are 12–15 kDa cytosolic proteins which are found in all eukaryotic cells and involved in actin cytoskeleton regulation (Radauer et al., 2006). They were identified for the first time in birch pollen grains and designated as *Bet v 2* allergens (Valenta et al., 1991). Although profilins are considered a minor apple allergen, profilin-allergic patients constitute about 20% of all pollen-allergic individuals (Valenta et al., 1992); moreover, the highly conserved aminoacidic sequence is responsible for multiple cross-reactivities among unrelated plant pollens, food and latex, and they are considered pan-allergens (Valenta et al., 1992). When subjected to heat treatment or enzymatic digestion, profilins are quite unstable, which explains the rather restricted but mild symptoms related to the OAS (Ma et al., 2006b).

High degree of sequence similarity has been demonstrated among profilins from different plant *spp.*, with more than 50% of aminoacidic similarity (Radauer et al., 2006). By comparison of all surface-exposed residues, eleven overlapping potential epitopes were identified. In particular, two high-conserved epitopes were assumed to be responsible for the high extent of cross-reactivity (Radauer et al., 2006). In apple, three profilins (*Mal d 4.01*, *4.02* and *4.03*) were fully sequenced, resulting in 78%–80% identity for coding DNA and amino acid sequences, respectively; they are located on LG 9, 2 and 8 (Gao et al., 2005b). A fourth complete isoform and pseudogenes were found in the apple genome sequence (Velasco et al., 2010).

Expressed sequences are found in young shoots, flowers and phloem, other than in fruit tissues. Cultivars with high content of transcripts are Golden Delicious,

Granny Smith and Rajca, while Fuji and Brina have lower quantities in fruits (Botton et al., 2008). Among isoforms, *Mal d 4.03* was clearly detected in fruits, concentrating in fruit maturity stages and in skin tissue, while the other two main isoforms were undetectable (Pagliarani et al., 2009).

Post-translational modifications may be responsible for introducing new epitopes and influence antibody binding and allergenicity: glycosylation sites are present on *Mal d 2* sequences (Breiteneder, 2004), disulphide bonds can dimerize or even polymerize *Mal d 2* and *3* proteins, as detected by immunoblot analysis of apple extracts (Krebitz et al., 2003), and recombinant *Mal d 1* can polymerize in *E. coli* cells (Roulias et al., 2014). Moreover, other than the above protein families, extracellular apple pollen transglutaminases (TGase) may be activated by environmental stress conditions and they could act as mediators of pollen allergy, by increasing the pro-inflammatory phospholipase A2 (sPLA2) activity present in human mucosa (Iorio et al., 2012). Finally, an additional allergen protein was found using serological screening of 2-DE apple fruit extracts: the protein was identified as a glyceraldehyde-3-phosphate dehydrogenase and reacted intensively with all patients' sera tested (Hernd et al., 2007).

In conclusion, even if apple allergens seem to be mainly restricted to four families, several additional proteins located in pollen and fruit tissues were reported as potential allergens.

3.2 Abiotic factors: Influence of environment and cultivation techniques

The impact of cultivation techniques and endogenous/environmental factors on apple fruit quality is commonly recognized, but few studies focused on their effect on allergens content in fruits.

No clear correlation was found between protein content of *Mal d 1* and 1-aminocyclopropane-l-carboxylic acid (ACC) oxidase, a key-enzyme of ethylene biosynthesis, during both ripening and under different storage conditions, suggesting that this allergen is more affected by factors related to resistance to several diseases than senescence processes occurring during storage (Hsieh et al., 1995). In contrast, a recent study demonstrated that the levels of *Mal d 1.01*, *Mal d 1b*, *Mal d 1d* and *Mal d 1e, Mal d 1*-associated protein (MdAP) and *Mal d 4.01* were up-regulated during ripening and enhanced by ethylene treatment; *Mal d 1.04*; *Mal d 2.01*, *2.02* and *2.03*; *Mal d 3.01* and *Mal d 4.02* and *4.03* were down-regulated. Treatment with 1-methylcyclopropene (1-MCP), an ethylene-action inhibitor, and ethylene generally produced opposite effects, which provides additional evidence that regulation of these genes is ethylene dependent (Yang et al., 2012).

The effect of growing site, cultivation and environmental factors are difficult to evaluate, and should be combined with varietal differences. Matthes and Schmitz-Eiberger (2009) found significant differences in *Mal d 1* protein content in fruit of the same cultivar, which was higher in integrated versus organic production, as pesticide

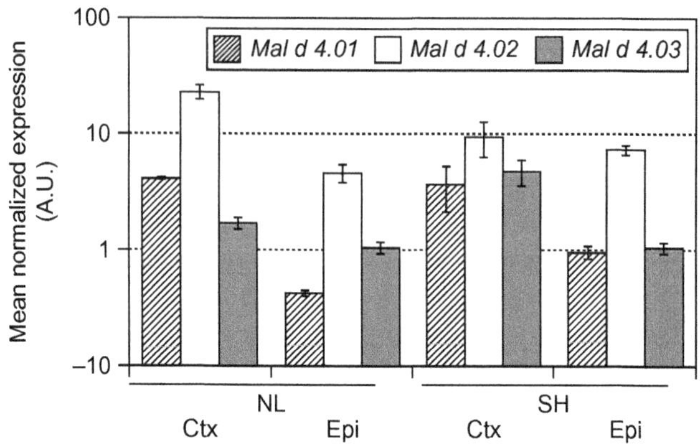

Figure 3.4 Effect of light growth conditions on the expression profile of three *Mal d 4* genes in Golden Delicious fruits. Plants were grown under normal light (NL) or shadowing (SH); fruits were divided into cortex (Ctx) and epidermis (Epi) tissues. Expression values were reported as arbitrary units (A.U.) after normalization to ubiquitin gene. With the exception of *Mal d 4.02* in cortex, shadowing has an irrelevant or up-regulation effect.

treatment leading to stronger responses than any biotic factors. Among the agronomical practices, shadowing and elevation significantly affected the transcription of the allergen-encoding genes, whereas water deprivation caused a minor effect, slightly influencing the expression of only two allergens, *Mal d 1.04* and *4.01* (Botton et al., 2008). In particular, shadowing has an important effect on *Mal d* transcript content in fruits: generally, *Mal d* genes were inhibited in the flesh and stimulated in the epidermal tissues. A statistically significant difference was observed for the isoforms *Mal d 1.03*, *Mal d 2.02*, *Mal d 3.01* and *3.02* and *Mal d 4.02* and *4.03* (Botton et al., 2008; Figure 3.4). Finally, fruits cultivated at low altitude showed a 50% reduction of global gene expression after 5-month storage, compared to high elevation orchards, which were unaffected (Botton et al., 2008).

3.3 Biotic factors: Pathogen infection and allergens content

Among plant allergens listed in the Official Allergen Database of the International Union of Immunological Societies, 25% are classified as PR proteins (Hoffmann-Sommergruber, 2002). Plant PR proteins are induced, both locally as systemically, in response to pathogen attack (viruses, bacteria and fungi), or by chemicals such as ethylene and SA that mimic the effects of pathogen infection (van Loon et al., 2006), as well as by physical stimuli, such as wounding, UV-B radiation, osmotic shock, low temperature and water deficit and -excess (Edreva, 2005). Presently, PRs are 17 families found in all plant organs as leaves, stems, roots, flowers and they are particularly abundant in leaves, where they can amount up to 5%–10% of the total proteins (van Loon et al., 2006).

In apple, three out of four allergen families belong to PR proteins: PR-10 (*Mal d 1*), PR-5 (*Mal d 2*) and PR-14 (*Mal d 3*). PR-10 proteins have a putative ribonuclease function with anti-viral activity: in grape they were expressed in leaves challenged with *P. viticola* (Kortekamp, 2006). PR-5 are supposed to be inserted directly into fungal membranes, forming a trans-membrane pore and causing influx of water followed by osmotic rupture; PR-5 have been detected in the leaves of young plants, but they accumulated rapidly to high levels upon biotic or abiotic stress (Hoffmann-Sommergruber, 2002). It has been demonstrated that PR-14 bind elicitins, among the most important elicitors in plant defence mechanism (Buhot et al., 2001). Besides, they exhibit antifungal and antibacterial activity, exerting their effect on plasma membrane of the target microorganism (Van Loon et al., 2006); finally, PR-14 overexpressing transgenic carrot plants showed higher resistance towards fungal pathogens (Jayaraj and Punja, 2007).

PR families encode genes constitutively expressed, to a variable extent, in all stages during plant development; as a consequence, the discrimination among isoforms up-regulated after a biotic challenge can be difficult. In apple few data are available about *Mal d* genes under biotic stress, mostly on challenged leaves. *Mal d 1* up-regulation in response to reduced glutathione (GSH) and SA treatments has been detected in leaves; in the same study, the *Mal d 1* promoter, co-transformed with the reporter gene b-glucoronidase (GUS) in tobacco, was highly induced by the same treatments and by virus infection (Puehringer et al., 2000). Moreover, genes encoding proteins related to plant defence were constitutively expressed at higher levels in healthy leaves of the scab-resistant cultivar Remo, in comparison to the susceptible cultivar Elstarand, among all, *Mal d 1* transcripts, indicating a putative protective role against biotic stress (Degenhardt et al., 2005). As a confirmation of the involvement of *Mal d 1* genes during pathogen infection, two ESTs, almost identical to previously found in senescent leaves (Beuning et al., 2004), were found in apple leaves challenged with *V. inaequalis*, the causal agent of apple scab: one of them was up-regulated and the other one was down-regulated, suggesting differences of functions among isoforms (Paris et al., 2009).

The involvement of *Mal d 2* in plant disease resistance can be confirmed by the numerous ESTs found in public database obtained from experiments on leaves challenged with apple scab (*V. inaequalis*) and leaf roller (*C. rosaceana*). Thaumatin-like proteins were constitutively highly expressed in the apoplastic fluid of healthy leaves of the scab-resistant cultivar Remo, suggesting a protective role against pathogens; moreover, they accumulated rapidly after the infection of *V. inaequalis* in the susceptible cultivar Elstar, confirming their involvement during the early stage of pathogen infection (Gau et al., 2004). Bonasera et al. (2006) detected enhanced expression of *Mal d 2* genes in leaves of 1-year-old apple plantlets, in response to inoculation with the pathogen *E. amylovora*. By contrast, thaumatin-like proteins isolated from apple fruit did not show any antifungal activity against *V. alboatrum*, even if they demonstrated endo-glucanase activity, suggesting distinct roles among isoforms expressed in different plant tissues inside this family (Menu-Bouaouiche et al., 2003).

Finally, the amount of *Mal d 3* protein content declined to a non-detectable level within the first week after infection by *V. inaequalis* (Gau et al., 2004), in agreement with the down-regulation of one *Mal d 3* EST recently found by Paris et al. (2009), using the same fungal spp. challenge.

3.4 Post-harvest, food processing and breeding strategies towards allergenic content decrease

3.4.1 Post-harvest and food processing

Apples are harvested at ripening time, but they are available for consumption for several months, due to their intrinsic preservation capabilities and to the controlled atmosphere in the conservation chambers. The controlled atmosphere (low O_2 and elevated CO_2 concentrations) and low temperature have the effect of reducing ethylene synthesis, extending the storage time (Gorny and Kader, 1996). However, post-harvest treatments may have an influence on fruit allergenicity. Both *Mal d 1* and *Mal d 2* protein levels increased during prolonged storage at 4°C, for commercial apple varieties as Granny Smith and Golden Delicious (Hsieh et al., 1995), unrelated to cultivation technique, orchard or variety (Matthes and Schmitz-Eiberger, 2009) and storage techniques (Sancho et al., 2006a; Figure 3.5). *Mal d 1* content, measured by ELISA test, increased also in low-allergenic varieties, such as Elise and Santana; furthermore, fruits treated with the addition to atmosphere of ethylene inhibitor 1-MCP had a significantly lower *Mal d 1* content than fruits without 1-MCP (Kiewning and Schmitz-Eiberger, 2014). In contrast, different storage conditions, in particular the controlled atmosphere, were proven to be efficient

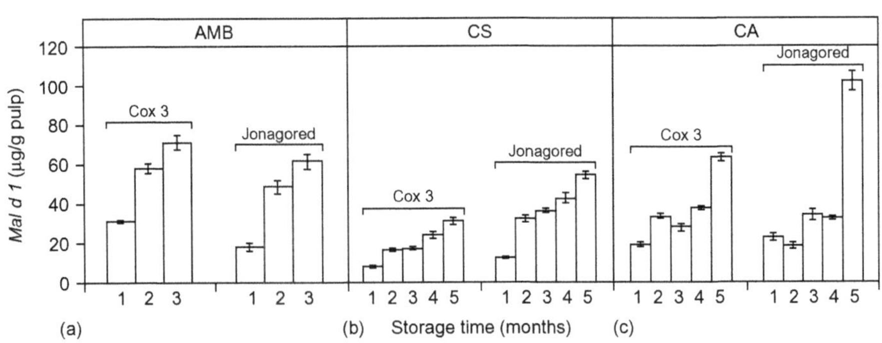

Figure 3.5 Effect of storage time and cultivar (Cox and Jonagored) on *Mal d 1* levels (μg/g pulp) in apples stored under (a) ambient conditions (AMB, 20°C), (b) cold with no controlled atmosphere (CS, 2°C) or (c) controlled atmosphere conditions (CA, <0.5% CO_2 and 1.25% O_2 at 3.8°C) for up to 5 months.

Source: From Sancho et al. (2006a).

post-harvest treatments in reducing *Mal d 3* content in apple fruits (Sancho et al., 2006b). After 5-month storage at low temperature, several isoforms of *Mal d 1* and *Mal d 2* genes were up-regulated, while *Mal d 3* and *Mal d 4* genes were down-regulated, confirming previously proteomic data (Botton et al., 2008). Thus, the allergenic content growth in post-harvest seems correlated to cold stress during storage in low temperature chambers (Matthes and Schmitz-Eiberger, 2009).

Even if apples are more suitable as fresh fruits, for allergenic people processed apples can be an alternative, but frequently apple food as puree, compote or chips still contain detectable allergens. In apple juice and puree, *Mal d 1* allergen can be effectively denatured and allergenic activity destroyed by pasteurization thermal processing. However, differences exist inside the same family, with apple *Mal d 1* totally denatured by heating, while other homologues from celery, peach and carrot were more thermostable (Wigotzki, 2001). By contrast, the ingestion of cooked apple foods did not induce OAS but probably late eczematous skin reactions (Bohle et al., 2006). A difference in thermostability was found among the four *Mal d* protein families, with *Mal d 2* and *Mal d 3* the most thermoresistant (Smole et al., 2008). As a consequence, the view that cooked pollen-allergenic-related foods can be consumed without allergologic consequences should be reconsidered. The addition of oxidizing enzymes could link phenolic compound to the allergen, reducing *Mal d 1* capability to react with the human IgE and the allergenic potential of apple extracts (Garcia et al., 2007). Recently, a novel interesting food processing technology, the combination of high pressure (HP) and mild thermal treatment, demonstrated to be effective in the inactivation of the main allergens *Mal d 1* and *Mal d 3*, opening new perspective for apple-sensitive consumers (Husband et al., 2011).

3.4.2 Breeding programs

About 1500 cultivars (Della Strada and Fideghelli, 2011) and several thousand accessions are available for breeders, dispersed in public centres and private collections, and the possibility to find differences in allergenic extent is reasonable. The importance of genotype on *Mal d* gene expression profile is determinant; the early-harvested apples show the highest allergenic potential, whereas those with the lowest amount of allergen-related transcripts are the late-ripening ones (Botton et al., 2008). On the other hand, discrepancies between genetic studies and clinical trials are common. Clinical studies advise that most of the commercialized apple cultivars are dangerous for allergic people. In particular, Golden Delicious, Elstar, Fiesta, Pinova and Jonathan appear to be among the most allergenic apple cultivars in several clinical surveys (Bolhaar et al., 2005a; Gao et al., 2008; Ricci et al., 2010; Vlieg-Boerstra et al., 2011). Nevertheless, traditionally bred apple cultivars with low-allergenic potential were identified and one of them, Santana, is now commercialized as hypoallergenic apple and consumed by people with mild allergenic history, generally suffering OAS syndrome (Bolhaar et al., 2005a). Even if Santana is still not well known among dealers and consumers, this variety seems to be well accepted to allergic people, who can for the first time approach the taste of this fruit. Moreover, its scab-resistance trait can be attractive to organic producers because of lower pesticides employed on this crop (Schenk et al., 2011).

The multi-genic complexity of the allergenic trait does not help plant geneticists in breeding programs. Reducing the target genes to a few ones can help to find suitable molecular markers for a fast screening of the germplasm, but the production of new variety by classical breeding has still the disadvantage to be time-consuming. Moreover, the function of several *Mal d* isoforms in plant defence, development, fruit quality traits and in still unknown pathways can produce unwanted characteristics and waste of time for plant selection. Recently, the feasibility of the RNA interference (RNAi) approach was successfully demonstrated by Gilissen et al. (2005). In this study the expression of *Mal d 1* gene family was drastically reduced in apple plantlets transformed with a hairpin structure designed on the *Mal d 1.02* sequence (NCBI AF020542). Moreover, the leaves of the transformed apple plantlets showed a consistent *in vivo* reduced allergenicity on patient's SPT. On the other hand, the cisgenic–transgenic approach, despite its utility in speed-up of the breeding time, may differ in consumer acceptance, which opinion should be considered by the breeder's choice before starting any long-lasting breeding programme (Schenk et al., 2011).

3.5 Conclusion

Apple allergy can be considered a new subject for breeders with a forwarded consumer-oriented approach, because it is highly related to fruit quality and health. Even if this allergy is not the most widespread and dangerous for human health, increasing the consumption of this fruit should be one of the first aims, in particular among children and young adults, who are more susceptible to food allergies. Despite the important steps achieved, we still need to improve our understanding of apple allergen biochemistry and interaction with the immune system, leading to more efficient diagnostic and immunotherapy techniques, other than to precise detection of these allergens in processed foods. Future studies should be organized in a multidisciplinary way, in order to achieve a complete picture and to find out the determinants of this allergy (Figure 3.1).

The complexity and the uncertainty of isoforms can be a disadvantage for starting a breeding programme designed to reduce the allergenic content; another disadvantage could be the long-lasting procedures for selection. The use of the most advanced techniques for accelerating the breeding procedures (i.e. transgenic, cisgenic or RNAi) should be accurately evaluated in accordance to the European Politics on GM and after a consumer screening for acceptance of GM varieties. On the other hand, there are many accessions and cultivars with still unknown allergenic potential, and some hypoallergenic varieties could be discovered and employed for a local commercialization, making apples a friendly fruit for people who never appreciated their taste before.

References

Asero, R., Mistrello, G., Roncarolo, D., de Vries Sacco, C., Gautier, M.F., Ciurana, C.L.F., et al., 2000. Lipid transfer protein: a panallergen in plant-derived foods that is highly resistant to pepsin digestion. Int. Arch. Allergy Immunol. 122, 20–32.

Atkinson, R.G., Perry, J., Matsui, T., Ross, G.S., MacRae, E.A., 1996. A stress, pathogenesis, and allergen-related cDNA in apple fruit is also ripening-related. New Zeal. J. Crop Hort. 24, 103–107.

Beuning, L.L., Bowen, J.H., Persson, H.A., Barraclough, D., Bulley, S., MacRae, E.A., 2004. Characterisation of Mal d 1-related genes in Malus. Plant Mol. Biol. 55, 369–388.

Bohle, B., Zwolfer, B., Heratizadeh, A., Jahn-Schmid, B., Dall Antonia, Y., Alter, M., et al., 2006. Cooking birch pollen–related food: divergent consequences for IgE- and T cell-mediated reactivity *in vitro* and *in vivo*. J. Allergy Clin. Immunol. 118, 1242–1249.

Bolhaar, S.T., van de Weg, W.E., van Ree, R., Gonzalez-Mancebo, E., Zuidmeer, L., Bruijnzeel-Koomen, C.A.F.M., et al., 2005a. *In vivo* assessment with prick-to-prick testing and double blind, placebo-controlled food challenge of allergenicity of apple cultivars. J. Allergy Clin. Immunol. 116, 1080–1086.

Bolhaar, S.T., Zuidmeer, L., Ma, Y., Ferreira, F., Bruijnzeel-Koomen, C.A., Hoffmann-Sommergruber, K., et al., 2005b. A mutant of the major apple allergen, *Mal d 1*, demonstrating hypo-allergenicity in the target organ by double-blind placebo-controlled food challenge. Clin. Exp. Allergy. 35, 1638–1644.

Bolhaar, S.T.H.P., Tiemessen, M.M., Zuidmeer, L., van Leeuwen, A., Hoffmann-Sommergruber, K., Bruijnzeel-Koomen, C.A.F.M., et al., 2004. Efficacy of birch-pollen immunotherapy on cross-reactive food allergy confirmed by skin tests and double-blind food challenges. Clin. Exp. Allergy. 34, 761–769.

Bonasera, J.M., Kim, J.F., Beer, S.V., 2006. PR genes of apple: identification and expression in response to elicitors and inoculation with *Erwinia amylovora*. BMC Plant Biol. 6, 23.

Borges, J.P., Jauneau, A., Brulé, C., Culerrier, R., Barre, A., Didier, A., et al., 2006. The lipid transfer proteins (LTP) essentially concentrate in the skin of Rosaceae fruits as cell surface exposed allergens. Plant Physiol. Biochem. 44, 535–542.

Borges, J.P., Barre, A., Culerrier, R., Granier, C., Didier, A., Rouge, P., 2008. Lipid transfer proteins from Rosaceae fruits share consensus epitopes responsible for their IgE-binding cross-reactivity. Biochem. Biophys. Res. Commun. 365, 685–690.

Botton, A., Lezzer, P., Dorigoni, A., Barcaccia, G., Ruperti, B., Ramina, A., 2008. Genetic and environmental factors affecting allergen-related gene expression in apple fruit (*Malus domestica* L. Borkh). J. Agric. Food Chem. 56, 6707–6716.

Breiteneder, H., 2004. Thaumatin-like proteins – a new family of pollen and fruit allergens. Allergy. 59, 479–481.

Breiteneder, H., Ebner, C., 2000. Molecular and biochemical classification of plant-derived food allergens. J. Allergy Clin. Immunol. 106, 27–36.

Buhot, N., Douliez, J.P., Jacquemard, A., Marion, D., Tran, V., Maume, B.F., et al., 2001. A lipid transfer protein binds to a receptor involved in the control of plant defence responses. FEBS Lett. 509, 27–30.

Chen, L., Zhang, S., Illa, E., Song, L., Wu, S., Howad, W., et al., 2008. Genomic characterization of putative allergen genes in peach/almond and their synteny with apple. BMC Genomics. 9, 543.

Cortegano, I., Civantos, E., Aceituno, E., del Moral, A., Lopez, E., Lombardero, M., et al., 2004. Cloning and expression of a major allergen from *Cupressus arizonica* pollen, Cup a 3, a PR-5 protein expressed under polluted environment. Allergy. 59, 485–490.

Cromwell, O., Hafner, D., Nandy, A., 2011. Recombinant allergens for specific immunotherapy. J. Allergy Clin. Immunol. 127, 865–872.

de Oliveira Carvalho, A., Moreira Gomes, V., 2007. Role of plant lipid transfer proteins in plant cell physiology – a concise review. Peptides. 1144–1153.

Degenhardt, J., Al-Masri, A.N., Kürkcüoglu, S., Szankowski, I., Gau, A.E., 2005. Characterization by suppression subtractive hybridization of transcripts that are differentially expressed in leaves of apple scab-resistant and susceptible cultivars of *Malus domestica*. Mol. Genet. Genomics. 273, 326–335.

Della Strada, G., Fideghelli, C., 2011. The fruit varieties released in the world from 1980 through 2008. CRA-Centro di Ricerca per la Frutticoltura- Ministero Politiche Agricole e Forestali (Ed).

Diaz-Perales, A., Lombardero, M., Sanchez-Monge, R., Garcia-Selles, F.J., et al., 2000. Lipid-transfer protein as potential plant-panallergens: cross-reactivity among proteins of Artemisia pollen, Castanea, nut and Rosaceae fruits, with different IgE-binding capacities. Clin. Exp. Allergy. 30, 1403–1410.

Diaz-Perales, A., Garcia-Casado, G., Sanchez-Monge, R., Garcia-Selles, F.J., Barber, D., Salcedo, G., 2002. cDNA cloning and heterologous expression of the major allergens from peach and apple belonging to the lipid-transfer protein family. Clin. Exp. Allergy. 32, 87–92.

Ebner, C., Birkner, T., Valenta, R., Rumpold, H., Breitenbach, M., Scheiner, O., et al., 1991. Common epitopes of birch pollen and apples studies by western and northern blot. J. Allergy Clin. Immunol. 88, 588–594.

Ebner, C., Hirschwehr, R., Bauer, L., Breiteneder, H., Valenta, R., Ebner, H., et al., 1995. Identification of allergens in fruits and vegetables – IgE cross-reactivities with the important birch pollen allergens *Bet V 1* and *Bet V 2* (birch profilin). J. Allergy Clin. Immunol. 95, 962–969.

Edreva, A., 2005. Pathogenesis-related proteins: research progress in the last 15 years. Gen. Appl. Plant Physiol. 31, 105–124.

Eriksson, N.E., Formgren, H., Svenonius, E., 1982. Food hypersensitivity in patients with pollen allergy. Allergy. 37, 437–443.

Fernandez-Rivas, M., Bolhaar, D., Gonzalez-Mancebo, E., Asero, R., Van Leeuwen, A., Bohle, B., et al., 2006. Apple allergy across Europe: how allergen sensitization profiles determine the clinical expression of allergies to plant foods. J. Allergy Clin. Immunol. 118, 481–488.

Ferreira, F., Ebner, C., Kramer, B., Casari, G., Briza, P., Kungl, A.J., et al., 1998. Modulation of IgE reactivity of allergens by site-directed mutagenesis: potential use of hypoallergenic variants for immunotherapy. FASEB J. 12, 231–242.

Fritsch, R., Bohle, B., Vollmann, U., Wiedermann, U., Jahn-Schmid, B., Krebitz, M., et al., 1998. *Bet v 1*, the major birch pollen allergen, and *Mal d 1*, the major apple allergen, cross-react at the level of allergen specific T helper cells. J. Allergy Clin. Immunol. 102, 679–686.

Gao, Z.S., van de Weg, W.E., Schaart, J.G., Schouten, H.J., Tran, D.H., Kodde, L.P., et al., 2005a. Genomic cloning and linkage mapping of the *Mal d 1* (PR-10) gene family in apple (*Malus domestica*). Theor. Appl. Genet. 111, 171–183.

Gao, Z.S., van de Weg, W.E., Schaart, J.G., Arkel, G., Breiteneder, H., Hoffmann-Sommergruber, K., et al., 2005b. Genomic characterization and linkage mapping of the apple allergen genes *Mal d 2* (thaumatin-like protein) and *Mal d 4* (profilin). Theor. Appl. Genet. 111, 1087–1097.

Gao, Z.S., van de Weg, W.E., Schaart, J.G., van der Meer, I.M., Kodde, L., Laimer, M., et al., 2005c. Linkage map positions and allelic diversity of two *Mal d 3* (nonspecific lipid transfer protein) genes in the cultivated apple (*Malus domestica*). Theor. Appl. Genet. 110, 479–491.

Gao, Z.S., van de Weg, W.E., Matos, C.I., Arens, P., Bolhaar, S.T.H.P., Knulst, A.C., et al., 2008. Assessment of allelic diversity in intron-containing *Mal d 1* genes and their association to apple allergenicity. BMC Plant Biol. 8, 116–128.

Garcia, A., Wichers, J.H., Wichers, H.J., 2007. Decrease of the IgE-binding by *Mal d 1*, the major apple allergen, by means of polyphenol oxidase and peroxidase treatments. Food Chem. 103, 94–100.

Garcia-Casado, G., Pacios, L.F., Diaz-Perales, A., Sanchez-Monge, R., Lombardero, M., Garcia-Selles, F.J., et al., 2003. Identification of IgE-binding epitopes of the major peach allergen Pru p 3. J. Allergy Clin. Immunol. 112, 599–605.

Garcia-Olmedo, F., Molina, A., Segura, A., Moreno, M., 1995. The defensive role of nonspecific lipid-transfer proteins in plants. Trends Microbiol. 3, 72–74.

Gau, A.E., Koutb, M., Piotrowski, M., Kloppstech, K., 2004. Accumulation of pathogenesis-related proteins in the apoplast of a susceptible cultivar of apple (*Malus domestica* cv. Elstar) after infection by *Venturia inaequalis* and constitutive expression of PR genes in the resistant cultivar Remo. Eur. J. Plant Pathol. 110, 703–711.

Gilissen, L.J.W.J., Bolhaar, S.T.H.P., Matos, C.I., Rouwendal, G.J.A., Boone, M.J., Krens, F.A., et al., 2005. Silencing the major apple allergen *Mal d 1* by using the RNA interference approach. J. Allergy Clin. Immunol. 115, 364–369.

Gorny, J.R., Kader, A.A., 1996. Controlled-atmosphere suppression of ACC synthase and ACC oxidase in 'Golden Delicious' Apples during long-term cold storage. J. Am. Soc. Hortic. Sci. 121, 751–755.

Hernd, A., Marzban, G., Kolarich, D., Hahn, R., Boscia, D., Hemmer, W., et al., 2007. Mapping of *Malus domestica* allergens by 2-D electrophoresis and IgE-reactivity. Electrophoresis. 28, 437–448.

Hoffmann-Sommergruber, K., 2002. Pathogenesis-related (PR)-proteins identified as allergens. Biochem. Soc. Trans. 30, 930–935.

Hoffmann-Sommergruber, K., Vanek-Krebitz, M., Radauer, C., Wen, J., Ferreira, F., Scheiner, O., et al., 1997. Genomic characterization of members of the Bet v 1 family: genes coding for allergens and pathogenesis-related proteins share intron positions. Gene. 197, 91–100.

Holm, J., Ferreras, M., Ipsen, H., Wurtzen, P.A., Gajhede, M., Larsen, J.N., et al., 2011. Epitope grafting, re-creating a conformational Bet v 1 antibody epitope on the surface of the homologous apple allergen Mal d 1. J. Biol. Chem. 286, 17569–17578.

Hsieh, L.S., Moos, M., Yuan Lin, Y., 1995. Characterization of apple 18 and 31 kd allergens by microsequencing and evaluation of their content during storage and ripening. J. Allergy Clin. Immunol. 96, 960–970.

Husband, F.A., Aldick, T., Van der Plancken, I., Grauwet, T., Hendrickx, M., Skypala, I., et al., 2011. High-pressure treatment reduces the immunoreactivity of the major allergens in apple and celery. Mol. Nutr. Food Res. 55, 1087–1095.

Iorio, R.A., Di Sandro, A., Paris, R., Pagliarani, G., Tartarini, S., Ricci, G., et al., 2012. Simulated environmental criticalities affect transglutaminase of Malus and Corylus pollens having different allergenic potential. Amino Acids. 42, 1007–1024.

Jayaraj, J., Punja, Z.K., 2007. Combined expression of chitinase and lipid transfer protein genes in transgenic carrot plants enhances resistance to foliar fungal pathogens. Plant Cell Rep. 26, 1539–1546.

Jung, H.W., Kim, W., Hwang, B.K., 2003. Three pathogen-inducible genes encoding lipid transfer protein from pepper are differentially activated by pathogens, abiotic, and environmental stresses. Plant Cell Environ. 26, 915–928.

Kader, J.C., 1996. Lipid-transfer proteins in plants. Ann. Rev. Plant Physiol. Plant Mol. Biol. 47, 627–654.

Kiewning, D., Schmitz-Eiberger, M., 2014. Effects of long-term storage on Mal d 1 contents of four apple cultivars with initial low Mal d 1 contents. J. Sci. Food Agric. 94, 798–802.

Kootstra, H.S., Vlieg-Boerstra, B.J., Dubois, A.E.J., 2007. Assessment of the reduced properties of the Santana apple. Ann. Allergy Asthma Immunol. 99, 522–525.

Kortekamp, A., 2006. Expression analysis of defence-related genes in grapevine leaves after inoculation with a host and a non-host pathogen. Plant Physiol. Biochem. 44, 58–67.

Krebitz, M., Wagner, B., Ferreira, F., Peterbauer, C., Campillo, N., Witty, M., et al., 2003. Plant-based heterologous expression of *Mal d 2*, a thaumatin-like protein and allergen of apple (*Malus domestica*) and its characterization as an antifungal protein. J. Mol. Biol. 329, 721–730.

Ma, Y., Gadermaier, G., Bohle, B., Bolhaar, S., Knulst, A., Markovic-Housley, Z., et al., 2006a. Mutational analysis of amino acid positions crucial for IgE binding epitopes of the major apple (*Malus domestica*) allergen, *Mal d 1*. Int. Arch. Allergy Immunol. 139, 53–62.

Ma, Y., Zuidmeer, L., Bohle, B., Bolhaar, S.T.H., Gadermaier, G., Gonzalez-Mancebo, E., et al., 2006b. Characterization of recombinant *Mal d 4* and its application for component resolved diagnosis of apple allergy. Clin. Exp. Allergy. 36, 1087–1096.

Marzban, G., Helene, H., Dey, R., Brynda, S., Ma, Y., Martinelli, A., et al., 2005. Localisation and distribution of the major allergens in apple fruits. Plant Sci. 169, 387–394.

Matthes, A., Schmitz-Eiberger, M., 2009. Apple (*Malus domestica* L. Borkh.) allergen *Mal d 1*: effect of cultivar, cultivation system, and storage conditions. J. Agr. Food Chem. 57, 10548–10553.

Menu-Bouaouiche, L., Vriet, C., Peumans, W.J., Barre, A., Van Damme, E.J.M., Rougé, P., 2003. A molecular basis for the endo-b1,3-glucanase activity of the thaumatin-like proteins from edible fruits. Biochimie. 85, 123–131.

Mills, E.N.C., Madsen, C., Shewry, P.R., Wichers, H.J., 2003. Food allergens of plant origin – their molecular and evolutionary relationships. Trends Food Sci. Technol. 14, 145–156.

Oh, D.H., Song, K.J., Shin, Y.U., Chung, W.I., 2000. Isolation of a cDNA encoding a 31-kDa, pathogenesis-related 5 thaumatin-like (PR5/TL) protein abundantly expressed in apple fruit (*Malus domestica* cv. Fuji). Biosci. Biotechnol. Biochem. 64, 355–362.

Pagliarani, G., Paris, R., Tartarini, S., Sansavini, S., 2009. Cloning and expression of the major allergen genes in apple fruit. J. Hortic. Sci. Biotechnol. 176–181, ISAFRUIT (special issue).

Pagliarani, G., Paris, R., Iorio, A.R., Tartarini, S., Del Duca, S., Arens, P., et al., 2012. Genomic organisation of the *Mal d 1* gene cluster on linkage group 16 in apple. Mol. Breed. 29, 759–778.

Pagliarani, G., Paris, R., Arens, P., Tartarini, S., Ricci, G., Smulders, M.J.M., et al., 2013. A qRT-PCR assay for the expression of all Mal d 1 isoallergen genes. BMC Plant Biol. 13, 51.

Paris, R., Cova, V., Pagliarani, G., Tartarini, S., Komjanc, M., Sansavini, S., 2009. Expression profiling in HcrVf2-transformed apple plants in response to *Venturia inaequalis*. Tree Genet. Genomes. 5, 81–91.

Pastorello, E.A., Rivolta, F., 2004. Clinical role of lipid transfer proteins in food allergy. Mol. Nutr. Food Res. 48, 356–362.

Pastorello, E.A., Pravettoni, V., Farioli, L., Ispano, M., Fortunato, D., Monza, M., et al., 1999. Clinical role of a lipid transfer protein that acts as a new apple-specific allergen. J. Allergy Clin. Immunol. 104, 1099–1106.

Puehringer, H., Moll, D., Hoffmann-Sommergruber, K., Watillon, B., Katinger, H., Laimer da Câmara Machado, M., 2000. The promoter of an apple Ypr10 gene, encoding the major allergen *Mal d 1*, is stress- and pathogen-inducible. Plant Sci. 152, 35–50.

Puehringer, H., Zinoecker, I., Marzban, G., Katinger, H., Laimer, M., 2003. MdAP, a novel protein in apple, is associated with the major allergen *Mal d 1*. Gene. 321, 173–183.

Radauer, C., Willerroiderw, M., Fuchs, H., Hoffmann-Sommergruber, K., Thalhamerw, J., Ferreira, F., et al., 2006. Cross-reactive and species-specific immunoglobulin E epitopes of plant profilins: an experimental and structure-based analysis. Clin. Exp. Allergy. 36, 920–929.

Ricci, G., Dondi, A., Belotti, T., Baldi, E., Tartarini, S., Paris, R., et al., 2010. Allergenicity of different apple cultivars assessed by means of skin prick test and sensitisation to recombinant allergens *Mal d 1* and *Mal d 3* in a group of Italian apple-allergic patients. Int. J. Food Sci. Technol. 45, 1517–1523.

Roulias, A., Pichler, U., Hauser, M., Himly, M., Hofer, H., Lackner, P., et al., 2014. Differences in the intrinsic immunogenicity and allergenicity of Bet v 1 and related food allergens revealed by site-directed mutagenesis. Allergy. 2014 (69), 208–215.

Rudeschko, O., Fahlbusch, B., Henzgen, M., Schlenvoigt, G., Herrmann, D., Vieths, S., et al., 1995. Investigation of the stability of apple allergen extracts. Allergy. 50, 575–580.

San Miguel-Moncin, M., Krail, M., Scheurer, S., Enrique, E., et al., 2003. Lettuce anaphylaxis: identification of a lipid transfer protein as the major allergen. Allergy. 58, 511–517.

Sancho, A.I., Foxall, R., Browne, T., Dey, R., Zuidmeer, L., Marzaban, G., et al., 2006a. Effect of postharvest storage on the expression of the apple allergen *Mal d 1*. J. Agric. Food Chem. 54, 5917–5923.

Sancho, A.I., Foxall, R., Rigby, N.M., Browne, T., Zuidmeer, L., van Ree, R., et al., 2006b. Maturity and storage influence on the apple (*Malus domestica*) allergen *Mal d 3*, a non-specific lipid transfer protein. J. Agric. Food Chem. 54, 5098–5104.

Schenk, M.F., van der Maas, M.P., Smulders, M.J.M., Gilissen, L.J.W.J., Fischer, A.R.H., van der Lans, I.A., et al., 2011. Consumer attitudes towards hypoallergenic apples that alleviate mild apple allergy. Food Qual. Preference. 22, 83–91.

Scheurer, S., Wangorsch, A., Nerkamp, J., Skov, P.S., Ballmer-Weber, B., Wuthrich, B., et al., 2001. Cross-reactivity within the profilin panallergen family investigated by comparison of recombinant profilins from pear (*Pyr c 4*), cherry (*Pru av 4*) and celery (*Api g 4*) with birch pollen profilin *Bet v 2*. J. Chromatogr. B Analyt. Technol. Biomed. Life Sci. 756, 315–325.

Schmidt-Andersen, M.B., Hall, S., Dragsted, L.O., 2011. Identification of European allergy patterns to the allergen families PR-10, LTP, and profilin from Rosaceae fruits. Clin. Rev. Allergy Immunol. 41, 4–19.

Smole, U., Bublin, M., Radauer, C., Ebner, C., Breiteneder, H., 2008. *Mal d 2*, the thaumatin-like allergen from apple, is highly resistant to gastrointestinal digestion and thermal processing. Int. Arch. Allergy Immunol. 147, 289–298.

Son, D.Y., Scheurer, S., Hoffmann, A., Haustein, D., Vieths, S., 1999. Pollen-related food allergy: cloning and immunological analysis of isoforms and mutants of *Mal d 1*, the major apple allergen, and *Bet v 1*, the major birch pollen allergen. Eur. J. Nutr. 38, 201–215.

Spangfort, M.D., Gajhede, M., Osmark, P., Poulsen, F.M., Ipsen, H., Larsen, J.N., et al., 1997. Three-dimensional structure and epitopes of Bet v 1. Int. Arch. Allergy Immunol. 113, 243–245.

Sterk, P., Booij, H., Schellekens, G.A., Van Kammen, A., De Vries, S.C., 1991. Cell-specific expression of the carrot EP2 lipid transfer protein gene. Plant Cell. 3, 907–922.

Togawa, A., Panzani, R.C., Garza, M.A., Kishikawa, R., Goldblum, R.M., Midoro-Horiuti, T., 2006. Identification of Italian cypress (*Cupressus sempervirens*) pollen allergen *Cup s 3* using homology and cross-reactivity. Ann. Allergy Asthma Immunol. 97, 336–342.

Valenta, R., Duchene, M., Pettenburger, K., Sillaber, C., Valent, P., Bettelheim, P., et al., 1991. Identification of profilin as a novel pollen allergen: IgE autoreactivity in sensitized individuals. Science. 253, 557–560.

Valenta, R., Duchene, M., Ebner, C., Valent, P., Sillaber, C., Deviller, P., et al., 1992. Profilins constitute a novel family of functional plant pan-allergens. J. Exp. Med. 175, 377–385.

van Loon, L.C., Rep, M., Pieterse, C.M., 2006. Significance of inducible defense-related proteins in infected plants. Annu. Rev. Phytopathol. 44, 135–162.

van Ree, R., Fernandez-Rivas, M., Cuevas, M., van Wijngaarden, M., Aalberse, R.C., 1995. Pollen-related allergy to peach and apple: an important role for profilin. J. Allergy Clin. Immunol. 95, 726–734.

Vanek-Krebitz, M., Hoffmann-Sommergruber, K., Laimer da Camara Machado, M., Susani, M., Ebner, C., Kraft, D., et al., 1995. Cloning and sequencing of *Mal d 1*, the major allergen from apple (*Malus domestica*), and its immunological relationship to *Bet v 1*, the major birch pollen allergen. Biochem. Biophys. Res. Commun. 214, 538–551.

Velasco, R., Zharkikh, A., Affourtit, J., Dhingra, A., Cestaro, A., Kalyanaraman, A., et al., 2010. The genome of the domesticated apple (*Malus* x *domestica* Borkh.). Nat. Genet. 42 (10), 833–841.

Vieths, S., Janek, K., Aulepp, H., Petersen, A., 1995. Isolation and characterization of the 18-kDa major apple allergen and comparison with the major birch pollen allergen (*Bet v* I). Allergy. 50, 421–430.

Vlieg-Boerstra, B.J., van de Weg, W.E., van der Heide, S., Kerkhof, M., Arens, P., Heijerman-Peppelman, G., et al., 2011. Identification of low allergenic apple cultivars using skin prick test and oral food challenges. Allergy. 66, 491–498.

Vlieg-Boerstra, B.J., van de Weg, W.E., van der Heide, S., Skypala, I., Bures, P., Ballmer-Weber, B.K., et al., 2013. Additional indications for the low allergenic properties of the apple cultivars Santana and Elise. Plant Foods Hum. Nutr. 68, 391–395.

Wigotzki, M., 2001. Einfluß der technologischen Verarbeitung auf das allergene Potential der birken pollen assoziierten Lebensmittel Haselnuß, Apfel und Karotte (PhD discussion thesis). <http://www.chemie.uni-hamburg.de/bibliothek/diss2001/DissertationWigotzki.pdf>.

Yang, X.T., Song, J., Campbell-Palmer, L., Walker, B., Zhang, Z., 2012. Allergen related gene expression in apple fruit is differentially controlled by ethylene during ripening. Postharvest Biol. Technol. 63, 40–49.

Ziadi, S., Poupard, P., Brisset, M.N., Paulin, J.P., Simoneau, P., 2001. Characterization in apple leaves of two subclasses of PR-10 transcripts inducible by acibenzolar-S-methyl, a functional analogue of salicylic acid. Physiol. Mol. Plant Pathol. 59, 33–43.

Zuidmeer, L., Goldhahn, K., Rona, R.J., Gislason, D., Madsen, M.D., Summers, C., et al., 2008. The prevalence of plant food allergies: a systematic review. J. Allergy Clin. Immunol. 121, 1210–1218.

Non-food interventions: Exploring plant biotechnology applications to therapeutic protein production

4

Matteo Busconi, Mariangela Marudelli and Corrado Fogher
Istituto di Agronomia, Genetica e Coltivazioni erbacee, Università Cattolica del Sacro Cuore, Piacenza, Italy

4.1 Introduction

The impressive advances of biotechnology in the last 20 years has greatly promoted the development of genetically modified (GM) crops worldwide. First-generation genetically modified organisms (GMOs), whose cultivation began in 1996, and second-generation GMOs were usually create in order to improve the agronomic properties and the nutritional value of major food crops. To obtain this, a large number of transgenes have been successfully transferred into crops through biotechnology approaches (mainly by *Agrobacterium tumefaciens* and gene gun-based methodologies) to induce, among the others: herbicide resistance (such as LibertyLink and Roundup Ready rice), insect (such as MON810 maize) and disease resistance, salt and drought tolerance, yield increase and quality improvement (Lu and Yang, 2009). The agronomic improved varieties with enhanced tolerance to herbicides and insects have been successfully cultivated by millions of farmers worldwide being planted in 2011 on at least 160 million of hectares in 29 countries (James, 2012). These transformations, sometimes involving the insertion of single genes (as for herbicide-resistant varieties) and other times involving the reconstruction of entire and novel metabolic pathways in the host plants (as for golden rice; Ye et al., 2000), were mainly thought to reduce the environmental impact of agriculture or to improve quality, quantity and safety of food at affordable price to meet the feeding demand of the increasing world population.

At the same time, alternative use of GM plants have been explored and, in the recent years, several works carried out by independent research teams throughout the world successfully resulted in the development of transformed plants for various applications that can be globally referred to as non-food intervention because these do not affect and have nothing to do with food. More in details in the last 10 years an increased number of papers, among the other, reported on:

- Development of GM plants with enhanced biodegradation and phytoremediation capacity of organic xenobiotics, heavy metals, metalloids, soil and sediment pollution (Abhilash et al., 2009; Kotrba et al., 2009)

Applied Plant Genomics and Biotechnology.

- Plant biotechnology solutions for bioenergy by means of increasing biomass production and yield, modifying lignin biosynthesis and pre-processing of biomass *in planta* by expressing cellulases and cellulosomes (Yuan et al., 2008)
- Use of plants as heterologous expression systems to obtain high levels of products with high commercial value, the application of biotechnology to this aim takes the name of molecular farming

Molecular farming, in principle, is a term referring to the use of GM crops (rather than animals) to produce highly priced compounds. There are two types of molecular farming: non-medical and medical molecular farming. The first one refers to the production of industrial enzymes and polymers, one of the greatest development in this field is represented by the expression of biopolymers as bacterial polyhydroxyalkanoates for bio-plastic production in high biomass crops (Tilbrook et al., 2011). Also the expression in plants of cellulosolytic enzymes for bioenergy production can be considered as a form of non-medical molecular farming. On the other hand, medical molecular farming refers to the use of transgenic plants to produce biopharmaceuticals (Daniell et al., 2009; Tacket, 2009; Paul et al., 2011; Thomas et al., 2011). Medical molecular farming represents an unprecedented opportunity to manufacture affordable medicines and make them available at a global scale especially in underdeveloped countries (Paul et al., 2011). This chapter will focus mainly on the applications in the field of medical molecular farming.

4.2 Plant as heterologous expression system for molecular farming

Concerning the evolution of plant-made pharmaceuticals (PMPs), the first feasibility studies and evidences that plants could be considered as alternative heterologous production systems for therapeutic proteins have been obtained in the early 1990s (Thomas et al., 2011). Among the first works referred to the possible use of plants to obtain plant-made vaccines and antibodies, Mason et al. (1992) expressed hepatitis B surface antigen in transgenic tobacco plants while few years later (Ma et al., 1994) the successful production and assembly of monoclonal antibodies in transgenic tobacco was reported. This last study demonstrated that transgenic plants possess the requested characteristics to produce and correctly assemble biologically active complex proteins with intricate quaternary structure. In synthesis, a functional antibody, also in its simplest form and not considering glycosylation, is made by two heavy and two light chains connected with disulphide bridges. According to the requirements, plants can be engineered to produce a wide range of compounds from simple molecules to complex therapeutic proteins.

In a recent report it has been reported that over 200 biopharmaceuticals had entered the market until 2010 (Walsh, 2010). All these pharmaceuticals have been produced by means of recombinant DNA technology with the following heterologous expression systems: bacterial fermentations, yeast cell cultures and mammalian cell cultures. Up to 2007, 57% of the recombinant biologics have been obtained in mammalian cell cultures (mainly CHO – Chinese Hamster Ovary cells) while the remaining 43% have been obtained in bacteria or yeasts (Boehm, 2007). Despite their widespread use, all these methods present, along with some advantages, several limitations mainly due to the need for (i) expensive infrastructure and media components (in particular for mammalian cells) that influence scalability form laboratory to industry; and (ii) highly accurate downstream processing to eliminate the risk of bacterial toxins, viral or oncogenic contamination of the final product that could pose safety issues for patients. Moreover, bacteria usually produce high level of protein aggregates as inclusion bodies, involving the use of laborious and expensive renaturation steps to recover the active proteins. Yeasts are characterized by high amount of protein loss because of the degradation of the produced compound (Boehm, 2007). All these limitations tend to influence the overall production costs. Daniell et al. (2001) reported values ranging between US$ g − 1, for mammalian cell culture, depending on the kind and on the amount of the desired protein.

With respect to these methods, transgenic plants offer several advantages due to high-scalability, cost-effectiveness and greater safety (no mammalian pathogen is conveyed by plants). In addition, plants can produce complex recombinant proteins with the correct form and post-translational modification, with full biological activity (Yusibov et al., 2011). Moreover, some organ of the plant, such as seeds, can provide a system to (i) accumulate high level of protein, (ii) store and extend protein stability over a long period even at room temperature. More detailed and clear lists of the advantages linked to the use of plants as heterologous expression system can be found in several different reviews dealing with this matter (Boehm, 2007; Daniell et al., 2009; Tacket, 2009; Wiktorek-Smagur et al., 2012). The possible use of transgenic plants, as a complement to the longer-established range of prokaryotic, yeast and mammalian cell-based recombinant protein production systems, it is now a reality and it has led to the definition of specific guideline on the approaches which should be employed to achieve satisfactory quality for biologically active substances proposed to be produced using the new technology (EMEA, 2006–2008).

4.3 Stable transformation

Concerning the choice of the host plant species, Boehm (2007) reported that, considering all the work published, there was not one clear advantageous, leading plant expression system, although well-established plant species are preferably used, including tobacco, rice and corn.

The biology of the production host needs to be considered from both the production perspective and according to how it impacts on the environment, food safety and human health. The best species for a particular application must satisfy and balance, as well as possible, all of the criteria required. Three 'classes' of plant species could potentially be used as non-cultivated species, non-food crops and food crops. The last can be further subdivided in (i) seed crops (i.e. rice, barley, maize), (ii) vegetable crops (potato, carrot) and (iii) fruit bearing/green leaf crops (lettuce, banana, tomato, etc.) (Sparrow et al., 2007).

There are two basic strategies for gene delivery and protein expression: stable and transient transformation (Jones and Sparks, 2009; Jones et al., 2009). Stable transformation involves the generation of transgenic plants with the transgene stably integrated in the nuclear or chloroplast DNA. The earliest research in this sector have been carried out by transformation of the nuclear genome of the host plant (usually tobacco; Mason et al., 1992; Ma et al., 1994) through *A. tumefaciens*. Subsequently, several experiments have considered also the transformation of edible plants (carrots, tomato, rice, potato, reviewed in Daniell et al., 2009) to support the oral administration of plant-made vaccines (with exclusion of tobacco, which is not an edible plant, and in which heterologous proteins must be purified before use as therapeutics). The target proteins can be expressed:

- In whole terrestrial plant (Rosales-Mendoza et al., 2008) by controlling the gene with strong constitutive promoters (e.g. CaMV35S)
- In specific plant organs (e.g. seeds) or time frames during the life of the plant by using inducible promoters (Reggi et al., 2005; Zhang et al., 2006)
- In plant culture cell suspensions (Kim et al., 2008; Aviezer et al., 2009)
- In aquatic higher plants (Boehm et al., 2001) and algae (Franklin and Mayfield, 2005; Specht et al., 2010)

To overcome some of the limits of nuclear transformation, mainly the relatively low levels of expression ($<1\%$ total soluble protein [TSP], which can vary from plant to plant and generation to generation, probably because of gene silencing or position effects) and the transgene spread in the environment (by vertical gene flow), the chloroplast transformation is increasingly used. Chloroplast transformation (Chebolu and Daniell, 2009) and the generation of transplastomic plants offer several advantages with respect to nuclear transformation: it allows for homologous recombination; no gene silencing and position effects have been yet reported; it achieves high production level (Oey et al. in 2009 reported a production level of approximately 70% TSP), it offers the possibility to co-express multiple genes in operons; and it provides enhanced confinement of the transgene because of maternal inheritance (Lössl and Waheed, 2011). Indeed, concerning this last point, a recent study has reported that there is a small degree (frequency smaller than 10^{-6}) of paternal transmission of plastid in tobacco (Ruf et al., 2007). On the other hand, chloroplast transformation present several limits, the main ones are:

- The protocols for transformation based on gene gun methodologies, *Agrobacterium* or plant viruses are not able to transform plastids, while initially only protocols for tobacco chloroplasts have been standardized for efficient transformation
- With the exception of disulphide bonds, chloroplasts are not able to perform other post-translational modification, like glycosylation, that are often crucial for the biologic activity of the expressed proteins (Daniell et al., 2009; Lössl and Waheed, 2011)

Concerning the first point, since tobacco being not suitable for oral delivery of therapeutic proteins, its proteins must to be purified. Alternatively, some encouraging results have been obtained with edible plants like carrots (which, however, have a slow regeneration time and consequently are not usable for the rapid production of high amount of proteins), tomato and lettuce. This last plant has shown characteristics similar to those of tobacco making them an optimum choice to develop oral delivery coupled with chloroplast transformation (reviewed in Daniell et al., 2009). Concerning the second limit, actually only proteins not requiring any post-translational modification can be expressed in chloroplast. One further concern linked to plastid transformation is the possible presence of pleiotropic effects, influencing the growth and morphology of the plant, following the insertion and high expression of the transgene in condition of homoplasy. Homoplasy is a term referring a situation in which all the chloroplast DNA molecules have inserted the transgene, a situation that is usually reached after some cycles of selection and regeneration. Few reports detected the presence of pleiotropic effects as male sterility, yellow leaves and stunted growth (Ruiz and Daniell, 2005; Oey et al., 2009). Similar to nuclear transformation, in chloroplasts the expression of transgenes can be constitutive or inducible by using appropriate promoters in order to express the heterologous protein at any specific developmental stage or eventually after harvesting (Lössl and Waheed, 2011). Foreign protein expression is actually higher using constitutive than inducible expression. A common limit of both nuclear and chloroplast transformation is that the time requested to pass from the first transformation event to the availability of a stable transformed line and to generate a sufficient amount of seed material is long.

4.4 Transient transformation

Transient transformation provides an alternative to the stable transformation procedures that does not involve the generation of transgenic plants. Transient expression does not depend on the integration of the heterologous DNA and it is not confounded by position effects. Transient expression is rapid: just 3 h after the delivery of DNA. The heterologous protein can be detected, reaching maximum levels between 18 and 48 h, and it can persist up to 10 days (Jones et al., 2009). To obtain transient expression of a desired protein, the DNA of interest is inserted in appropriate viral vectors and usually introduced by vacuum infiltration of whole plants or plant parts with *Agrobacterium*. Among the most used viruses are tobacco mosaic virus (TMV), alfalfa mosaic virus (AMV) and potato virus X (PVX). The most used plant hosts are tobacco and *Nicotiana benthamiana*, consequently the produced protein must be purified to remove Agrobacteria and toxic host plant compounds (Daniell et al., 2009). By using these methodologies, the amount of protein that can be obtained after few days from the infection can be considerable, between 0.5 and 1 g per kg fresh weight. Some systems provide lab-scale quantity of proteins usable in feasibility or preclinical studies. Other approaches, like magnIcon technology, used altered viruses that can be used also for commercially large = scale protein production in open field plants (Huang et al., 2008). Rapid manufacturing of therapeutic proteins in plants is especially promising for therapeutic application in those situations when a rapid supply of the pharmaceutical is needed. Bendandi et al. (2010) proved that by using transient expression approaches, it was possible to obtain high yield production of individualized idiotype vaccines for non-Hodgkin's lymphoma. The manufacturing time from biopsy to vaccine is less than 12 weeks. The recent advances in transient plant transformation systems resulted in the production and accumulation of the recombinant protein up to 80% of TSP in tobacco (Marillonnet et al., 2004).

4.5 Limits on the use of plants as expression systems for molecular farming

Until recently the major limitation of plant as heterologous expression systems for pharmaceutical was linked to the low accumulation levels. Several works reported that the production level should be higher than 1% TSP to render the systems usable for commercial exploitation (Lössl and Waheed, 2011). In the last years, several advances have been made to produce an increase of the accumulation level, mainly with chloroplast expression, transient expression and subcellular targeting. The different approaches adopted and studied to overcome these limitations have been reviewed by Obembe et al. in 2011.

Most pharmaceutical proteins are complex proteins requiring multiple co- or post-translational modifications to show biological activity. Although plants can perform most of these modifications including *N*-glycosylation (Gomordand Faye, 2004), differences between plant and mammalian glycosylation are actually the major limitation to the commercial production of glycosylated PMP in human therapy, especially for those proteins considered for parenteral administration (Gomord et al., 2005; Walsh and Jefferis, 2006). Glycosylation is one of the principal post-translational modifications of eukaryotic proteins. More than 50% of the proteins in eukaryotes, as well as one-third of approved biopharmaceuticals are glycoproteins (Gomord et al., 2010). Glycosylation is often fundamental to allow a protein to assume the correct structure and functionality. Carbohydrates added to proteins can be divided into two categories: *N*-glycans, linked to asparagine (N) residues mainly at the sites NXT/S (Asn (N) - any aminoacids except Pro (X) - Thr (T)/Ser (S)). where X could be any amino acid except proline (P) and *O*-glycans, linked to the hydroxyl group of serine (S), threonine (T), hydroxylysine or hydroxyproline residues in the protein chain. The biology and immunogenicity of several proteins is influenced by the glycosylation status. Plant complex *N*-glycans were shown indeed to be immunogenic, as indicated by the detection of antibodies specific for plant glyco-epitopes in laboratory mammals and human sera (Gomord et al., 2005). This finding implies that glycosylation is a problematic modification to be faced with all the heterologous expression systems and it is one of the main reasons why mammalian cells are actually the most used systems to produce human therapeutic proteins: mammalian cells have a glycosylation system that is the highest similarity to human glycosylation.

N-glycosylation in eukaryotic cells is a very complex reaction starting inside the endoplasmic reticulum (ER) and continuing in the Golgi apparatus. While the steps in the ER are similar in all the eukaryotic organisms, the subsequent step of maturation in the Golgi reticulum is responsible for the differences among glycosylation of different organisms. Plant glycoproteins are characterized by: (i) a $\beta(1,2)$ xylose on the first mannose of the core that is never found in mammals and (ii) an $\alpha(1,3)$ fucose instead of α(1,6) fucose on the proximal *N*-acetylglucosamine of the core, which is present in mammals. Other differences can be found also at the level of the terminal *N*-acetylglucosamine that is usually combined with different residues in plants (β[1,3] galactose and α[1,4] fucose) and mammals (sialic acid) (Gomord et al., 2010).

To solve this problem great progress has been made in the field of *N*-glycosylation engineering to redesign the *N*-glycan structure of proteins synthesized in plant systems (Gomord et al., 2010). The most used strategies are:

- Targeting of the recombinant protein in the ER by adding at the C-terminal of the protein the ER retention signals H/KDEL (His (H)/Lys (K), Asp (D), Glu (E), Leu (L)), this strategy often resulted in the production of PMP with non-immunogenic high-mannose-type *N*-glycans
- knockout of host plant's endogenous glycosyltransferases to prevent the addition of unwanted and potentially immunogenic, sugar residues; and
- insertion of heterologous (human) glycosyltransferases for the addition of mammalian-type sugars to complex glycan structures (production of galactosylated and sialylated *N*-glycans) (Saint-Jore-Dupas et al., 2007; Gomord et al., 2010).

N-glycosylation has been deeply investigated; by contrast very little attention has been paid until now to the *O*-glycosylation of plant-made therapeutic proteins. In humans, *O*-glycosylation is essential for the expression or biological activity of many proteins and hundreds of human proteins represent good candidates for *O*-glycosylation when produced in a plant expression system. *O*-glycosylation is still less known compared to *N*-glycosylation, and no approaches to engineer plants in order to produce human like *O*-glycosylation has been proposed and carried out. Some works reported that plant-specific *O*-glycosylation can have a positive impact on half-life and pharmacodynamics of PMP (Xu et al., 2007); on the other hand, a high immunogenicity and allergenicity provided by the presence of arabinogalactan and arabinan *O*-glycans, as shown in several studies, could be a limit to the use of PMP for human therapy (Gomord et al., 2010).

4.6 Plant-made recombinant pharmaceuticals

The therapeutic molecules expressed in plants can be mainly divided into three groups: (i) plant-made vaccines; (ii) plant-made antibodies; (iii) plant-made therapeutics.

4.6.1 Plant-made vaccines

Several published works have dealt with the development of plant-made vaccines and the references to these works can be found in many recent reviews (Chebolu and Daniell, 2009; Daniell et al., 2009; Lössl and Waheed, 2011; Thomas et al., 2011; Wiktorek-Smagur et al., 2012). The idea behind this sector of biotechnology is the possibility to develop inexpensive and immunogenic edible vaccines for oral immunization as alternative to immunization by injection. The area of greatest potential is in the prevention of infectious diseases, particularly in underdeveloped countries where access to vaccines is very limited (Paul et al., 2011). To this aim, the host plant transformed should be an edible plant that can be consumed uncooked (cooking may denature the antigen). To induce oral immunization, a vaccine should be resistant to the enzymatic degradation in the stomach and arrive in a sufficient amount in the gut lumen. It has been shown in several studies that plant-made vaccines delivered orally are naturally protected from degradation in the stomach very likely because of plant cell wall, which can provide a protection to the antigen that is bioencapsulated inside plant cells. From the literature it is clearly visible that plants can be easily transformed to express antigens. Several antigens have been obtained by stable nuclear or chloroplast transformation and by transient transformation. This last approach is considered of particular interest for the rapid production of high amount of vaccines: a level of antigens up to 80% of TSP in transgenic tobacco leaves has been obtained. In the production of vaccines by means of transient transformation, a systems attractive is the so-called 'epitope presentation'. In this system, the expression of the vaccine is obtained using viral vector designed so that short antigenic peptides fused to the coat protein (CP) are displayed on the surface of assembled viral particles without interfering with the ability of the modified CP to assemble. These modified viruses – cowpea mosaic virus (CPMV), TMV, plum pox virus (PPV) – are attractive as vaccines, because the modified particles can be readily purified and the presentation of multiple copies of an antigenic peptide on the surface of a single virus particle can significantly increase their immunogenicity (Canizares et al., 2005).

4.6.2 Plant-made antibodies

Antibodies are important tools in research, as well as in therapy and diagnosis. In recent years, the demand for antibodies has constantly increased and to overcome the overloading of current production facilities, alternative and less expensive expression systems have been developed, of which plants appear the most promising (De Muynck et al., 2010). The first antibody was produced in tobacco plants (Hiatt et al., 1989). From then on, several antibodies have been produced in plant systems. Functional and correctly assembled antibodies have been obtained against human/animal viruses and bacteria (rabies, hepatitis B, anthrax, *Clostridium botulinum* neurotoxins) and against different cancer antigens (skin, breast and colorectal cancer) (Wiktorek-Smagur et al., 2012). Because of their complex glycosylation pattern and their assembly from four polypeptides linked by disulphide bridges, antibodies require an appropriate expression host. The most used expression systems for antibodies have been *N. tabacum* and *N. benthamiana* (De Muynck et al., 2010). This is justified with the necessity that antibodies need to be highly pure for administration to patients. Consequently, it is possible to choose hosts with a preference for those species with well-established transformation protocols. The most used promoters are strong constitutive viral promoters such as CaMV35S or CPMV. Other works have been carried out by using eukaryotic or inducible promoters. Concerning the organs of expression, some works reported the expression in whole plants, while other focused on specific plant organs. Also the expression in callus and suspension cell cultures has been considered. Both stable and transient transformation methods have been used, and the expression levels are usually below the 1% of TSP; only for some reports a level greater than 1% has been reported (De Muynck et al., 2010). Plantibodies have been developed by Biolex Therapeutics, the Research Triangle of North Carolina. In 2012, Biolex sold the LEX System to Synthon, a Netherlands-based specialty pharmaceutical company. The sale included two preclinical biologics made with the LEX System, proprietary technology that genetically transforms the aquatic plant Lemna, in the duckweed family (Cox et al., 2006). Several works reported the functionality of the plant-made antibodies; this could reflect the fact that, at least for antibodies, the glycosylation state does not necessarily determine the biologic activity of the antibodies itself.

4.6.3 Plant-made therapeutics

Plants, as demonstrated in the production of antibodies, can be used to produce therapeutic proteins with both simple and complex quaternary structures. Several reports on the expression of therapeutics in plants can be found in literature and several reviews report the bibliographic references to these articles. Edible/not edible plants and suspension cell cultures have been transformed to produce:

- Small, bioactive peptides such as lactostatin and novokinin which have hypocholesterolemic and antihypertensive activity, respectively (Wakasa et al., 2006, 2011)
- Antimicrobial nutraceutics and enzymes such as human lactoferrin, human lysozyme, trypsin (Wiktorek-Smagur, 2012)
- Secondary metabolites with high therapeutic and commercial value such as artemisinin, used in malaria therapy or its precursor (van Herpen et al., 2010)
- Therapeutic proteins including blood proteins and substitutes, and protein for prevention and treatment of diseases such as Gaucher's disease (Reggi et al., 2005), atherosclerosis (Fogher et al., 2008) and glycogen storage disease type II/maltase deficiency (Martiniuk et al., 2013)

More detailed descriptions of these last cases will be reported subsequently. Both stable (nuclear and chloroplast) and transient transformation have been used to produce therapeutics. The expression has been obtained both constitutively in the whole plant by using strong viral promoters and in specific organs (such as seeds) by using inducible promoters. Compared with the constitutive expression, the expression in seeds present several advantages and seed-based expression systems for plant molecular farming have been established:

- The proteins expressed in seeds can be stably stored for several months without losing their properties at room temperature; it is not required the availability of a cold chain to maintain protein stability before and after the purification (if needed)
- Seeds can be efficiently sterilized at the surface before protein extraction reducing bio-loads to industrially accepted standard
- Seeds accumulate high level of proteins with percentage of proteins varying between 10% and 40%, several reports suggest that recombinant protein accumulation in seed mimes the situation of naturally occurring proteins
- Seeds have a reduced content of proteases that make easy the recovery step (Boothe et al., 2010)

Moreover, the accumulation of the therapeutic protein in the edible part of the plant makes the development of direct oral administration possible, providing a safe and reliable delivery system which contributes to promote human health. A further advantage that must be confirmed by further studies is that the recombinant protein synthesized in seeds seem to lack potentially immunogenic glycans (Reggi et al., 2005).

4.6.4 Production of recombinant human acid β-glucosidase stored in tobacco seed

Gaucher disease is the most frequent human sphingolipidosis caused by the quantitative/qualitative deficiency of the lysosomal hydrolytic enzyme, acid β-glucosidase (GCase) and is currently treated at a very high cost by enzyme replacement therapy. At present, the enzyme used in therapy is a recombinant, processed form of human acid β-glucosidase (imiglucerase, Genzyme Corporation) obtained from cultured CHO cells. In Reggi et al., 2005, nuclear stable transformed tobacco plants have been obtained by cloning the gene for the functional human GCase recovered from human placenta. The expression of the protein has been directed into the seeds under the control of the soybean basic 7S globulin promoter. A second transformation event has been carried out by using the constitutive promoter CaMV35S. It was observed that in transgenic tobacco, GCase accumulation and stability cannot be achieved with a constitutive expression system. In spite of promoter strength, the amount of GCase mRNA was very low; a further part of the enzyme was cleaved by proteases. On the basis of these results, it was hypothesized that large quantities of recombinant GCase could interfere with the plant metabolism, hindering regeneration/survival of the best transformants. In contrast, the expression in seeds resulted in higher levels of functionally active recombinant protein.

The protein has been purified from the seeds of the elite stable transformed lines. Follow-up experiments have shown that plant-derived recombinant GCase was found to be enzymatically active, uptaken by human fibroblasts and free of immunogenic xylose and fucose residues.

Tobacco plants represent a safe and cost-effective production system for human placental GCase compared to established expression systems and can be used as bioreactors in the large-scale production of injectable proteins required for lifelong therapy.

4.6.5 Transgenic expression, accumulation and recovery of dimer human apolipoprotein A-I$_{Milano}$ in rice seeds (Fogher et al., 2008)

Apolipoprotein A-I is the major component of circulating high-density lipoprotein (HDL) also known as 'good cholesterol' and has been extensively studied in the last 20 years because of its anti-atherogenic properties. Apolipoprotein A-I$_{Milano}$ (ApoA-I$_{Milano}$) is a naturally occurring variant of ApoA-I, first described in an Italian population, characterized by a cysteine-to-arginine substitution at position 173 leading to the formation of disulphide-linked homodimers and heterodimers with ApoA-II. Several observations suggest that variant could have peculiar properties preserving from cardiovascular disease (Franceschini et al., 1980; Chiesa et al., 2002).

Concerning the mode of action, although there are no certainties, it seems likely that a crucial role is played by the spontaneous formation of homodimers. In fact ApoA-I_{Milano} homodimers show peculiar structure and function possibly responsible of the anti-atherogenic effects (Sirtori et al., 1999; Calabresi et al., 1999). In this application nuclear stable transgenic lines accumulating high percentage of dimeric ApoA-I_{Milano} (85%–90%) have been obtained in the seeds. The transformation has been obtained placing the expression of the gene under the control of the rice prolamin promoter to drive the expression and accumulation of the protein directly in the seeds. In addition the coding sequence corresponding to the mature protein has been fused with the signal peptide of prolamin that targets the nascent protein to the ER where it undergoes spontaneous dimerization producing the dimer directly and removing the need for further processing after protein purification. In the state of the art this is the first study reporting a high production of the dimeric form of ApoA-I_{Milano} in the seeds of transgenic plants.

In rice seed, more than 80% of the total seed proteins are stored in protein bodies (PB) that act as protein storage organelles. In these organelles a protein is protected by membranes from proteases during seed maturation until the protein is required. There are two types of organelles: protein bodies I (PB-I) derived from ER containing mainly prolamin, and protein bodies II (PB-II) that are storage vacuole protein bodies storing mainly glutelin and globulin. The recombinant Apo has been expressed as a promoter and signal peptide of prolamin and so it is possible, but it should be demonstrated that seed ApoA-I_{Milano} can be stored in PB-I. Some works suggest that rice seed ER-derived protein bodies (PB-I) act as an efficient delivery vehicle for oral-administrated peptides (Takagi et al., 2010); it should be tested if the same is true also for more complex proteins than peptides.

4.6.6 Production of a functional human acid maltase in tobacco seeds

Genetic deficiency of lysosomal acid alpha glucosidase (GAA) or acid maltase results in glycogen storage disease type II (GSDII) or acid maltase deficiency, encompassing at least five clinical subtypes of varying severity. Currently, there is no treatment or cure for GSDII. Enzyme and gene replacement therapies are being developed. Enzyme therapy using a recombinant human GAA produced in a CHO cell line (Genzyme Corporation) has shown moderate success in patients using a biweekly infusion regimen. In this work, nuclear stable transgenic tobacco plants have been obtained to provide a cheaper alternative to established production systems.

The expression of the recombinant GAA has been driven to the seeds and regulated under the control of the promoter for soybean β-conglycinin, a seed protein synthesized in huge amounts (Martiniuk et al., 2013). Further the expression of β-conglycinin is highly regulated being restricted to the embryo during the mid-maturation phase of embryogeny.

The protein has been extracted from the elite lines and first analysis demonstrated that tobacco GAA had size similar to human GAA. Further, several *in vitro* and *in vivo* tests showed that the recombinant GAA from plant is biologically active, is readily taken up by GSDII fibroblasts and it is able to reverse the enzyme defect in tissues at 7 days after a single dose following administration in GAA−/− mice. These data demonstrate indirectly that the plant-made GAA is fully functional; proteolytically cleaved and contains the minimal phosphorylation and mannose-6-phosphate residues to not disrupt activity. Data on *E. coli* and unpublished data on yeast show that both have altered glycosylation and lost substantial activity. Additionally, the purified tobacco recombinant GAA has high specific activity, similar to the native human placental GAA making it ideal for enzyme replacement therapy.

4.7 Regulatory aspects and clinical status of PMPs

The use of plants to produce pharmaceutical and industrial proteins is likely to be one of the next major commercial developments in biotechnology. Starting from the first feasibility studies, hundreds of papers actually report also that PMPs are often functional with biologic activity similar or identical to the correspondent recombinant pharmaceutical produced by well-established and accepted expression systems. These results have provided the foundations for following experiments and, in recent years, several PMPs have moved from the exploratory research phase towards clinical trials. This transition is very important to demonstrate that the production of pharmaceutics in plants is not just a proof of concept or ability but could also provide significant practical applications. At the same time the use of GM plants to synthesize proteins that will be subsequently processed, regulated and sold as pharmaceuticals, challenges two very different established regulatory frameworks, one concerning GM plants and the other covering the development of biotechnology-derived drugs (Spök, 2006; Spök et al., 2008). Within these regulatory systems, specific regulations and guidelines for PMPs are still evolving. As reported before, in 2006 and 2008 the EMEA provided some guidelines to the development of PMP under good manufacturing practice (GMP). Thus far there has been no commercial application of PMP technology. The products nearing commercial viability will ultimately help to road test and fine-tune these regulations, and might help to reduce regulatory uncertainties. The EMEA guidelines also highlight the fact that different countries have different approaches to the GMO problems and different regulation in this field (Spök et al., 2008). The use of plants to produce biologics raised several concerns. Sparrow et al. (2007) reported the role of consortium like Pharma-Planta (EU-funded academic research consortium established in 2004). The objectives of this consortium are to address to these concerns: (i) to build a plant-based production platform and to help the development of appropriate regulatory oversight in the EU; (ii) to produce recombinant antibodies in transgenic plants that will be developed through regulatory requirements; and

(iii) to demonstrate a practical commitment to the humanitarian use of PMPs and the associated production technology in developing countries.

Besides regulatory aspects, ethical arguments should also be considered for humanitarian purposes in developing countries, e.g. by establishing licensing strategies for humanitarian purposes (Spök et al., 2008; Paul et al., 2011).

Concerning the actual clinical status, several PMPs are currently in clinical phases. The majority of PMP are in phase I or II and just few of them are moved beyond into phase III and are expected to be commercialized within a short time. Detailed lists of the current PMP status can be found in reviews by Tacket (2009); Thomas et al. (2011) and Yusibov et al. (2011).

Relevant for plant-made vaccines, the first world licensing for an injectable vaccine against Newcastle disease virus (NDV) occurred in 2006. The vaccine has been produced to protect birds which were exposed to NDV. While this vaccine was never brought forward and turned into a commercially available product, the formulation was advanced through the USDA Center for Veterinary Biologics' regulatory approval, demonstrating that plant-made vaccines could be developed within the existing regulatory framework (Thomas et al., 2011). Other vaccines that have been tested on human volunteers (Yusibov et al., 2011) have been: (i) enterotoxigenic *E. coli* vaccine; (ii) Norwalk virus vaccine; (iii) hepatitis B vaccines (Yusibov et al. [2011] reported that a vaccine against hepatitis B virus (HBV) is already marketed in Cuba) and (iv) rabies vaccine.

Relative to plant-made antibodies, Thomas et al. (2011) reported that two plant-made antibody products have made it to advanced human clinical trials: (i) CaroRX a topic sIgA anti-*Streptococcus mutans* for dental caries; (ii) a plant-made antibody directed against cancer (non-Hodgkin's lymphoma). Other antibodies at the early stages of clinical investigation are two antibodies (anti-αCCR5 and anti-gp120) directed against HIV and thought to provide topical protection against the virus.

It is relevant to plant-made therapeutics the case of Taliglucerase alfa, a recombinant glucocerbrosidase for Gaucher disease produced in carrot cell cultures. It was shown to be as safe and functional as the recombinant proteins currently used in enzyme replacement therapy (ERT) in different clinic trials. It was approved by the United States (U.S.) Food and Drug Administration (FDA) on May 1, 2012 for long-term ERT for adults with a confirmed diagnosis of Type 1 Gaucher disease (http://www.gaucherdisease.org). At the same time the enzyme is in final clinical trials for the treatment of pediatric patients with Gaucher's disease (Haddley, 2012). The enzyme is marketed under the name ELELYSO™ (http://www.elelyso.com).

In addition to therapeutic proteins, dietary proteins and nutraceutics produced in plants have been evaluated in phase I and II clinical trials (Yusibov et al., 2011).

References

Abhilash, P.C., Jamil, S., Singh, N., 2009. Transgenic plants for enhanced biodegradation and phytoremediation of organic xenobiotics. Biotechnol. Adv. 27, 474–488.

Aviezer, D., Brill-Almon, E., Shaaltiel, Y., Hashmueli, S., Bartfeld, D., Mizrachi, S., et al., 2009. A plant-derived recombinant human glucocerebrosidase enzyme – a preclinical and phase I investigation. PLoS One. 4, e4792.

Bendandi, M., Marillonnet, S., Kandzia, R., Thieme, F., Nickstadt, A., Herz, S., et al., 2010. Rapid, high-yield production in plants of individualized idiotype vaccines for non-Hodgkin's lymphoma. Ann. Oncol. 21, 2420–2427.

Boehm, R., 2007. Bioproduction of therapeutic proteins in the 21st century and the role of plants and plant cells as production platforms. Ann. NY Acad. Sci. 1102, 121–134.

Boehm, R., Kruse, C., Voeste, D., Barth, S., Schnabl, H.A., 2001. A transient transformation system for duckweed (*Wolffia columbiana*) using *Agrobacterium*-mediated gene transfer. J. Appl. Bot. 75, 107–111.

Boothe, J., Nykiforuk, C., Shen, Y., Zaplachinski, S., Szarka, S., Kuhlman, P., et al., 2010. Seed-based expression systems for plant molecular farming. Plant Biotechnol. J. 8, 588–606.

Calabresi, L., Canavesi, M., Bernini, F., Franceschini, G., 1999. Cell cholesterol efflux to reconstituted high-density lipoproteins containing the Apolipoprotein A-I_{Milano} dimer. Biochemistry. 38, 16307–16314.

Canizares, M.C., Nicholson, L., Lomonossoff, G.P., 2005. Use of viral vectors for vaccine production in plants. Immunol. Cell Biol. 83, 263–270.

Chebolu, S., Daniell, H., 2009. Chloroplast-derived vaccine antigens and biopharmaceuticals: expression, folding, assembly and functionality. Curr. Top. Microbiol. Immunol. 332, 33–54.

Chiesa, G., Monteggia, E., Marchesi, M., Lorenzon, P., Laucello, M., Lorusso, V., et al., 2002. Recombinant apolipoprotein A-I_{Milano} infusion into rabbit carotid artery rapidly removes lipid from fatty streaks. Circ. Res. 90, 974–980.

Cox, K.M., Sterling, J.D., Regan, J.T., Gasdaska, J.R., Frantz, K.K., Peele, C.G., et al., 2006. Glycan optimization of a human monoclonal antibody in the aquatic plant *Lemna minor*. Nat. Biotechnol. 24, 1591–1597.

Daniell, H., Streatfield, S.J., Wycoff, K., 2001. Medical molecular farming: production of antibodies, biopharmaceuticals and edible vaccines on plants. Trends Plant Sci. 6, 219–226.

Daniell, H., Singh, N.D., Mason, H., Streatfield, S.J., 2009. Plant-made vaccine antigens and biopharmaceuticals. Trends Plant Sci. 14, 669–679.

De Muynck, B., Navarre, C., Boutry, M., 2010. Production of antibodies in plants: status after twenty years. Plant Biotechnol. J. 8, 529–563.

EMEA (European Medicines Agency), 2006–2008. Guideline on the quality of biological active substances produced by stable transgene expression in higher plants. EMEA/CHMP/BWP/48316/2006.

Fogher, C., Reggi, S., Perfanov, K., 2008. In-plant production of dimeric and/or oligomeric (comprising three or more units) forms of human Apo A-1 protein muteins. World Intellectual Property Organization, WO2008/017906.

Franceschini, G., Sirtori, C.R., Capurso, A., Weisgraber, K.H., Mahley, R.W., 1980. A-I_{Milano} apoprotein. Decreased high density lipoprotein cholesterol levels with significant lipoprotein modifications and without clinical atherosclerosis in an Italian family. J. Clin. Invest. 66, 892–900.

Franklin, S.E., Mayfield, S.P., 2005. Recent developments in the production of human therapeutic proteins in eucaryotic algae. Expert Opin. Biol. Ther. 5, 225–235.

Gomord, V., Faye, L., 2004. Post-translational modification of therapeutic proteins in plants. Curr. Opin. Plant Biol. 7, 171–181.

Gomord, V., Chamberlain, P., Jefferis, R., Faye, L., 2005. Biopharmaceutical production in plants: problems, solutions and opportunities. Trends Biotechnol. 23, 559–565.

Gomord, V., Fitchette, A.C., Menu-Bouaouiche, L., Saint-Jore-Dupas, C., Plasson, C., Michaud, D., et al., 2010. Plant-specific glycosylation patterns in the context of therapeutic protein production. Plant Biotechnol. J. 8, 564–587.

Haddley, K., 2012. Taliglucerase alfa for the treatment of Gaucher's disease. Drugs Today (Barc). 48 (8), 525–532. http://dx.doi.org/10.1358/dot.2012.48.8.1844808.

Hiatt, A., Cafferkey, R., Bowdish, K., 1989. Production of antibodies in transgenic plants. Nature. 342, 76–78.

Huang, Z., Le Pore, K., Elkin, G., Thanavala, Y., Mason, H.S., 2008. High-yield rapid production of hepatitis B surface antigen in plant leaf by a viral expression system. Plant Biotechnol. J. 6, 202–209.

James, C., 2012. 2011 ISAAA Report on Global Status of Biotech/GM Crops. <http://www.isaaa.org>.

Jones, H.D., Sparks, C.A., 2009. Stable transformation of plants. Methods Mol. Biol. 513, 111–130.

Jones, H.D., Doherty, A., Sparks, C.A., 2009. Transient transformation of plants. Methods Mol. Biol. 513, 131–152.

Kim, T.G., Baek, M.Y., Lee, E.K., Kwon, T.H., Yang, M.S., 2008. Expression of human growth hormone intransgenic rice cell suspension culture. Plant Cell Rep. 27, 885–891.

Kotrba, P., Najmanova, J., Macek, T., Ruml, T., Mackova, M., 2009. Genetically modified plants in phytoremediation of heavy metal and metalloid soil and sediment pollution. Biotechnol. Adv. 27, 799–810.

Lössl, A.G., Waheed, M.T., 2011. Chloroplast-derived vaccines against human diseases: achievements, challenges and scopes. Plant Biotechnol. J. 9, 527–539.

Lu, B.R., Yang, C., 2009. Gene flow from genetically modified rice to its wild relatives: assessing potential ecological consequences. Biotechnol. Adv. 27, 1083–1091.

Ma, J.K.C., Lehner, T., Stabila, P., Fux, C.I., Hiatt, A., 1994. Assembly of monoclonal-antibodies with IgG1 and IgA heavy chain domains in transgenic tobacco plants. Eur. J. Immunol. 24, 131–138.

Marillonnet, S., Giritch, A., Gils, M., Kandzia, R., Klimyuk, V., Gleba, Y., 2004. In planta engineering of viral RNA replicons: Efficient assembly by recombination of DNA modules delivered by *Agrobacterium*. Proc. Natl. Acad. Sci. USA. 101, 6852–6857.

Martiniuk, F., Reggi, S., Tchou-Wong, K.-M., Rom, W.N., Busconi, M., Fogher, C., 2013. Production of a functional human acid maltase in tobacco seeds; biochemical analysis, uptake by human GSDII cells and in vivo studies in GAA knockout mice. Appl. Biochem. Biotechnol. 171, 916–926.

Mason, H.S., Man-Kit Lam, D., Arntzen, C.J., 1992. Expression of hepatitis B surface antigen in transgenic plants. Proc. Natl. Acad. Sci. USA. 89, 11745–11749.

Obembe, O.O., Popoola, J.O., Leelavathi, S., Reddy, S.V., 2011. Advances in plant molecular farming. Biotechnol. Adv. 29, 210–222.

Oey, M., Lohse, M., Kreikemeyer, B., Bock, R., 2009. Exhaustion of the chloroplast protein synthesis capacity by massive expression of a highly stable protein antibiotic. Plant J. 57, 436–445.

Paul, M., van Dolleweerd, C., Drake, P.M.W., Reljic, R., Thangaraj, H., Barbi, T., et al., 2011. Molecular pharming – future targets and aspirations. Hum. Vaccin. 7, 375–382.

Reggi, S., Marchetti, S., Patti, T., De Amicis, F., Cariati, R., Bembi, B., et al., 2005. Recombinant human acid β-glucosidase stored in tobacco seed is stable, active and taken up by human fibroblasts. Plant Mol. Biol. 57, 101–113.

Rosales-Mendoza, S., Soria-Guerra, R.E., López-Revilla, R., Moreno-Fierros, L., Alpuche-Solís, A.G., 2008. Ingestion of transgenic carrots expressing the *Escherichia coli* heat-labile enterotoxin B subunit protects mice against cholera toxin challenge. Plant Cell Rep. 27, 79–84.

Ruf, S., Karcher, D., Bock, R., 2007. Determining the transgene containment level provided by chloroplast transformation. Proc. Natl. Acad. Sci. USA. 104, 6998–7002.

Ruiz, O.N., Daniell, H., 2005. Engineering cytoplasmic male sterility via the chloroplast genome by expression of β-ketothiolase. Plant Physiol. 138, 1232–1246.

Saint-Jore-Dupas, C., Faye, L., Gomord, V., 2007. From planta to pharma with glycosylation in the toolbox. Trends Biotechnol. 25, 317–323.

Sirtori, C.R., Calabresi, L., Franceschini, G., 1999. Recombinant apolipoproteins for the treatment of vascular diseases. Atherosclerosis. 142, 29–40.

Sparrow, P.A.C., Irwin, J.A., Dale, P.J., Twyman, R.M., Ma, J.K.C., 2007. Pharma-Planta: road testing the developing regulatory guidelines for plant-made pharmaceuticals. Transgenic Res. 16, 147–161.

Specht, E., Miyake-Stoner, S., Mayfield, S., 2010. Micro-algae come of age as a platform for recombinant protein production. Biotechnol. Lett. 32, 1373–1383.

Spök, A., 2006. Molecular farming on the rise – GMO regulators still walking a tightrope. Trends Biotechnol. 25, 74–82.

Spök, A., Twyman, R.M., Fischer, R., Ma, J.K.C., Sparrow, P.A.C., 2008. Evolution of a regulatory framework for pharmaceuticals derived from genetically modified plants. Trends Biotechnol. 26, 506–517.

Tacket, C.O., 2009. Plant-based oral vaccines: results of human trials. Curr. Top. Microbiol. Immunol. 332, 103–117.

Takagi, H., Hiroi, T., Hirose, S., Yang, L., Takaiwa, F., 2010. Rice seed ER-derived protein body as an efficient delivery vehicle for oral tolerogenic peptides. Peptides. 31, 1421–1425.

Thomas, D.R., Penney, C.A., Majumder, A., Walmsley, A.M., 2011. Evolution of plant-made pharmaceuticals. Int. J. Mol. Sci. 12, 3220–3236.

Tilbrook, K., Gebbie, L., Schenk, P.M., Poirier, Y., Brumbley, S.M., 2011. Peroxisomal polyhydroxyalkanoate biosynthesis is a promising strategy for bioplastic production in high biomass crops. Plant Biotechnol. J. 9, 958–969.

van Herpen, T.W.J.M., Cankar, K., Nogueira, M., Bosch, D., Bouwmeester, H.J., Beekwilder, J., 2010. *Nicotiana benthamiana* as a production platform for artemisinin precursors. PLoS One. 5, e14222.

Wakasa, Y., Yasuda, H., Takaiwa, F., 2006. High accumulation of bioactive peptide in transgenic rice seeds by expression of introduced multiple genes. Plant Biotechnol. J. 4, 499–510.

Wakasa, Y., Zhao, H., Hirose, S., Yamauchi, D., Yamada, Y., Yang, L., et al., 2011. Antihypertensive activity of transgenic rice seed containing an 18-repeat novokinin peptide localized in the nucleolus of endosperm cells. Plant Biotechnol. J. 9, 729–735.

Walsh, G., 2010. Biopharmaceutical benchmarks 2010. Nat. Biotechnol. 28, 917–924.

Walsh, G., Jefferis, R., 2006. Post-translational modifications in the context of therapeutic proteins. Nat. Biotechnol. 24, 1241–1252.

Wiktorek-Smagur, A., Hnatuszko-Konka, K., Gerszberg, A., Kowalczyk, T., Luchniak, P., Kononowicz, A.J., 2012. Green way of biomedicine – how to force plants to produce new important proteins. In: Çiftçi, Y.Ö. (Ed.), Transgenic Plant – Advances and Limitations. InTech Europe, Janeza Trdine, Rijeka, Croatia, pp. 63–90 (Chapter 3).

Xu, J., Tan, L., Goodrum, K.J., Kieliszewski, M.J., 2007. High yields and extended half-life of human interferon ά 2 b expressed in tobacco cells as arabinogalactan-protein fusions. Biotechnol. Bioeng. 97, 997–1008.

Ye, X., Al-Babili, S., Kloti, A., Zhang, J., Lucca, P., Beyer, P., et al., 2000. Engineering the provitamin A (b-carotene) biosynthetic pathway into (carotenoid-free) rice endosperm. Science. 287, 303–305.

Yuan, J.S., Tiller, K.H., Al-Ahmad, H., Stewart, N.R., Stewart Jr., C.L., 2008. Plants to power: bioenergy to fuel the future. Trends Plant Sci. 13, 421–429.

Yusibov, V., Streatfield, S.J., Kushnir, N., 2011. Clinical development of plant-produced recombinant pharmaceuticals – vaccines, antibodies and beyond. Hum. Vaccin. 7, 313–321.

Zhang, X., Buehner, N.A., Hutson, A.M., Estes, M.K., Mason, H.S., 2006. Tomato is a highly effective vehicle for expression and oral immunization with Norwalk virus capsid protein. Plant Biotechnol. J. 4, 419–432.

In planta produced virus-like particles as candidate vaccines

5

Slavica Matić and Emanuela Noris
Istituto di Virologia Vegetale, CNR, Torino, Italy

5.1 Introduction

Plants have been proposed as alternative biotechnological means for the production of pharmaceutical proteins almost 25 years ago. The idea to use plants as an alternative protein-producing source is based on advantages such as relative ease and reduced cost of production and high scalability. Virus-like particles (VLPs) are one of the most attractive pharmaceuticals produced in plants, as they offer the possibility to obtain plant-made vaccines against human and animal diseases. VLPs do not necessitate long purification steps and the cold chain which are required in the production of commercial vaccines. The first concept of plant-made vaccines appeared in the nineties and these were planned as 'edible vaccines' where fruits and vegetables containing antigens would be eaten and the 'person' would become vaccinated in economical and safe way. However, few drawbacks of this concept have arisen such as the difficult estimation of the amount of antigen delivered to the immune system due to the oral administration, possible digestion of protein in the gastrointestinal tract and associated risks of contamination of the food chain regarding the cultivation of transgenic plants carrying the antigens. This influenced the moving of plant-based vaccine research towards non-food crops such as *Nicotiana* species and *Arabidopsis thaliana* which allowed quick purification and non-oral administration of the proteins.

VLPs can be produced in plants by transient expression systems (replicative viral vectors or non-replicative binary vectors) and by stable transformation (binary vectors with the T-DNA region). Although stable transformation offers constitutive expression of foreign gene in transgenic plants (with gene of interest introduced into nucleus or chloroplast DNA), a major disadvantage of this system is the time required to obtain a stable transgenic line producing the protein of interest (from 18 to 24 months). A great facilitation of this aspect was reached through transient expression systems which allow the expression of proteins in plants in one to few weeks by commonly using the agroinfiltration method. Agroinfiltration is based on the delivery of *Agrobacterium tumefaciens* strains into the lower leaf part, forcing the replacement of the intercellular air with a bacterial suspension containing the antigen-expressing plasmid. Infiltration is routinely performed using the positive pressure of a needle-less syringe. This delivery is suitable for few plants, but when huge quantity of protein and high number of plants are required, a vacuum infiltration method should be used.

Applied Plant Genomics and Biotechnology.

Different VLP-producing proteins were obtained in transgenic plants or experimental plants by transient expression systems. Thus, single proteins, multiple proteins forming the higher-order structures, enveloped and naked particles and protein-carriers of heterologous epitopes have been successfully produced in plants. This has been achieved by using native or synthetic genes of interest expressed in plants. The expression of native human or animal genes may result in low-yield protein accumulation which can require the use of synthetic genes and the optimization of the coding sequence. For successful expression in the plants, a foreign gene requires proper codon usage, especially when codons rarely used and less preferred by plants are present. Gene optimization can also mean to use an appropriate GC content, eliminate gene portions prone to gene silencing or forming secondary structures, stabilize mRNA, eliminate cryptic splicing signals and premature PolyA sites. The site of gene expression within the plant cell with regard to post-translational modification and glycosylation potentials are also important factors regulating protein yield; these can be modulated by 'targeting' the proteins to different cell compartments, such as endoplasmic reticulum, chloroplast and apoplast.

In this chapter, we will show the progress in VLP production in plants considering the few most important human and animal viral diseases for which alternative candidate vaccines are being sought.

5.2 Papillomaviruses

Papillomaviruses (PVs) are tumorigenic, non-enveloped viruses with an 8-kbp circular double-stranded DNA genome. Human papillomaviruses (HPVs) are classified into mucosal and cutaneous viruses. High-risk mucosal HPVs, of which HPV-16 and HPV-18 are the most important, cause cervical cancer in women (Bosch et al., 1995). HPV-8, a high-risk cutaneous HPV, is one of the first HPV suspected of human tumorigenesis due to its association with skin epidermodysplasia verruciformis (Pfister et al., 1981) and non-melanoma skin cancer (Weissenborn et al., 2005). Current effective prophylactic vaccines, produced in yeast or insect cells, include the high-risk types HPV-16 and HPV-18 alone, or in combination with the genital wart types HPV-6 and HPV-11.

VLPs of HPVs can assemble *in vivo* and *in vitro* from the major capsid protein L1 expressed alone or together with the L2 (minor) capsid protein (Hagensee et al., 1993). The L1 protein can form two types of VLPs, either an icosahedral lattice with a $T=7$ symmetry, composed of 72 pentamers and morphologically identical to the native virions, or a smaller $T=1$ particle composed of 12 pentamers (Baker et al., 1991; Chen et al., 2000).

The first HPV VLP assembly in transgenic plants was obtained in 2000 using the L1 of HPV-16 and HPV-18 which resulted in a patent (Sohn et al., 2002). The next step was the demonstration that HPV VLPs were immunogenic; this was achieved by Varsani et al. (2003) through the expression of a native HPV-16 L1 gene in transgenic *Nicotiana tabacum* plants and the demonstration of a weak immunogenicity after immunization of rabbits with plant-produced particles, and by Biemelt et al. (2003) through the expression of a human codon optimized HPV-16 L1 in transgenic potato and tobacco resulting in immunogenicity of mice after treatment with VLPs and in anti-L1 response after feeding mice with transgenic tubers. However, initial yields ranged only from 4 μg to 20 mg per kg of fresh leaf weight (FLW), which was lower than the basic amount required for economical vaccine production, i.e. 40 mg of purified protein per kg (Fischer et al., 2004). More recently, L1 yields were highly improved by targeting the gene to the chloroplasts. Thus, yields reached 0.53 g per kg in the case of a chloroplast-targeted HPV-16 L1 protein transiently expressed from a human codon-optimized gene (Maclean et al., 2007), and 3 g per kg of FLW with a native HPV-16 L1 gene in the chloroplast transgenic tobacco (Fernández-San Millán et al., 2008). These L1 yields resulted in a much higher immunogenicity obtained in both studies, creating promising perspectives for developing a candidate plant-based vaccine against HPV-16. Improvements of the HPV-16 L1 expression in the cytoplasm of the plant cells were also achieved through the use of a novel autonomously replicating geminivirus shuttle vector, reaching up to 0.55 g per kg of FLW (Regnard et al., 2010).

The HPV-16 L1 protein self-assembles into VLPs in plant cells not only in its native form, but also in chimeric forms including either fusions or insertions of other polypeptides. Translational fusion of HPV-16 L1 with the plastid photosynthetic proteins in transplastomic tobacco increased antigen accumulation in plants (Lenzi et al., 2008); L1 fusion with the adjuvant '*Escherichia coli* heat-labile enterotoxin subunit B' enhanced L1 immunogenicity and decreased the costs of separate co-administration of adjuvants (Waheed et al., 2011).

Further developments were based on the concept that due to the VLP-assembly potential L1 may be used as carrier of heterologous epitopes to design more efficient vaccines with multiple (prophylactic and therapeutic) characteristics, or with wider range of viral targets. In the first case, chimeric HPV-16 L1s were designed with epitopes of other proteins of the same virus: (i) chimeric L1 was fused to epitopes of the E6 and E7 oncoproteins and expressed in transgenic tomato, resulting in both neutralizing antibodies against viral particles and cytotoxic T-lymphocytes activity against the epitopes after mice immunization with chimeric VLPs (Paz De la Rosa et al., 2009); and (ii) chimeric L1 containing cross-protective epitopes from the L2 protein was transiently expressed in *Nicotiana benthamiana* with a chloroplast targeting signal, giving a yield of 1.2 g per kg of FLW and self-assembling into small immunogenic VLPs (Pineo et al., 2013). In the second case, HPV-16 L1 acted as a carrier of the epitopes derived from a heterologous virus: the $M2e_{2\text{-}24}$ epitope, ectodomain of the M2 protein (M2e) of the *Influenza A virus* (IAV), that is highly conserved among all influenza A isolates, and the $M2e_{2\text{-}9}$ epitope, which is common for both M1 and M2 influenza proteins. All chimeric constructs were transiently expressed in *N. benthamiana* plants (Matić et al., 2011) using the non-replicating *Cowpea mosaic virus*-derived expression vector pEAQ-HT (Sainsbury et al., 2009). Plant-made chimeric proteins spontaneously assembled in higher-order structures (Figure 5.1), such as VLPs of $T = 1$ or $T = 7$ symmetry that were recognized by linear and conformation-specific anti-HPV-16 L1 MAbs; two of the

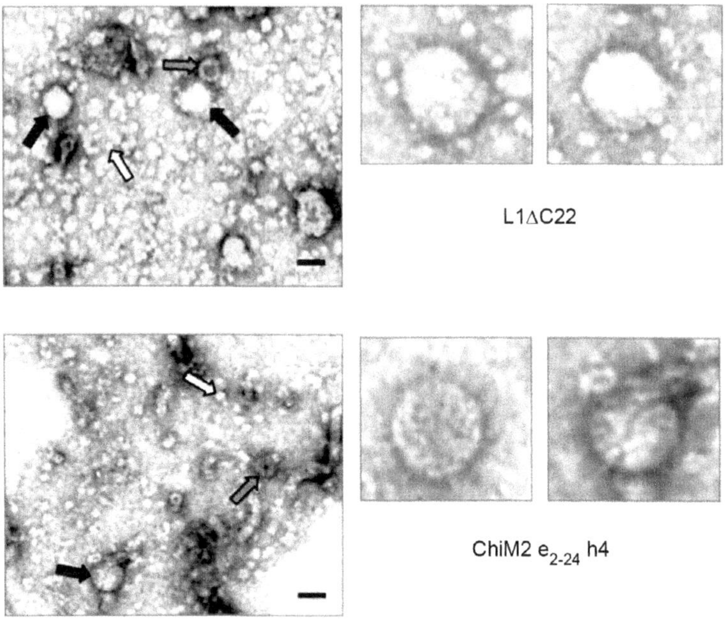

Figure 5.1 Electron micrographs of plant-produced HPV-16 L1 (L1ΔC22) and chimeric construct (ChiM2 $e_{2\text{-}24}$ h4) (Matić et al., 2011). Black arrows = 55-nm T7 VLPs; grey arrows = 30-nm T1 VLPs; white arrows = capsomers. Assembled L1 structures for each construct are enlarged on the right side. Bar = 50 nm.

chimeras also reacted with the anti-influenza antibodies. The results obtained by these studies opened perspectives for producing multivalent vaccines in plants.

Recent studies on cross-neutralizing properties of the capsid proteins of cutaneous HPVs have also raised an interest in vaccine production against these viruses. Successful transient expression of L1 protein in *N. benthamiana* plants was also obtained with the high-risk cutaneous HPV-8 (Matić et al., 2012) using the non-replicating pEAQ-HT vector (Sainsbury et al., 2009) and the replicating *Tobacco mosaic virus* (TMV)-based vectors (Icon Genetics, Halle, Germany) with different targeting signals. The highest L1 yield (240 mg per kg of FLW) was achieved with the pEAQ-HT vector when 22 C-terminal amino acids of L1 were deleted (possibly eliminating a nuclear localization signal), which presents a fourfold yield increase compared to the full-length L1. HPV-8 L1 assembled in plants in appropriate $T = 1$ or $T = 7$ VLPs. Ultra-thin sections of L1ΔC22-expressing plant cells showed that VLPs accumulated in the cytoplasm. This combined with the immunogenicity results of plant-produced HPV-8 L1 (Noris and Meyes, unpublished) open perspectives for the production of another HPV candidate vaccine in plants.

There is still one report about the expression of the L1 protein of another PV, HPV-11, which is associated with anogenital warts (Lowy and Howley, 2001). L1 was expressed in plants from a plant codon-optimized gene in transgenic potato (Warzecha et al., 2003). The L1 protein localized in plant cell nuclei and self-assembled into appropriate VLPs, but a low yield was reached (20 μg per kg of plant material) and only limited immunogenicity after oral delivery of transgenic L1 potato to mice was obtained.

Beside structural HPV protein, there are reports of expression of non-structural HPV proteins in plants, such as the E7 oncoprotein, which modifies the cell cycle control and is the main regulator of HPV-induced oncogenesis (Hebner and Laimins, 2006). Vaccines against this antigen have therapeutic purposes, and their use is intended for people with already established HPV infections. Successful transient expression of HPV-16 E7 (mediated by a *Potato virus X* [PVX]-vector; Chapman et al., 1992) targeted to the secretory system of *N. benthamiana* plants yielded 20 mg of protein per kg of FLW (Franconi et al., 2006). Mice immunized by plant-made E7 showed high stimulation of cell-mediated immune response and tumour growth inhibition. Another HPV from which E7 was successfully expressed in plants is HPV-8. HPV-8 E7 was transiently expressed using the replicating TMV-based vectors (Icon Genetics, Halle, Germany). E7 yielded up to 4.6 g per kg of FLW, and the highest yields were obtained when the antigen was targeted to the apoplast (Noris et al., 2011). HPV-8 E7 protein produced in *N. benthamiana* plants also triggered the mouse immune system, delaying the development of skin papillomatous lesions.

Bovine papillomavirus (BPV) induces benign tumours of cutaneous or mucosal epithelia in cattle, and fibroblastic tumours in equids (Nasir and Campo, 2008). BPV cannot be grown in culture for the preparation of traditional virus killed or attenuated vaccines (Campo, 2003), and thus yeast or insect cell-expression systems are used for producing L1- and L2-based vaccines, but these vaccines are rather expensive for cattle. Similarly to HPVs, BPV L1 self-assembles into VLPs of $T=1$ or $T=7$ symmetry (Kirnbauer et al., 1992; Campo, 2003). Love et al. (2012) reported for the first time transient expression of BPV-1 L1 in plants using the non-replicating pEAQ-HT vector (Sainsbury et al., 2009). Plant codon-optimized BPV-1 L1 gave higher yields (0.22 g per kg of FLW of *N. benthamiana*) compared to the native L1 gene, it self-assembled into $T=1$ particles, which induced strong immune response in rabbits after immunization. This study represents the first plant-produced candidate vaccine against BPV-1.

5.3 Hepatitis B virus

Hepatitis B virus (HBV) is an enveloped virus of the family *Hepadnaviridae*, with a genome of approximately 3 kbp. The viral particle (42 nm) is composed of an outer lipoprotein coat containing the surface antigen (HBsAg) and an inner nucleocapsid core containing the core antigen (HBcAg) protecting a dsDNA molecule (Gitlin, 1997; Torresi, 2002). A commercial efficient HBV vaccine contains 22-nm VLPs derived from the HBsAg gene produced in yeast cells. However, HBsAg and HBcAg, both capable of self-assembling into VLPs, are of considerable interest for the production of alternative vaccines of plant origin.

The first expression of HBsAg in plants was achieved in transgenic tobacco (Mason et al., 1992). HBsAg, derived from a native gene, self-assembled in plants into appropriate 22-nm VLPs with physical characteristics similar to those of the human serum-derived HBsAg. The HBsAg yield was 66 mg per kg of soluble protein. Subsequently, the immunogenicity of a plant-derived HBsAg from the same transgenic tobacco was demonstrated (Thanavala et al., 1995). An anti-hepatitis B response was obtained after intraperitoneal immunization in mice using HBsAg similar to that achieved with a yeast-derived HBsAg of the commercial vaccine. The transgenic tobacco-derived HBsAg showed immunological properties necessary to elicit HBsAg-specific B- and T-cell responses. Afterwards, important results were obtained following oral immunization with the plant-made antigen obtained from a native HBsAg gene (Kong et al., 2001). Comparison of the oral immunogenicity of HBsAg obtained from yeast (purified product) or transgenic potato (uncooked sample), using a cholera toxin as an oral adjuvant, resulted in higher efficiency of transgenic plants in inducing a primary immune response and priming mice to respond to successive parenteral HBsAg injections. Plant-derived HBsAgs were found localized within specific membrane-bound vesicles of the plant cells. Mice fed with HBsAg-potatoes generated HBsAg-specific serum antibodies at levels higher than the necessary protective level (in humans) and produced a high secondary antibody response, with a single parenteral injection. Another important goal was achieved by improving HBsAg yield using a transient expression system (Huang et al., 2008). A high HBsAg yield (295 mg per kg of *N. benthamiana* FLW) was reached at 10 days post-infection (dpi) using the replicating TMV vector (Icon Genetics, Halle, Germany). Transiently expressed HBsAg formed immunologically important disulphide-linked dimers; it preserved the conformational 'a' antigenic determinant and assembled into recognizable VLPs. Intraperitoneal immunization of mice with partially purified HBsAg elicited also HBsAg-specific antibodies. Recently, a candidate oral HBV vaccine suitable for human immunization was developed from lyophilized plant material derived from herbicide-resistant transgenic lettuce carrying small, medium or large antigens of HBV (S-, M- or L-HBsAg) (Pniewski et al., 2011, 2012a). Lyophilized plant tissue converted into tablets maintained HBsAg content for at least 1 year of storage at room temperature. Plant-expressed S-HBsAg accumulated in the plant cells within the endoplasmic reticulum in the form of vesicles containing VLPs with 17–22 nm in size. Immune response in mice after oral delivery of lyophilized lettuce containing S-HBsAg was not strong, but the cut-off value considered as the minimal protective level for humans (10 mIU/ml) was reached. In spite of its feasible formulation, oral immunization by a plant lyophilized tissue should be improved in terms of efficiency, since oral tolerance appeared to be the main drawback for the production of anti-HBV plant-based vaccine (Pniewski, 2012b).

HBcAg was also successfully expressed in plants by stable or transient expression means (Tsuda et al., 1998; Huang et al., 2006; Mechtcheriakova et al., 2006; Zahmanova et al., 2008; Thuenemann et al., 2013). The highest yield (2.38 g of HBcAg per kg of FLW) was reached in *N. benthamiana* at 7 dpi by transient expression using the replicating TMV vector (Icon Genetics, Halle, Germany).

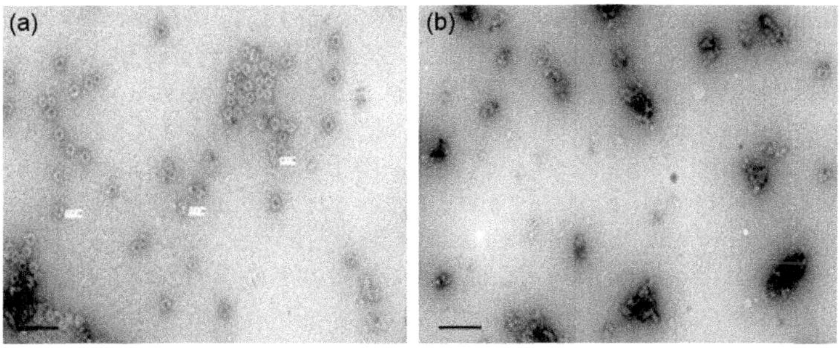

Figure 5.2 Electron micrographs of plant-derived full-length HBcAg (a) and the C-terminal deleted HBcAg (b) (Thuenemann et al., 2013). Bar = 100 nm.

Plant-derived HBcAg self-assembled into 30 nm VLPs and elicited a stronger immune response in mice upon intraperitoneal injection compared to the *E. coli*-derived HBcAg; it developed also HBcAg-specific serum IgG in mice upon mucosal immunization in the absence of adjuvants (Huang et al., 2006). HBcAg was also produced in plants from a plant codon-optimized gene in a form of different C-terminal deleted HBcAg mutants which were capable to form VLPs, having the advantage to be free of nucleic acid due to their C-terminal truncation (Figure 5.2; Thuenemann et al., 2013).

HBcAg self-assembles into VLPs not only from a native gene, but also from chimeric genes, similar to the HPV-16 L1 protein. HBcAg was proposed as a carrier for the insertion of the VP21 epitope of the capsid protein of *Foot-and-mouth disease virus* (FMDV), which was successfully expressed in transgenic tobacco (Huang et al., 2005). Chimeric HBcAg could form VLPs in plants, and induced immune responses in mice upon intraperitoneal immunization with crude plant extracts. Specific antibody responses were raised by both HBcAg and FMDV VP1. The HBcAg carrier was also used for inserting the M2e epitope of *Influenza virus A* (discussed below in the specific section).

5.4 Human immunodeficiency virus-1

Human immunodeficiency virus-1 (HIV-1), the causal agent of AIDS, belongs to the genus *Lentivirus* (family *Retroviridae*). It is composed of two copies of positive ssRNA enclosed by a conical capsid composed of the viral protein p24 and surrounded by the viral lipid envelope (Chiu et al., 1985; Karlsson Hedestam et al., 2008). Virus assembly is mainly driven by the viral Gag multi-domain polyprotein ($Pr55^{gag}$) whose expression is sufficient for VLP formation (Freed, 1998; Briggs and Kräusslich, 2011). These VLPs can be safely delivered to animals and humans since they are non-infectious and can elicit potent cellular and humoral immune responses (Doan et al., 2005). $Pr55^{gag}$, one of the promising candidates for the development of an HIV-1 vaccine, has also been studied and expressed in plants.

The transient expression of $Pr55^{gag}$ and truncated Gag versions (p17/24 and 24) was achieved by Meyers et al. (2008) in *N. benthamiana* plants by using the pTRA vectors with different targeting signals (Maclean et al., 2007) and by stable transformation in tobacco. A plant codon-optimized $Pr55^{gag}$ gave a low yield with both systems (44 μg per kg of FLW in the case of transient expression of the cytoplasm-targeted $Pr55^{gag}$, and 48 μg per kg of FLW in the case of stable transformation of the chloroplast-targeted $Pr55^{gag}$). Ten-fold increase in yield was obtained for both Gag-truncated constructs, p24 and p17/p24, when expressed transiently and targeted to the chloroplasts (4.0 and 4.8 mg per kg of FLW, respectively). The plant-derived p17/p24 was not immunogenic in mice when administered alone, but it considerably boosted humoral and cellular responses in mice primed by a gag DNA vaccine. A similar study of $Pr55^{gag}$ expression was performed by Scotti et al. (2009), where the gag protein was expressed in plants transiently or by stable nuclear and plastid transformation. The stable transformation improved importantly the $Pr55^{gag}$ yield, and a much higher yield was obtained in transplastomic tobacco plants (363 mg per kg of FLW) then in transgenic plants (28 mg per kg of FLW), although in both systems the protein was addressed to the chloroplasts. Yield differences were possibly due to different promoters and protein fusion partners used in two systems. In the case of the plastid transformation, the strong plastid *rrn* promoter was used for the expression and the $Pr55^{gag}$ was translationally fused to the plastid photosynthetic RbcL protein. Plant-derived $Pr55^{gag}$ assembled into spherical VLPs of 80–100 nm in size (Figure 5.3), resembling those produced in baculovirus/insect and *E. coli* expression systems.

Recently, Kessans et al. (2013) improved the yield of the full-length gag protein in the cytoplasm, a compartment in which VLP formation naturally initiates in animal cells. This was achieved by co-expressing the plant codon-optimized gag and gp41 genes, where the latter is associated with HIV-1 mucosal transmission and CD4+ cell infection. The Gag/gp41 co-expression was obtained in *N. benthamiana* plants combining stable transformation and transient expression by a TMV vector (Icon Genetics, Halle, Germany). Thus, protein yield in the cytoplasm was 400-fold

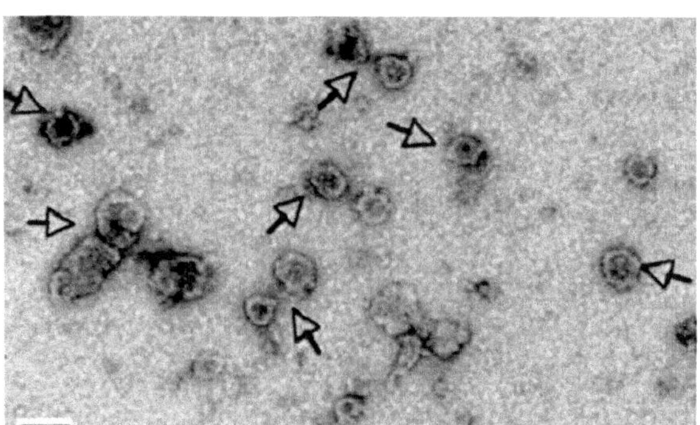

Figure 5.3 Immuno-sorbent electron microscopy of leaf tissue extracts from transplastomic plants (Scotti et al., 2009). Arrows indicate plant-produced HIV-1 Gag VLPs. Bar = 100 nm.

increased, reaching more than 20 mg per kg of FLW. Furthermore, the assembly of enveloped gag VLPs with 100 nm diameter was observed in the apoplast and in intracellular compartment within cytoplasmic membrane vesicles and these VLPs were able to bud across cell membranes, a process similar to VLP budding occurring in animal cells.

5.5 Influenza A virus

IAV is an enveloped virus of the family Orthomyxoviridae with the negative-sense RNA genome consisting of eight segments which encode 12 proteins (Hutchinson et al., 2010). IAV causes acute respiratory diseases in humans; efficient preventive treatments have not yet been obtained against this virus because of its high variability and broad host specificity. The continuous concern of emerging new influenza variants is induced by avian viruses which can be transmitted between different species and may cause devastating diseases on humans due to the lack of the specific immunity (Varečková et al., 2013). Current vaccines against influenza virus are based on the highly immunogenic surface proteins such as hemagglutinin (HA) and neuraminidase (NA).

The first report of influenza HA production in tobacco plants was documented in a patent by Cardineau et al. (2004). Then, HA was successfully expressed in a transient way by agroinfiltration in *N. benthamiana* plants, using a binary vector with the plastocyanin promoter (D'Aoust et al., 2008). HA was able to assemble into VLPs resembling those of influenza native virions. Plant-derived HA-based VLPs budded from the plasma similarly to the budding of influenza virions occurring in animal cells. Intramuscular immunization of mice with two doses of purified influenza HA-based VLPs showed a strong immunogenicity against the homologous virus. Due to a high variability of HA and NA, a broad range vaccine against influenza virus has not been developed using these proteins.

As an alternative, the use of the highly conserved minor M2 protein was proposed. M2e, the ectodomain of the M2 protein is a promising vaccine candidate since it is rather conserved among all influenza human strains, while the M2e of influenza pandemic animal strains have few different amino acids. Due to its low size, M2e can not be expressed alone and requires the fusion to carrier proteins. The first M2e transient expression in *N. benthamiana* plants was reported using HPV-16 L1 protein as carrier (Figure 5.1) (Matić et al., 2011) and this has been already discussed in PVs section. M2e expression was also achieved through the use of HBcAg as carrier. Ravin et al. (2012) expressed a chimeric HBcAg fused to M2e using a recombinant PVX vector in *N. benthamiana* plants. Plant-produced HBcAg-chimeric particles were highly immunogenic in mice, and mice were protected against influenza challenges. Two approaches for transient M2e expression using HBcAg as carrier were further reported in *N. benthamiana* by Thuenemann et al. (2013) using a replicating PVX vector (Mardanova et al., 2009) and the non-replicating pEAQ-HT system (Sainsbury et al., 2009). Using a PVX vector, a

M2eHBc chimeric construct was successfully expressed in plants (30–40 mg per kg of FLW) and assembled into spherical VLPs with 30–35 nm diameter similar to the native HBc antigen. Intraperitoneal immunization of mice with the plant-derived M2eHBc showed immune response against influenza virus. Another chimeric M2eHBc construct was well expressed in plants using the pEAQ-HT vector, showing slightly higher yield (48 mg per kg of FLW) and inducing antibodies specific to both the HBcAg carrier and the inserted M2e epitope in the immune sera, confirming the immunogenicity of the chimeras. Recently, the M2e epitope was successfully expressed in *N. benthamiana* plants using the structural coat protein (CP) of the plant virus (TMV) as a carrier (Petukhova et al., 2013). This approach adopted a recombinant TMV designed to deliver the protein by agroinfiltration in the plant cell nucleus. TMV-M2e chimeric viruses were able to systemically infect plants. Antisera produced against TMV-M2e chimeras had a higher content of antibodies specific to M2e compared to those specific to TMV CP. Immunogold electron-microscopy assays showed the M2e exposure on the surface of chimeric TMV virions. Mice immunized with TMV-M2e chimeras were resistant to five lethal doses of the homologous influenza virus strain, and partially resistant after challenge with a heterologous influenza virus. The results obtained with M2e chimeric constructs in these studies sustain the idea that a novel candidate universal vaccine against influenza could be produced in plants.

References

Baker, T.S., Newcomb, W.W., Olson, N.H., Cowsert, L.M., Olson, C., et al., 1991. Structures of bovine and human papillomaviruses. Analysis by cryoelectron microscopy and three-dimensional image reconstruction. Biophys. J. 60, 1445–1456.

Biemelt, S., Sonnewald, U., Galmbacher, P., Willmitzer, L., Müller, M., 2003. Production of human papillomavirus type 16 virus-like particles in transgenic plants. J. Virol. 77, 9211–9220.

Bosch, F.X., Manos, M.M., Muñoz, N., Sherman, M., Jansen, A.M., et al., 1995. Prevalence of human papillomavirus in cervical cancer: a worldwide perspective. International biological study on cervical cancer (IBSCC) Study Group. J. Natl. Cancer Inst. 87, 796–802.

Briggs, J.A., Kräusslich, H.G., 2011. The molecular architecture of HIV. J. Mol. Biol. 410, 491–500.

Campo, M.S., 2003. Vaccination against papillomavirus in cattle. Clin. Dermatol. 15, 275–283.

Cardineau, G., Mason, H.S., Vaneck, J.M., Kirk, D.D., Wamsley, A.M., 2004. Vectors and cells for preparing immunoprotective compositions derived from transgenic plants. International Patent application WO2004/098533.

Chapman, S., Kavanagh, T., Baulcombe, D., 1992. Potato virus X as a vector for gene expression in plants. Plant J. 2, 549–557.

Chen, X., Garcea, R., Goldberg, I., Casini, G., Harrison, S.C., 2000. Structure of small virus-like particles assembled from the L1 protein of human papillomavirus 16. Mol. Cell. 5, 557–567.

Chiu, I.M., Yaniv, A., Dahlberg, J.E., Gazit, A., Skuntz, S.F., et al., 1985. Nucleotide sequence evidence for relationship of AIDS retrovirus to lentiviruses. Nature 317, 366–368.

D'Aoust, M.A., Lavoie, P.O., Couture, M.M., Trépanier, S., Guay, J.M., et al., 2008. Influenza virus-like particles produced by transient expression in *Nicotiana benthamiana* induce a protective immune response against a lethal viral challenge in mice. Plant Biotechnol. J. 6, 930–940.

Doan, L.X., Li, M., Chen, C., Yao, Q., 2005. Virus-like particles as HIV-1 vaccines. Rev. Med. Virol. 15, 75–88.

Fernández-San Millán, A., Ortigosa, S.M., Hervás-Stubbs, S., Corral-Martínez, P., Seguí-Simarro, J.M., et al., 2008. Human papillomavirus L1 protein expressed in tobacco chloroplasts self-assembles into virus-like particles that are highly immunogenic. Plant Biotechnol. J. 6, 427–441.

Fischer, R., Stoger, E., Schillberg, S., Christou, P., Twyman, R.M., 2004. Plant-based production of biopharmaceuticals. Curr. Opin. Plant Biol. 7, 152–158.

Franconi, R., Massa, S., Illiano, E., Mullar, A., Cirilli, A., 2006. Exploiting the plant secretory pathway to improve the anticancer activity of a plant-derived HPV16 E7 vaccine. Int. J. Immunopathol. Pharmacol. 19, 187–197.

Freed, E.O., 1998. HIV-1 Gag proteins: diverse functions in the virus life cycle. Virology. 251, 1–15.

Gitlin, N., 1997. Hepatitis B: diagnosis, prevention, and treatment. Clin. Chem. 43, 1500–1506.

Hagensee, M.E., Yaegashi, N., Galloway, D.A., 1993. Self-assembly of human papillomavirus type 1 capsids by expression of the L1 protein alone or by coexpression of the L1 and L2 capsid proteins. J. Virol. 67, 315–322.

Hebner, C.M., Laimins, L.A., 2006. Human papillomaviruses: basic mechanisms of pathogenesis and oncogenicity. Rev. Med. Virol. 16, 83–97.

Huang, Y., Liang, W., Wang, Y., Zhou, Z., Pan, A., et al., 2005. Immunogenicity of the epitope of the foot-and-mouth disease virus fused with a hepatitis B core protein as expressed in transgenic tobacco. Viral Immunol. 18, 668–677.

Huang, Z., LePore, K., Elkin, G., Thanavala, Y., Mason, H.S., 2008. High-yield rapid production of hepatitis B surface antigen in plant leaf by a viral expression system. Plant Biotechnol. J. 6, 202–209.

Huang, Z., Santi, L., LePore, K., Kilbourne, J., Arntzen, C.J., et al., 2006. Rapid, high-level production of hepatitis B core antigen in plant leaf and its immunogenicity in mice. Vaccine. 24, 2506–2513.

Hutchinson, E.C., von Kirchbach, J.C., Gog, J.R., Digard, P., 2010. Genome packaging in influenza A virus. J. Gen. Virol. 91, 313–328.

Karlsson Hedestam, G.B., Fouchier, R.A., Phogat, S., Burton, D.R., Sodroski, J., et al., 2008. The challenges of eliciting neutralizing antibodies to HIV-1 and to influenza virus. Nat. Rev. Microbiol. 6, 143–155.

Kessans, S.A., Linhart, M.D., Matoba, N., Mor, T., 2013. Biological and biochemical characterization of HIV-1 Gag/dgp41 virus-like particles expressed in *Nicotiana benthamiana*. Plant Biotechnol. J. 11, 681–690.

Kirnbauer, R., Booy, F., Cheng, N., Lowy, D.R., Schiller, J.T., 1992. Papillomavirus L1 major capsid protein self-assembles into virus-like particles that are highly immunogenic. Proc. Natl. Acad. Sci. USA. 89, 12180–12184.

Kong, Q., Richter, L., Yang, Y.F., Arntzen, C.J., Mason, H.S., et al., 2001. Oral immunization with hepatitis B surface antigen expressed in transgenic plants. Proc. Natl. Acad. Sci. USA. 98, 11539–11544.

Lenzi, P., Scotti, N., Alagna, F., Tornesello, M.L., Pompa, A., et al., 2008. T: Translational fusion of chloroplast-expressed human papillomavirus type 16 L1 capsid protein enhances antigen accumulation in transplastomic tobacco. Transgenic Res. 17, 1091–1102.

Love, A.J., Chapman, S.N., Matic, S., Noris, E., Lomonossoff, G.P., et al., 2012. *In planta* production of a candidate vaccine against bovine papillomavirus type 1. Planta. 236, 1305–1313.

Lowy, D.R., Howley, P.M., 2001. Papillomaviruses. In: fourth ed. Knipe, D.M., Howley, P. M. (Eds.), Fields Virology, vol. 2. Lippincott Williams & Wilkins, Philadelphia, PA, pp. 2231–2264.

Maclean, J., Koekemoer, M., Olivier, A.J., Stewart, D., Hitzeroth, I.I., et al., 2007. Optimization of human papillomavirus type 16 (HPV-16) L1 expression in plants: comparison of the suitability of different HPV-16 L1 gene variants and different cell-compartment localization. J. Gen. Virol. 88, 1460–1469.

Mardanova, E.S., Kotliarov, R. Yu, Ravin, N.V., 2009. Increased efficiency of recombinant proteins production in plants due to optimised translation of RNA of viral vector. Mol. Biol. (Mosk). 43, 568–571.

Mason, H.S., Lam, D.M., Arntzen, C.J., 1992. Expression of hepatitis B surface antigen in transgenic plants. Proc. Natl. Acad. Sci. USA. 89, 11745–11749.

Matić, S., Rinaldi, R., Masenga, V., Noris, E., 2011. Efficient production of chimeric *Human papillomavirus* 16 L1 protein bearing the M2e influenza epitope in *Nicotiana benthamiana* plants. BMC Biotechnol. 11, 106.

Matić, S., Masenga, V., Poli, A., Rinaldi, R., Milne, R.G., et al., 2012. Comparative analysis of recombinant Human Papillomavirus 8 L1 production in plants by a variety of expression systems and purification methods. Plant Biotechnol. J. 10, 410–421.

Mechtcheriakova, I.A., Eldarov, M.A., Nicholson, L., Shanks, M., Skryabin, K.G., et al., 2006. The use of viral vectors to produce hepatitis B virus core particles in plants. J. Virol. Methods. 131, 10–15.

Meyers, A., Chakauya, E., Shephard, E., Tanzer, F.L., Maclean, J., et al., 2008. Expression of HIV-1 antigens in plants as potential subunit vaccines. BMC Biotechnol. 8, 53.

Nasir, L., Campo, M.S., 2008. Bovine papillomaviruses: their role in the aetiology of cutaneous tumours of bovids and equids. Vet. Dermatol. 19, 243–254.

Noris, E., Poli, A., Cojoca, R., Rittà, M., Cavallo, F., et al., 2011. A human papillomavirus 8 E7 protein produced in plants is able to trigger the mouse immune system and delay the development of skin lesions. Arch. Virol. 156, 587–595.

Paz De la Rosa, G., Monroy-García, A., Mora-García Mde, L., Peña, C.G.R., Hernández-Montes, J., et al., 2009. An HPV 16 L1-based chimeric human papilloma virus-like particles containing a string of epitopes produced in plants is able to elicit humoral and cytotoxic T-cell activity in mice. Virol J. 6, 2.

Petukhova, N.V., Gasanova, T.V., Stepanova, L.A., Rusova, O.A., Potapchuk, M.V., et al., 2013. Immunogenicity and protective efficacy of candidate universal influenza A nanovaccines produced in plants by tobacco mosaic virus-based vectors. Curr. Pharm. Des. 19, 5587–5600.

Pfister, H., Nurnberger, F., Gissmann, L., zur Hausen, H., 1981. Characterization of a human papillomavirus from epidermodysplasia verruciformis lesions of a patient from Upper-Volta. Int. J. Cancer. 27, 645–650.

Pineo, C.B., Hitzeroth, I.I., Rybicki, E.P., 2013. Immunogenic assessment of plant-produced human papillomavirus type 16 L1/L2 chimaeras. Plant Biotechnol. J. 11, 964–975.

Pniewski, T., 2012b. Is an oral plant-based vaccine against hepatitis B virus possible? Curr. Pharm. Biotechnol. 13, 2692–2704.

Pniewski, T., Kapusta, J., Bociąg, P., Wojciechowicz, J., Kostrzak, A., et al., 2011. Low-dose oral immunization with lyophilized tissue of herbicide-resistant lettuce expressing hepatitis B surface antigen for prototype plant-derived vaccine tablet formulation. J. Appl. Genetics. 52, 125–136.

Pniewski, T., Kapusta, J., Bociąg, P., Kostrzak, A., Fedorowicz-Strońska, O., et al., 2012a. Plant expression, lyophilisation and storage of HBV medium and large surface antigens for a prototype oral vaccine formulation. Plant Cell Rep. 31, 585–595.

Ravin, N.V., Kotlyarov, R.Y., Mardanova, E.S., Kuprianov, V.V., Migunov, A.I., et al., 2012. Plant-produced recombinant influenza vaccine based on virus-like HBc particles carrying an extracellular domain of M2 protein. Biochemistry (Mosc). 77, 33–40.

Regnard, G.L., Halley-Stott, R.P., Tanzer, F.L., Hitzeroth, I.I., Rybicki, E.P., 2010. High level protein expression in plants through the use of a novel autonomously replicating geminivirus shuttle vector. Plant Biotechnol. J. 8, 38–46.

Sainsbury, F., Thuenemann, E.C., Lomonossoff, G.P., 2009. pEAQ: versatile expression vectors for easy and quick transient expression of heterologous proteins in plants. Plant Biotechnol. J. 7, 682–693.

Scotti, N., Alagna, F., Ferraiolo, E., Formisano, G., Sannino, L., et al., 2009. High-level expression of the HIV-1 Pr55gag polyprotein in transgenic tobacco chloroplasts. Planta. 229, 1109–1122.

Sohn, U., Nam, H.G., Park, D.H., Kim, K.H., 2002. Recombinant human papillomavirus vaccine expressed in transgenic plants. U.S. patent 6,444,805.

Thanavala, Y., Yang, Y.F., Lyons, P., Mason, H.S., Arntzen, C., 1995. Immunogenicity of transgenic plant-derived hepatitis B surface antigen. Proc. Natl. Acad. Sci. USA. 92, 3358–3361.

Thuenemann, E.C., Lenzi, P., Love, A.J., Taliansky, M., Bécares, M., et al., 2013. The use of transient expression systems for the rapid production of virus-like particles in plants. Curr. Pharm. Des. 19, 5564–5573.

Torresi, J., 2002. The virological and clinical significance of mutations in the overlapping envelope and polymerase genes of hepatitis B virus. J. Clin. Virol. 25, 97–106.

Tsuda, S., Yoshioka, K., Tanaka, T., Iwata, A., Yoshikawa, A., 1998. Application of the human hepatitis B virus core antigen from transgenic tobacco plants for serological diagnosis. Vox Sang. 74, 148–155.

Varečková, E., Mucha, V., Kostolanský, F., 2013. HA2 glycopolypeptide of influenza A virus and antiviral immunity. Acta Virol. 57, 247–256.

Varsani, A., Williamson, A.L., Rose, R.C., Jaffer, M., Rybicki, E.P., 2003. Expression of human papillomavirus type 16 major capsid protein in transgenic *Nicotiana tabacum* cv. Xanthi. Arch. Virol. 148, 1771–1786.

Waheed, M.T., Thönes, N., Müller, M., Hassan, S.W., Gottschamel, J., et al., 2011. Plastid expression of a double-pentameric vaccine candidate containing human papillomavirus-16 L1 antigen fused with LTB as adjuvant: transplastomic plants show pleiotropic phenotypes. Plant Biotechnol. J. 9, 651–660.

Warzecha, H., Mason, H.S., Lane, C., Tryggvesson, A., Rybicki, E., et al., 2003. Oral immunogenicity of human papillomavirus-like particles expressed in potato. Virol. 77, 8702–8711.

Weissenborn, S.J., Nindl, I., Purdie, K., Harwood, C., Proby, C., et al., 2005. Human papillomavirus-DNA loads in actinic keratoses exceed those in non-melanoma skin cancers. J. Invest. Dermatol. 125, 93–97.

Zahmanova, G., Falzarano, D., Naimov, S., Kostova, M., Boncheva, R., et al., 2008. Oral immunization with truncated Hepatitis B virus nucleocapsid expressed in transgenic potatoes. C. R. Acad. Sci. Comptes rendus de l'Acade'mie bulgare des Sciences. 61, 1293–1300.

Biotechnology of Euphorbiaceae (*Jatropha curcas, Manihot esculenta, Ricinus communis*)

6

Fatemeh Maghuly[1], *Johann Vollmann*[2] *and Margit Laimer*[1]
[1]Plant Biotechnology Unit (PBU), Department of Biotechnology, BOKU VIBT, Vienna, Austria; [2]Plant Breeding Division, Department of Crop Sciences, BOKU, Tulln an der Donau, Austria

6.1 Euphorbiaceae crops

Euphorbiaceae or the spurge family is one of the largest and genetically diverse plant family with nearly 322 genera and 8,910 species, ranging from large woody trees to simple weeds (Mwine and Van Damme, 2011) and includes economically important species like *Jatropha curcas, Manihot esculenta* (cassava), *Ricinus communis* (castor bean) and *Hevea brasiliensis* (rubber tree) (Mwine and Van Damme, 2011).

Jatropha curcas or physic nut, a drought resistant shrub with a height of 3–5 m and an average life of 50 years is grown for harvesting the 30%–40% semi-dry toxic oil contained in its seeds, which can be used for the production of biofuel (Vollmann and Laimer, 2013; Maghuly and Laimer, 2013; Makkar et al., 2009). *J. curcas*, native to Central America, was transported by Portuguese in the eighteenth century via Cabo Verde to Africa and Asia (Valdes-Rodriguez et al., 2013). Interestingly, all accessions were reported to be toxic, while non toxic plants were found only in Southern Mexico (Ramkat, 2013; Valdes-Rodriguez et al., 2013).

Jatropha is interesting, due to its ability to grow under rainfall regimes and tropical humid conditions (Makkar et al., 2009) and its capacity of reclaiming marginal soils and degraded land by exploring the soil through an adequate root system and providing shadow to the soil as well as reducing risks of erosion and desertification (Openshaw, 2000). *Jatropha* is planted on an estimated 1.5 mio ha in Indonesia, on 100–200,000 ha in China, 20,000 ha in Brazil, 120,000 ha in Africa (mainly in Madagascar and Zambia) and has a huge potential in India and other tropical countries (Carels, 2013; Brittaine and Lutaladio, 2010).

Although *Jatropha* is toxic, it is one of the most valuable crude drugs of primitive times and is still widely used in medicine (Soomro and Memon, 2007) as a remedy for alopecia, burns, eczema, inflammation, paralysis and yellow fever.

Applied Plant Genomics and Biotechnology.

In addition, the oil is used for burning and spinning in the manufacture of hard soaps, candles, paints and lubricants (Roy, 1990). The seed cake is nutrient rich and therefore suitable as fertilizer (Ramkat, 2013) and has a crude protein content of 68%, which is far higher than in soybean (45.7%) (Aregheore et al., 2003).

However, various antinutritional factors (trypsin inhibitors, phytate, saponins, lectins and phorbol esters) are present in the kernel meal of *Jatropha*, which make the seed cake and oil unsuitable for animal feed (Martinez-Herrera et al., 2010). The major fatty acids in *Jatropha* oil consist of monounsaturated oleic (44%), polyunsaturated linoleic (33.3%) and saturated palmitic (14.7%) and stearic (6.7%) acids (Devappa et al., 2010).

Manihot esculenta or cassava is the only member of family cultivated for food (Patil and Fauquet, 2009). It is the fourth most important source of carbohydrates for human consumption in the tropics, after rice, sugarcane and maize (Bellotti et al., 1999), and is also a cash crop, feed crop and a raw material for industrial uses like starch, alcohol and biofuel production (Jansson et al., 2009). Cassava is grown as a staple starch food crop on over 20 mio ha worldwide; major producers of cassava include Nigeria, Congo, Angola, Tanzania, Ghana and Mozambique in Africa; Thailand, Indonesia, Vietnam and Cambodia in Asia and Brazil in South America (FAOSTAT, 2014). Current consensus based on botanical, genetic and archaeological evidence supports that the southern border of the Amazon region was the first centre of cassava domestication (Olsen and Schaal, 2001). It was spread by Portuguese seafarers to West Africa in the form of flour and first cultivated in 1558 in the Congo basin (Carter et al., 1995), from where it spread to other countries like Guinea, Sierra Leone, Angolan Coast and Congo River, Mozambique and Kenya (Carter et al., 1995).

Cassava is adapted to warm humid lowland tropics and grows in areas where the mean annual temperature is higher than 20°C with annual rainfall between 500 and 800 mm (Puonti-Kaerlas, 1998). Its ability to thrive on degraded soils, tolerate drought and produce acceptable yield makes it of economic importance for tropical agriculture (Thresh and Cooter, 2005).

Ricinus communis or castor bean is an agricultural crop and garden ornamental that is widely cultivated and has been introduced worldwide (Foster et al., 2010). Historically castor bean has been cultivated as an agricultural crop for the oil derived from its seeds containing 40%–60% toxic oil with numerous industrial and cosmetic uses (Foster et al., 2010). Seeds are rich in triglycerides, mainly ricinoleic acid (12-hydroxy-cis-9-octadecenoic acid) and contain the poison ricin.

Castor is grown on 1.5–1.6 mio ha annually with India, China, Mozambique and Brazil being the most significant producers during the last decade (FAOSTAT, 2014). It is a seed oil plant that can vary greatly in its growth habit and appearance. Castor is indigenous to the Southeastern Mediterranean Basin, Eastern Africa and India, but is widespread throughout tropical and sub-tropical regions (Roger and Martyn, 1999). Castor bean originated in Eastern Africa based on the high diversity of plants found in Ethiopia (Foster et al., 2010), however this has not been directly tested. Plants are self- or cross-pollinated by wind, with outcrossing as predominant mode of reproduction (Foster et al., 2010, Brigham, 1968).

6.2 Genetic diversity

Genetic variability is the pre-requisite for any plant breeding programme and the basis for adaptability, stability and evolution of species and tree populations (Müller-Starck et al., 1992). The identification of taxonomic units and the determination of the uniqueness of many tree species, primitive cultivars, landraces, elite breeding lines and wild relatives of crops is the focus of conservation strategies. Populations with a narrow genetic basis are widely thought to be more sensitive to environmental changes or disease, leading to a decrease in productivity (Maghuly et al., 2006).

For crop improvement, genetic diversity is an important basis to identify genetically distinct plants for various breeding purposes (Brummer et al., 1995). Identification of genetic diversity based on morphological characteristics can be easily monitored, however, it is highly influenced by the environment and genetically-based characters (environment × genotype interaction) (Keivani et al., 2010). Biochemical and molecular traits were considered to detect and assess genetic diversity (Muthusamy et al., 2008). Unlike morphological markers, DNA markers are stable, and can provide direct access to the genomic regions where they are located (Jarne and Lagoda, 1996).

6.2.1 Genetic diversity in Jatropha

The genetic diversity of *Jatropha* is considered rather low (Jongschaap et al., 2007), particularly in Asia and Africa, whereas it appears larger at the site of origin, i.e. in South America. In Brazil, about 200 *Jatropha* accessions have been characterized in the Embrapa genebank both on a phenotypic and molecular genetic level. While phenotypic variation appears sufficient for breeding purposes (Laviola et al., 2012), genetic diversity in random amplified polymorphic DNA (RAPD) and microsatellites or simple sequence repeat (SSR) markers was found to be rather low (Rosado et al., 2010). In contrast to the outcrossing nature of *Jatropha* flowering biology, high levels of homozygosity were found; moreover, clustering of accessions based on molecular information was not related to the geographic origin, which indicates widespread seed or vegetative dispersion of particular genotypes across regions within Brazil. Thus, a need for introduction of novel diversity for breeding purposes was identified (Rosado et al., 2010). In contrast to the low *Jatropha* diversity of Brazil, much higher diversity has been reported from Central and Southeastern Mexico, considered as possible centre of *Jatropha* origin (Pecina-Quintero et al., 2014). Mexican *Jatropha* genotypes revealed a high genetic diversity between different geographic regions, and large diversity was also reported for seed oil and protein content and for the level of toxicity (Pecina-Quintero et al., 2014). However, standard genotypes and the utilization of the same marker system would be necessary for fully comparing genetic diversity between different countries towards an overall judgement of *Jatropha* variability. Apart from molecular data, information on genetic variability in yield and seed quality characters, heritability and correlations among traits is most valuable for selection of germplasm in plant-breeding programs, as illustrated for a base population of *Jatropha* accessions in India (Tripathi et al., 2013). Similarly, Montes et al. (2013a) investigated seed quality characters of a worldwide sampling of *Jatropha* germplasm; while they reported a large variation in seed and oil traits, their estimates of variation are affected by the environment, as their sampling of accessions was from different geographic locations. The genome of *J. curcas* was sequenced, using Sanger and multiplex sequencing methods, which 285,858,490 bp consisting of 120,586 contigs and 29,831 singlets was obtained. This resulted in 40,929 complete and partial structures of protein encoding genes (Sato et al., 2013). A large living reference collection of 1,400 *J. curcas* accessions and related species from 16 different countries has been established at the Plant Biotechnology Unit, BOKU University, Vienna, Austria.

6.2.2 Genetic diversity in cassava

From the time cassava was introduced to Africa from its centre of origin, selection for adaptation to African conditions has contributed to considerable diversity within the African cassava accessions (Beeching et al., 1993). Cassava varieties with high hydrogen cyanide (HCN) concentrations are referred to as bitter and are mainly consumed as flour, starch or glucose. Accessions with low HCN levels (<100 ppm in fresh roots), known as sweet cassava (Vieira et al., 2011). Colombo et al. (1998) showed that cultivated varieties for flour production are grouped together and separated from those destined to *in natura* consumption. Elias et al. (2000) using amplified fragment length polymorphism (AFLP) on cassava accessions grown by farmers in Guyana and wild accessions, found a high level of genetic variability. They described that the interaction between human and ecological factors could have facilitated the dynamics of diversity of local varieties (Elias et al., 2000). Second et al. (1997) analysed accessions from South America and observed introgression into cassava from *M. glaziovii*. The diversity of cassava itself was high, but the diversity was narrow in a single Amazonian field (Second et al., 1997). Although domestication appeared to have evolved primarily from *M. esculenta* ssp. *flabellifolia* and *peruviana*, it seemed that some other species also contributed (Second et al., 1997).

Using molecular markers in cassava was useful to illustrate the role played by farming practice in maintaining and/or adding to cassava genetic diversity. Montero-Rojas et al. (2011) developed SSR markers to evaluate genetic diversity of samples from Puerto Rico and samples with unknown genetic background. The obtained results showed higher genetic diversity in unknown cassava samples compared to samples originating from Puerto Rico. Traditional practices like intercropping and incorporation of volunteer seedlings contributes in recombination of genotypes to the cultivated stocks not only allows for selection and adaptation in cassava but also result to high levels of genetic diversity (Montero-Rojas et al., 2011).

Natural and artificial selection act on these seedlings leading to new accessions of cassava in the field (Montero-Rojas et al., 2011). Furthermore it is possible that whilst selecting, utilizing and distributing landraces with their preferred agronomic and quality traits, farmers have inadvertently added useful cassava mosaic disease (CMD) resistant accessions to the available germplasm (Lokko et al., 2006). The mapping of two CMD resistance genes *CMD1* (recessive) and *CMD2* (dominant) on the cassava genetic linkage map (Akano et al., 2002) and the identification of a sequence characterized region (SCAR) marker (RME1) and SSR markers (SSRY28 and NS158) associated with *CMD2* allowed the fast tracking of CMD-resistant germplasm (Bi et al., 2010) and marker assisted breeding in cassava (Bi et al., 2010).

The expressed sequence tags (ESTs) of cassava reveal a high degree of genetic diversity between cassava genotypes, because of its allotetraploid nature and lack of intense domestication (Anderson et al., 2004).

Since the content of provitamin A carotenoids in roots of commercial cassava cultivars is low, conventional breeding and genetic modification have been attempted to increase the level of provitamin A carotenoids in cassava (Rojas et al., 2009; Welsch et al., 2010). For conventional breeding, the heterozygous nature of the crop renders varietal recovery difficult, and long breeding cycles slow down the progress of this endeavour (Rojas et al., 2009). Advancement has shown that a single nucleotide polymorphism (SNP) present only in yellow-rooted cultivars cosegregates with coloured roots in a breeding pedigree. This newly characterized phytoene synthase allele will further provide means to improve cassava provitamin A content (Welsch et al., 2010).

A draft genome sequence of cassava has been generated using a 454-based whole genome shotgun strategy (Prochnik et al., 2012), which covers 69% of the predicted whole genome size (770 Mb) and 96% of protein-coding gene space (Prochnik et al., 2012). However, known SSR and SNP markers are sparsely distributed across the cassava genome which may make it not ideal for fine-mapping (Prochnik et al., 2012). Cassava ESTs could not only clarify the distribution and divergence of orthologous genes and their allelic diversity, but also can be used for mapping and breeding (Zou et al., 2011).

The International Institute of Tropical Agriculture (CIAT) in Columbia is holding most of the cassava germplasm, which is maintained *in vitro* (Ceballos et al., 2010). Apart from the primary gene pool of cassava comprising three subspecies within *M. esculenta*, the secondary gene pool consists of over 10 *Manihot* species. These species could be utilized for widening cassava diversity and for transferring useful traits such as improved nutritional quality or increased protein content into the primary gene pool (Nassar and Ortiz, 2009).

6.2.3 Genetic diversity in castor bean

Because of limited genetic diversity, understanding population structure and the distribution of castor bean cultivars is a challenge (Foster et al., 2010). Using AFLP markers to assess the genetic diversity of 200 castor beans from 35 countries worldwide showed a low level of genetic diversity (HE = 0.126) (Allan et al., 2008). Further, Qiu et al. (2010) designed EST-SSR markers to characterize 24 castor bean accessions from different countries, which showed a moderate level of genetic diversity (HE = 0.41). Foster et al., (2010) determined the population genetic structure of 676 castor beans using SNPs at 48 loci. Results showed five main groups and mixed genotypes. High levels of population differentiation occurred between most populations without geographic base (Foster et al., 2010). Further, using these SNPs revealed a low level of genetic diversity and minimal geographic structuring of populations worldwide due to mixing of genotypes (Foster et al., 2010). However, knowing the sequence of an entire plant genome (approximately 320 megabases, organized in 10 chromosomes) allows developing markers across the genome, which will facilitate the characterization of genotypes (Chan et al., 2010; Prochnik et al., 2012). The currently published castor bean genome is based on a 4.6× genome coverage from Solexa sequencing. Genes involved in the biosynthesis of fatty acids and triacylglycerols (TAGs), corresponding mainly to ricinoleic acid and triricinolein, were identified in the draft genome sequences of castor bean (Chan et al., 2010).

The chloroplast genome has been sequenced, from which sequence data of genes for structural and functional organellar components were assembled (Rivarola et al., 2011). This data was used to generate SNP markers to analyze the genetic diversity of castor bean from five geographical locations which revealed low levels of genetic diversity (Rivarola et al., 2011).

Breeding work has attempted to develop castor beans with reduced levels of ricin, *R. communis* agglutinin (RCA_{120}) toxins and dwarf-internode growth habit (Ramkat, 2013; Auld et al., 2003). Texas Tech University developed and released in 2002 an open-pollinated germplasm population of castor bean, TTU-LRC (Reg. no. GP-3, PI 631156, Auld et al., 2003). In addition, through induced mutations and subsequent selection and identification of pistillate variants, the longevity in maturity time of castor bean has been reduced from being a perennial to a high yielding annual domesticated crop (Divakara et al., 2009).

Castor genetic resources are available to a much wider extent in germplasm repositories as compared to *Jatropha* (Table 6.1). Most accessions are held by gene banks in India, China and Brazil, but the use of germplasm in castor breeding was reported to be rather limited due to the lack of characterization of accessions (Severino et al., 2012). Apart from castor genetic resources preserved in gene banks, Fernandez-Martinez and Velasco (2012) emphasize the presence of *in situ* castor resources such as landraces, ornamental castor and semi-wild weedy populations which might bear local adaptation traits in specific environments.

Table 6.1 Comparative listing of *Jatropha*, cassava and castor characteristics relevant to plant breeding

Characteristic	*Jatropha*	Cassava	Castor bean
Origin	Mexico, North-East South America	South America	Eastern Africa, Ethiopia
Chromosome number	$2n = 22$	$2n = 36$	$2n = 20$
Ploidy level	Diploid	Functional diploid	Diploid
Life cycle	Perennial	Perennial, handled as annual	Annual (perennial)
Flowering biology	Monoecious, protandry, insect cross pollination	Monoecious, protogynous	Monoecious, mainly cross pollination by wind
Commercial propagation	Seed or cutting	Stem cuttings	Seed
Cultivar types	Semi-wild local provenances in different growing regions	Clonal varieties	Open-pollinated varieties or hybrids
Genetic resources	About 200 described (Brazil)	>6,000	Approx. 11,300

6.3 Genetic improvement

Genetic improvement of *Jatropha*, cassava and castor bean is carried out at different levels due to considerable differences in domestication status, cultivation history and crop utilization pattern. *Jatropha* has been distributed from its centre of origin throughout the tropics and sub-tropics of Africa, Asia as well as South and Central America since centuries. However, systematic attempts to domesticate *Jatropha* as an industrial oilseed crop are very recent, dating back to the last two decades only. Therefore, *Jatropha* is still considered as a wild plant with large variation in seed and oil yield between genotypes (Brittaine and Lutaladio, 2010). In contrast, cassava has been domesticated several thousands of years ago in South America, cultivated since ancient times as a food crop and has a long breeding history (Ceballos et al., 2010). Similarly, castor bean is also considered an ancient oilseed crop with medical and various non-food uses throughout its cropping history (Fernandez-Martinez and Velasco, 2012). However, understanding population structure and the distribution of *Jatropha* and castor bean cultivars have been challenging because of limited genetic variability (Foster et al., 2010).

6.3.1 Breeding strategies

Although *Jatropha*, cassava and castor have similarities in their flowering biology, i.e. monoecious flowers which make it easy to carry out hybridization, differences in life cycle and mode of propagation (Table 6.1) require different breeding strategies.

6.3.1.1 Conventional breeding

Jatropha

The reproductive biology of *Jatropha* (Table 6.1) would suggest clonal breeding similar to cassava as a breeding method. However, the low domestication status of *Jatropha*, the lack of well-characterized genetic resources and the utilization pattern of long-term plantations for harvesting over decades are different from cassava thus prompting for more focused approach integrating both conventional and biotechnology methods. While vegetative propagation of superior elite genotypes is the method of choice for immediate action in order to generate planting material, mid- to long-term strategies should foresee recurrent selection for population improvement, utilization of heterosis and marker-assisted and genomic selection approaches. For hybrid breeding, superior clones from different world regions of *Jatropha* production could be tested for genetic distance using molecular marker technologies and test crosses for combining ability; similarly to oil palm (Soh et al., 2009), superior hybrids could be propagated vegetatively using tissue culture techniques for rapid propagation of uniform planting material, as seed propagation would generate heterogenous planting material due to heterozygosity of the parental material. In hybrid breeding programs, inbreeding might also be practised to obtain homozygous materials for test crosses, for identifying useful recessive traits and – as suggested in cassava (Ceballos et al., 2010) – for reducing the genetic load of undesirable negative alleles probably present in semi-wild *Jatropha* populations. As plants from vegetative propagation are weaker than seedlings due to the lack of tap root formation (Brittaine and Lutaladio, 2010), tissue culture protocols such as somatic embryogenesis procedures should be refined to ensure the production of vigorous planting materials.

A range of important breeding objectives has been listed for *Jatropha* improvement including agronomic, seed quality and various stress tolerance characters (Brittaine and Lutaladio, 2010). For seed quality screenings, high-throughput methods such as near-infrared reflectance spectroscopy have been partly implemented in selection for oil and protein content, individual fatty acids and for determination of the toxic phorbol ester content (Montes et al., 2013b). Further, highly efficient marker-assisted selection as presented for the quantitative trait locus (QTL) controlling phorbol ester synthesis (King et al., 2013) could be applied for mapping oil content or other qualities as well. The availability of genetic markers for selection purposes has recently increased in *Jatropha* (Maurya et al., 2013; Kumari et al., 2013), and partial synteny between *Jatropha* and other Euphorbiaceae genomes has been utilized (e.g. King et al., 2013). Thus, genome-wide selection approaches as discussed by Laviola et al. (2012) might be implemented in breeding programs for raising productivity, while metabolomic and more functional approaches highlighting fatty acid or terpenoid biosynthesis pathways (Maghuly and Laimer, 2013) represent key resources for tailoring *Jatropha* oil and meal quality features.

Cassava

In cassava as a vegetatively propagated crop, clonal breeding is the conventional method of choice for variety development. This involves crosses between elite clones to obtain recombinant botanical seed. Seedlings are planted in nurseries, and both selection and vegetative propagation of superior clones take place over several years. As the classical clonal breeding method is time-consuming, cassava multiplication rates are low and heterosis can only partially be utilized in clonal varieties, a number of additional strategies such as rapid cycling recurrent selection or inbreeding for better exploitation of heterosis have been proposed (Ceballos et al., 2010). Moreover, genomics-based techniques such as marker-assisted selection procedures are presently integrated into cassava breeding both on the level of breeding strategies and on the individual trait level due to the progress in cassava genome sequences (Ceballos et al., 2012). As cassava field experimentation is considerably more laborious than for small-grain crops, the successful application of genomic selection could shorten the lengthy cycles of phenotypic selection while increasing the efficiency of the programme. De Oliveira et al. (2012) evaluated 358 cassava genotypes for agronomic characters while genotyping them with 390 SNPs; using a random regression-best linear unbiased prediction model, they reported increased selection accuracy in high heritability characters such as fresh cassava root yield, dry matter content and starch yield.

Castor bean

In castor bean, breeding methods and techniques developed for annual seed propagated crops can be applied. As reviewed by Auld et al. (2009), segregating populations after hybridizations can be treated using pedigree or bulk methods. In addition, the single seed descent method could be applied for rapid generation advance, while backcrossing is utilized to transfer individual traits such as shattering or disease resistance. In contrast to predominantly self-pollinated crops, castor bean varieties would never be fully homogeneous and homozygous pure-line varieties, as outcrossing rates of about 50%–90% have been reported, which would enable recurrent selection for population improvement without hand pollination. Moreover, hybrid breeding is also practised since male sterile plants carrying only pistillate flowers could be identified (for details see recent reviews by Severino et al., 2012 and Fernandez-Martinez and Velasco, 2012). As heterosis for grain yield is significant in superior crosses, hybrid varieties have been developed which are planted on more than 50% of the castor bean acreage of India according to Severino et al. (2012).

6.3.1.2 Tissue culture and gene transfer technology

Plant tissue cultures represent the basis for many applications, such as the conservation of valuable genetic resources and biotechnological breeding, pathogen elimination through *in vitro* thermotherapy and meristem culture of selected genotypes (Laimer, 2006). In the frame of breeding programs involving mutagenesis (Mukherjee et al., 2011), *in vitro* culture can also be used to sort out chimeras (which represent the major challenge in these treatments), since only fragments of a plant body can be maintained and regenerated into entire plants (Laimer, 2006). Further, gene transfer technology for all crop species rely on tissue culture systems capable of generating totipotent cells and tissues (Taylor et al., 2004) to be used for transgene insertion, selection of successful integration and transformed plant regeneration (Sudhakar Johnson et al., 2011). The most widely used methods in Euphorbiaceae so far are the direct gene method using microprojectile bombardment and *Agrobacterium tumefaciens*-mediated transformation, each having their own advantages and limitations (Nyaboga et al., 2013; Sailaja et al., 2008). The transformation efficiency in microprojectile bombardment may be lower than with *Agrobacterium*-mediated transformation and the device and consumables are costly (Nyaboga et al., 2013). *Agrobacterium*-mediated transformation offers several advantages, e.g. the transfer of one or few copies of DNA fragments, large DNA fragments with minimal rearrangement, higher efficiencies and lower costs (Nyaboga et al., 2013; Sailaja et al., 2008). So far, both methods were used for transformation of Euphorbiaceae (Sailaja et al., 2008).

Jatropha

Jatropha in vitro regeneration was reported from different tissues (da Câmara Machado et al., 1997; Sudhaker Johnson et al., 2011; Kajikawa et al., 2012), e.g. leaves, cotyledon, petioles, hypocotyls, epicotyls, nodal segments, axillary nodes, shoot tips and stems (reviewed by Sudhaker Johnson et al., 2011).

Stable transformation of *Jatropha* using *Agrobacterium* or particle bombardment for modifying oil and toxin biosynthesis was reported by several groups (Ye et al., 2013; Kajikawa et al., 2012; Ceasar and Ignacimuthu, 2011; Mukherjee et al., 2011). The first *Agrobacterium*-mediated transformation from cotyledon explants was published by Li et al. (2008) using the herbicide phosphinothricin as selectable marker gene. Other groups followed by using hygromycin and kanamycin as resistance marker genes (Kajikawa et al., 2012). Recently, novel protocols to generate transgenic plants and hairy roots for *Jatropha* were established by Kajikawa et al. (2012). However, it is important to optimize regeneration and transformation procedures for each genotype of *Jatropha* (Basa and Sujatha, 2009).

Increasing oil yield and improving the FA composition are the obvious goals to achieve a premium sustainable bioenergy crop (Sato et al., 2013). Transgenic technology will help to improve *Jatropha* as a crop in a more desirable way to serve as a best feedstock for the efficient biodiesel production, e.g. modification of lipid profile, low toxin levels and resistance to biotic and abiotic stresses (Ceasar and Ignacimuthu, 2011). Therefore, to improve oil accumulation in *Jatropha* seeds by genetic modification, genes involved in the biosynthesis pathway of TAGs are of great interest (Sato et al., 2013). Qu et al. (2012) increased the olcic acids concentration in *Jatropha* by downregulating the conversion of oleic to linoleic acid using RNA interference technology.

Cassava

The potential transgenic technology for improving cassava was first reported in late 1980s (Taylor et al., 2004). In the mid-1990s Li et al. (1996) used *Agrobacterium*-mediated transformation of cotyledons to regenerate transgenic shoots by organogenesis, while Schöpke et al. (1996), demonstrated regeneration of transgenic plantlets through microparticle bombardment of embryogenic cells (Nyaboga et al., 2013). For the generation of transgenic cassava, *Agrobacterium*-mediated transformation of friable embryogenic callus (FEC) has become the most efficient and preferred strategy (Nyaboga et al., 2013; Zainuddin et al., 2012). FEC-based cassava transformation allowed the presentation of important agronomic traits, like resistance or tolerance to *Cassava mosaic virus* (CMV) and *Cassava brown streak virus* (CBSV) using RNAi-based approaches and *African cassava mosaic virus* (ACMV) using antisense RNA interference (Nyaboga et al., 2013; Zainuddin et al., 2012; Yadav et al., 2011; Zhang et al., 2005). Comparison between transgenic and non-transgenic cassava showed a reduction of viral replication activity by 86%–99% (Taylor et al., 2004). A procedure to transform landraces, which was used successfully to transform an industry-preferred cultivar in South Africa was developed (Zainuddin et al., 2012).

It is known that most cassava farmers are resource-poor, lacking the ability to purchase and apply agrochemicals, therefore, development and deployment of enhanced germplasm remain the most important method for ensuring improved cassava production (Taylor et al., 2004). Transgenic technology allows beneficial traits to be transferred from one cassava cultivar to another and from wild relatives to cultivated ones, and furthermore the integration of genetic material from exotic sources such as viruses for pathogen-derived resistance strategies, and bacterial genes for insect resistance become possible (Taylor et al., 2004). However, transformation of the cassava cultivars particularly preferred by African farmers remains challenging because FEC production tends to be genotype dependent and regeneration of transformed FEC is often problematic (Nyaboga et al., 2013).

Castor bean

Castor bean tissue culture provides an effective method for the conservation of genetic resources and for genetic transformation (Tan et al., 2013; Sailaja et al., 2008). Regeneration is reported mainly from explants with meristematic explants such as shoot apices and embryonic axes (Tan et al., 2013). Shoot regeneration from explants devoid of meristematic tissues was sporadic and the frequency of morphogenesis was low and genotype dependent (Sarvesh et al., 1992; Reddy and Bahadur, 1989). Most studies investigated callus initiation and plantlet regeneration from seedling explants (Tan et al., 2013; Kansara et al., 2013), while a few used meristematic tissues of shoot tips as explants (Sujatha and Reddy, 2007), and axillary buds of pistillate plants (Tan et al., 2013).

Genetic transformation studies relied on meristem-based proliferation systems developed by Sujatha and Reddy (1998). The first successful stable transformation of meristematic tissues of castor bean through *Agrobacterium*-mediated transformation has been reported (Sujatha and Sailaja, 2005) and transformation by vacuum infiltration of wounded flower buds was patented (US Patent No. 6,620,986) by McKeon and Chen (2003). Transgene technology to develop resistant cultivars in castor bean mainly focused on insect resistance (Sujatha et al., 2009). Transgenic castor beans resistant to castor semilooper through deployment of *Cry1Ab* gene have been developed (Malathi et al., 2006).

6.4 Phytosanitary improvement

In order to control the impact of pathogens, constituting important biotic limitation for many crop species, the detection and elimination of the most important viruses from planting material is a pre-requisite.

6.4.1 *Viral diseases of* Jatropha, *cassava and castor bean*

Jatropha curcas, cassava and castor bean are susceptible to infections by several viruses (Table 6.2), including cassava mosaic geminiviruses (CMG) leading to cassava mosaic disease (CMD), CBSV, *Cassava common mosaic virus* (CsCMV) and *Cucumber mosaic virus* (CMV).

Although not all regions are equally impacted, in sub-Saharan Africa, CMD is estimated to be responsible for a 30%–40% harvest loss, ranking it as the most damaging plant virus disease in the world (Patil and Fauquet, 2009). Legg et al. (2006) reported that the CMD pandemic has affected cassava in at least nine East and Central African countries, covering an area of 2.6 million square kilometres and causing an estimated annual economic loss of US$1.9–2.7 billion (Patil and Fauquet, 2009). The International Institute of Tropical Agriculture (IITA) has succeeded in producing varieties resistant to CMD by conventional breeding (Taylor et al., 2004). It is considered, however, that transgenic strategies also have an important role to play in combating CMD by introducing new sources of resistance into cassava germplasm. This can occur through integration of resistance genes directly into susceptible farmer-preferred landraces, by stacking transgene-derived resistance into existing conventionally improved varieties, and through genetic transformation of breeding material (Taylor et al., 2004).

6.4.2 *Virus detection*

Enzyme-linked immunosorbant assay (ELISA) with monoclonal antibodies (Mabs) was first used to distinguish ACMV and *East African cassava mosaic virus* (EACMV) in cassava (Harrison et al., 1997). ELISA offers the advantage of relative simplicity; however, it requires fresh samples with clear disease symptoms, since it cannot detect viruses at low titres (Aloyce et al., 2013).

To increase sensitivity, degenerate and virus specific primers were developed and polymerase chain reaction (PCR) used subsequently to detect the CMGs

Table 6.2 **Viruses infecting *Jatropha*, cassava and castor bean (Ramkat, 2013)**

Host plant	Virus	Genus/family	Symptoms	Vector	Distribution	References
Cassava, *Jatropha*, castor bean	*African cassava mosaic virus*	*Begomovirus/ Geminiviridae*	Mosaic, leaf distortion and stunting	Whitefly	Africa	Thottappilly et al., 2003; Ramkat et al., 2011
Cassava	*East African cassava mosaic virus*	*Begomovirus/ Geminiviridae*	Mosaic, leaf distortion and stunting	Whitefly	East Africa	Alabi et al., 2008; Pita et al., 2001; Thottappilly et al., 2003
Cassava	*East African cassava mosaic, Cameroon virus*	*Begomovirus/ Geminiviridae*	Mosaic, leaf distortion and stunting	Whitefly	West Africa, Tanzania	Fondong et al., 2000
Cassava	*East African cassava mosaic, Kenya virus*	*Begomovirus/ Geminiviridae*	Mosaic, leaf distortion and stunting	Whitefly	East Africa	Bull et al., 2006
Cassava	*East African cassava mosaic, Malawi virus*	*Begomovirus/ Geminiviridae*	Mosaic, leaf distortion and stunting	Whitefly	Malawi	Zhou et al., 1997
Cassava	*East African mosaic, Zanzibar virus*	*Begomovirus/ Geminiviridae*	Mosaic, leaf distortion and stunting	Whitefly	Zanzibar, Madagascar	Maruthi et al., 2004
Cassava, *Jatropha*	*East African cassava mosaic virus-Uganda*	*Begomovirus/ Geminiviridae*	Mosaic, leaf distortion and stunting	Whitefly	Africa	Pita et al., 2001; Ramkat et al., 2011
Jatropha	*Jatropha mosaic Nigeria virus*	*Begomovirus/ Geminiviridae*	Mosaic, leaf blistering and mottling	Whitefly	Nigeria	Kashina et al., 2013
Cassava	*Indian cassava mosaic virus*	*Begomovirus/ Geminiviridae*	Mosaic, leaf distortion and stunting	Whitefly	Indian, Sri Lanka	Hong et al., 1993; Gao et al., 2010

(*Continued*)

Table 6.2 (Continued)

Host plant	Virus	Genus/family	Symptoms	Vector	Distribution	References
Jatropha	*Jatropha Mosaic virus*	*Begomovirus/ Geminiviridae*	Mosaic, blistering on leaf surface	Whitefly	Puerto Rico, Jamaica, India	Gao et al., 2010
Cassava	*South African cassava mosaic virus*	*Begomovirus/ Geminiviridae*	Mosaic, leaf distortion and stunting	Whitefly	South Africa, Zanzibar, Madagascar, Zimbabwe	Berrie et al., 2001
Cassava	*Cassava brown streak virus*	*Ipomovirus/ Potyviridae*	Brown, elongate necrotic stem lesions, secondary and tertiary vein chlorosis, corky brown necrosis in tuberous roots	Whitefly	Africa	Monger et al., 2001
Cassava	*Uganda Cassava brown streak virus*	*Ipomovirus/ Potyviridae*	Brown, elongate necrotic stem lesions on, secondary and tertiary vein chlorosis, corky brown necrosis in tuberous roots	Whitefly	Africa	Winter et al., 2010
Cassava	*Cassava Ivorian bacilliform virus*	Unassigned/ *Ourmiavirus*	Symptomless	Unknown	Cote d' Ivoire	Thottappilly et al., 2003
Cassava	*Cassava virus C*	*Ourmiavirus/ unassigned*	Pronounced leaf fleck	Unknown	Cote d' Ivoire	Thottappilly et al., 2003

Jatropha castor bean	*Cucumber mosaic virus*	*Cucumovirus/ Bromoviridae*	Mosaic	Aphids	India	Raj et al., 2008, 2010
Cassava	*Sri Lankan cassava mosaic virus*	*Begomovirus/ Geminiviridae*	Mosaic, leaf distortion and stunting	Whitefly	India, Sri Lanka and India	Rothenstein et al., 2006
Cassava	*Cassava American latent virus*	*Nepovirus/ Comoviridae*	symptomless	Unknown	Brazil and Guyana	Thottappilly et al., 2003
Cassava	*Cassava vein mosaic virus*	*Cavemovirus/ Caulmoviridae*	Vein mosaic	Unknown	Brazil	Thottappilly et al., 2003
Cassava	*Cassava Colombian symptomless virus*	Tentative *Potexvirus/ Flexiviridae*	Symptomless	Unknown	Columbia	Thottappilly et al., 2003
Cassava	*Cassava virus X*	*Potexvirus/ Flexiviridae*	Symptomless	Unknown	Columbia	Thottappilly et al., 2003
Cassava castor bean	*Cassava common mosaic virus*	*Potexvirus/ Flexiviridae*	Mild mosaic	Unknown	South and North America, Africa and Asia	Kitajima et al., 1965; http://pvo.bio-mirror.cn/descr171.htm
Cassava	*Cassava frogskin-associated virus*	Tentative *Oryzavirus*	Frog skin symptoms in tubers	Unknown	South America	Thottappilly et al., 2003
Cassava castor bean	*Cassava green mottle virus*	*Nepovirus/ Comoviridae*	Local and systemic mottle	Unknown	Australasia and Pacific islands, Solomon islands	Thottappilly et al., 2003; http://ictvdb.bio-mirror.cn/ICTVdB
Cassava	*Cassava symptomless virus*	Unassigned *Nucleorbhabdo virus/ Rhabdoviridae*	Symptomless	Unknown	Unknown	Thottappilly et al., 2003

occurring in several African countries, including Uganda, South Africa, Rwanda, Senegal, Kenya, Tanzania and Nigeria (reviewed by Aloyce et al., 2013). For the first time, multiplex PCR combining multiple primer sets enabled the detection of ACMV and EACMCV in cassava (Alabi et al., 2008). Reverse transcriptase (RT) multiplex PCR was developed for the simultaneous detection of RNA and DNA viruses co-infecting cassava (Abarshi et al., 2012). Also, rolling cycle amplification was used to amplify full length DNA-A sequences from EACMV-UG and ACMV in *Jatropha* and cassava (Ramkat et al., 2011). RT multiplex PCR could detect the two cassava brown streak associated viruses, CBSV and *Uganda Cassava brown streak virus* (UCBSV), however, it distinguished RNA and DNA cassava viruses weakly (Aloyce et al., 2013). Recently, a more specific, sensitive and reliable single-tube duplex and multiplex PCR tool for the simultaneous detection of ACMV, EACMCV, EACMMV and EACMZV, prevalent in cassava in Kenya, Malawi, Mozambique, Rwanda, Tanzania and Zambia was developed (Aloyce et al., 2013).

Diverse symptoms, ranging from mild to severe leaf distortions, are reported on CMD-affected plants in the field. This could be either explained by plant varietal differences (host resistance), virus species/strains or by the nature of infection (single or mixed) (Gibson and Otim-Nape, 1997). Mixed infections of ACMV and EACMV-UG increase the accumulation in the host plant, facilitate replication and movement and result in more severe symptoms compared to single infections, as reported in Eastern and Central Africa (Aloyce et al., 2013).

6.4.3 Host–pathogen interactions

Only a thorough understanding of host–virus interaction mechanisms allows to develop resistance strategies against viruses. Plants have the ability to activate gene silencing pathways upon virus infection directed against the pathogen (Eckardt, 2011). On the other hand, viruses are in a position to trigger infection if they acquired virulence factors to counteract basal defence and suppress gene silencing by the host by producing silencing suppressors (Pallas and García, 2011; Eckardt, 2011; Bisaro, 2006). Silencing suppressors not only affect antiviral defence but also interfere with the plant physiological processes that depend on RNA silencing, an interference that may contribute to the pathogenesis of different viruses (Pallas and García, 2011).

Besides its regulatory role in plant development, RNA silencing also functions as a natural antiviral defence mechanism, a process also known as virus-induced gene silencing (VIGS) (Dunoyer and Voinnet, 2005). The virus-derived small RNAs (vsRNAs) are then recruited to host RNA-induced silencing complexes (RISCs), to target and inhibit gene expression and protein translation in the viral genome (Dunoyer and Voinnet, 2005; Baulcombe, 2004). On the other hand, many plant viruses have evolved as viral suppressors of RNA silencing to counteract antiviral silencing (Bisaro, 2006; Voinnet et al., 1999). Viruses can also exploit RNA silencing to modify host gene expression directly by homologies between vsRNAs and host transcripts, which may lead to development of viral disease symptoms (Dunoyer and Voinnet, 2005).

6.4.3.1 microRNAs

MicroRNAs (miRNAs) are small non-coding RNA molecules, of approximately 21–24 nt in length acting as key regulators of gene expression (Anselmo et al., 2011, Baulcombe, 2004). After the discovery of the first family of miRNAs from *Caenorhabditis elegans* (Reinhart et al., 2000), several miRNAs have been identified in plants, animals, viruses and green algae by molecular cloning, next generation sequencing and *in silico* approaches (Anselmo et al., 2011; Grundhoff and Sullivan, 2011; Wang et al., 2005; Bartel, 2004).

During biogenesis, plant miRNA genes are transcribed by Pol II to generate a stem loop containing primary miRNA (pri-miRNA), which can range in size from several hundred bps to tens of kbs (Lee et al., 2004). Primary miRNA (pri-miRNA) transcripts are first cleaved by the nuclear-based Dicer-like enzyme, release short stem loop precursor miRNA (pre-miRNA) (Bartel, 2004). MiRs are processed from stem loop regions by a Dicer-like enzyme and loaded into the RISC, where they directly cleave messenger RNAs (mRNAs) (Maghuly et al., 2014; Bartel, 2004).

Virus miRNA

Since miRNA-mediated gene silencing serves as a general defence mechanism against viruses, it would not be surprising that viruses also employ miRNAs to circumvent the host plant's defence system (Maghuly et al., 2014). While there is limited information on miRNA encoded by plant viruses, some human viruses encode miRNAs, which they utilize to modulate both their own gene expression and that of their host cells (Pfeffer et al., 2004). miRNAs have several features that make them useful to viruses. Furthermore, the down-regulation of specific genes allows the virus to establish a favourable environment for its own replication while attenuating or avoiding the host immune response (Umbach and Cullen, 2009). On the other hand, viruses have suppressors of gene silencing that can interact with the plant miRNA pathways. Examples are AC4 (pathogenicity-enhancer protein) encoded by ACMV and AC2 (Transcriptional Activator Protein) encoded by EACMV, which can bind to single-stranded miRNA and inhibit miRNA-mediated negative regulation of gene expression in plants, thus playing important roles in pathogenicity and post-transcriptional gene silencing (Maghuly et al., 2014; Chellappan et al., 2005).

miRNAs of Euphorbiaceae

miRNAs in plants have been found to regulate genes involved in plant growth and development, biotic and abiotic responses (Bazzini et al., 2007; Chellappan et al., 2005). Although a number of miRNAs have been identified from diverse plants, information on identification and characterization of miRNAs in the family Euphorbiaceae is very limited (Xu et al., 2013).

Plant miRNAs can accumulate to a higher percentage in virus-infected plants (Tagami et al., 2007). Begomoviruses generally increase the accumulation of miRNA, which leads to a decreased translation of genes involved in the development of plants (Amin et al., 2011). *Nicotiana benthamiana* plants infected by ACMV like *Cabbage leaf curl virus, Tomato yellow leaf curl virus* and *Cotton leaf curl Multan virus* increased the level of miR159, miR164, miR165/166, miR167 and miR168 (Amin et al., 2011).

Wang et al. (2012) found 52 putative miRNAs, and detected differential expression patterns of 15 miRNAs in root, stem, leaf, fruit and seed by real-time PCR. Among them, 10 miRNAs were highly expressed in fruit or seed, indicating that they may be involved in seed development or fatty acids synthesis. Silencing of JcumiR004 primary miRNA increased the expression of four target genes and oil composition was modulated significantly, indicating diverse functions of JcumiR004 (Wang et al., 2012).

One hundred and fifty-three cassava miRNAs available in miRBase (V.20) (Griffiths-Jones et al., 2008) were obtained by Patanun et al. (2013) using mostly computational techniques. Amiteye et al. (2011) revealed 17 conserved miRNA families and their target genes in cassava, Pérez-Quintero et al. (2012) identified small RNA from cassava tissues infected (and non-infected) with *Xanthomonas axonopodis* and Zeng et al. (2009) studied conserved miRNAs in the Euphorbiaceae family (Ballén-Taborda et al., 2013).

Sixty miRNAs conserved in other plant species and 821 potential cassava-specific miRNAs were identical (Ballén-Taborda et al., 2013). Identification of comprehensive sets of miRNAs in cassava is a critical step to facilitate our understanding of regulatory mechanisms or networks (Patanun et al., 2013).

The miRNA database (miRBase) contains 153 miRNAs identified from cassava, 63 from castor bean, 28 from rubber tree (*Hevea brasiliensis*) and no miRNAs identified from *Jatropha*, so far. Using high-throughput sequencing, 86 conserved miRNAs were identified, including 63 known and 23 newly identified. Sixteen isomiRs were found from the conserved miRNAs, which displayed diverse organ-specific expression levels. A total 72 novel miRNAs and their potential precursors were annotated and 20 miRNAs were validated (Xu et al., 2013).

6.4.4 Utilization of viral sequences to generate virus resistant Euphorbiaceae crops

The breeding of CMD-resistant cassava has been attempted through classical breeding. However, this has been quite difficult due to high heterozygosity of cassava and strong inbreeding depression of elite varieties (Zhang et al., 2005). A further limitation to the process is the rapid evolution of CMGs in the field leading to new aggressive strains (Legg and Thresh, 2003; Akano et al., 2002; Zhou et al., 1997). However, the advent of biotechnological approaches has opened the possibility to support the plant's defence system against viruses by generating virus-resistant plants (Vanderschuren et al., 2007; Zhang et al., 2005). Furthermore, it can be used to introduce new resistant traits or genes, which are currently absent within the cassava germplasm (Prins et al., 2008; Zhang et al., 2005). Vanderschuren et al. (2007) used sequences from almost the entire common region of DNA A and B including the bidirectional promoter of the ACMV-Kenya isolate (Genbank NC 001467) to design constructs, which were mobilized into *A. tumefaciens* and used for transformation of cassava plants. The transgenic plants generated expressed small interfering RNAs (siRNA) leading to the attenuation of CMD symptoms following inoculation with

ACMV-N*Og* infectious clones (Vanderschuren et al., 2007). Other attempts involved improved antisense RNA technology by targeting the ACMV viral mRNAs of three open reading frames (ORFs) AC1-3 of DNA-A (Zhang et al., 2005). The full coding sequences of these ORFs were inserted separately in antisense orientation to the 3′ untranslated region (UTR) of the hygromycin phosphotransferase gene under the control of the *Cauliflower mosaic virus* (CaMV) 35S promoter. Analyses performed on leaf discs showed that viral DNA accumulation was strongly decreased and the resistant cassava plants showed delayed and attenuated CMD symptoms depending on the viral titre (Zhang et al., 2005). The ability to resist ACMV infection was concluded to be via post transcriptional gene silencing (PTGS). Similarly, broad spectrum resistant cassava plants towards ACMV, EACMV and *Sri Lankan cassava mosaic virus* (SLCMV) were generated by using constructs from the entire AC1 gene of a Kenyan isolate of ACMV (Chellappan et al., 2004). The levels of AC1 mRNA were suppressed upon challenging with CMGs and the viral DNA accumulation was reduced to 98% compared to the controls. AC1 integration was found to initiate cross protection against several CMGs via PTGS. This robust cross protection has an implication towards field deployment of transgenic plants as an alternative to control CMGs (Chellappan et al., 2004).

The transgenic approach was further applied to develop CBSV-resistant plants. The CBSV (GenBank JN091565) coat protein region conserved for both CBSV and UCBSV (positions 538–1,063) was placed into a binary expression vector, electroporated into *A. tumefaciens* and used to transform cassava plants co-infected with CBSV and EACMV (Vanderschuren et al., 2012). All transgenic lines did not support virus replication, even under high viral pressure. Furthermore, these transgenic lines were consistently resistant to UCBSV (Vanderschuren et al., 2012).

6.5 Concluding remarks

The current chapter discussed different strategies, which may contribute towards the improvement of Euphorbiaceae with the focus on three economically important crops. Investigations of genetic variation, adaptive processes and populations' history showed, that genetic resources are rather limited in *Jatropha*, while sufficient germplasm resources appear to be available in cassava and castor bean (Table 6.1). Different molecular detection techniques led to the identification of viruses infecting these crops, which reduce and endanger their productivity. The transgenic strategy using miRNAs might help to improve and develop Euphorbiaceae cultivars with important traits of interest. The understanding of the complex plant–virus interaction at the molecular level opens new avenues to better understanding of the mechanisms of plant defence or virus pathogenesis. Due to the huge importance of these crop species, major efforts should be undertaken to ensure their productivity and quality for future generations.

Acknowledgement

The financial support of The Austrian Science Fund (FWF), The Austrian Research Promotion Agency (FFG) and Bioplant R&D is kindly acknowledged.

References

Abarshi, M.M., Mohammed, I.U., Jeremiah, S.C., Legg, J.P., Lava Kumar, P., Hillocks, R.J., et al., 2012. Multiplex RT-PCR assays for the simultaneous detection of both RNA and DNA viruses infecting cassava and the common occurrence of mixed infections by two cassava brown streak viruses in East Africa. J. Virol. Methods. 179, 176–184.

Akano, O., Dixon, O., Mba, C., Barrera, E., Fregene, M., 2002. Genetic mapping of a dominant gene conferring resistance to cassava mosaic disease. Theor. Appl. Genet. 105, 521–525.

Alabi, O.J., Ogbe, F.O., Bandyopadhyay, R., Lava Kumar, P., Dixon, A.G., Hughes, J.A., et al., 2008. Alternate hosts of *African cassava mosaic virus* and *East African cassava mosaic Cameroon virus* in Nigeria. Arch. Virol. 153, 1743–1747.

Allan, G., Williams, A., Rabinowicz, P.D., Chan, A.P., Ravel, J., Keim, P., 2008. Worldwide genotyping of castor bean germplasm (*Ricinus communis* L.) using AFLPs and SSRs. Genet. Resour. Crop Evol. 55, 365–378.

Aloyce, R.C., Tairo, F., Sseruwagi, P., Rey, M.E.C., Ndunguru, J., 2013. A single-tube duplex and multiplex PCR for simultaneous detection of four Cassava mosaic begomovirus species in cassava plants. J. Virol. Meth. 189, 148–156.

Amin, I., Patil, B.L., Briddon, R.W., Mansoor, S., Fauquet, C.M., 2011. A common set of developmental miRNAs are upregulated in *Nicotiana benthamiana* by diverse begomoviruses. Virol. J. 8, 143.

Amiteye, S., Corral, J.M., Sharbel, T.F., 2011. Overview of the potential of microRNAs and their target gene detection for cassava (*Manihot esculenta*) improvement. Afri. J. Biotech. 10, 2562–2573.

Anderson, J.V., Delseny, M., Fregene, M.A., Jorge, V., Mba, C., Lopez, C., et al., 2004. An EST resource for cassava and other species of Euphorbiaceae. Plant Mol. Biol. 56, 527–539.

Anselmo, A., Flori, L., Jaffrezic, F., Rutigliano, T., Cecere, M., Cortes-Perez, N., et al., 2011. Co-expression of host and viral microRNAs in porcine dendritic cells infected by the *Pseudorabies virus*. PLoS One. 6, e17374.

Aregheore, E.M., Becker, K., Makkar, H.P.S., 2003. Detoxification of a toxic variety of *Jatropha curcas* using heat and chemical treatments, and preliminary nutritional evaluation with rats. South Pac. J. Nat. Appl. Sci. 21, 51–56.

Auld, D.L., Pinkerton, S.D., Boroda, E., Lombard, K.A., Murphy, C.K., Kenworthy, K.E., et al., 2003. Registration of TTU-LRC castor germplasm with reduced levels of ricin and RCA120. Crop. Sci. 43, 746–747.

Auld, D.L., Zanotto, M.D., McKeon, T., Morris, J.B., 2009. Castor. In: Vollmann, J., Rajcan, I. (Eds.), Oil Crops, Handbook of Plant Breeding, 4. Springer, New York, NY, pp. 317–332.

Ballén-Taborda, C., Plata, G., Ayling, S., Rodríguez-Zapata, F., Lopez-Lavalle, L.A.B., Duitama, J., et al., 2013. Identification of cassava MicroRNAs under abiotic stress. Int. J. Geno. 2013, 857986.

Bartel, D.P., 2004. MicroRNAs: genomics, biogenesis, mechanism, and function. Cell. 116, 281–297.

Basa, S.D., Sujatha, M., 2009. Genetic analysis of *Jatropha* species and interspecifc hybrids of *Jatropha curcas* using nuclear and organelle specific markers. Euphytica. 168, 197–214.

Baulcombe, D., 2004. RNA silencing in plants. Nature. 431, 356–363.

Bazzini, A.A., Hopp, H.E., Beachy, R.N., Asurmendi, S., 2007. Infection and coaccumulation of tobacco mosaic virus proteins alter microRNA levels, correlating with symptom and plant development. Proc. Natl. Acad. Sci. USA. 104, 12157–12162.

Beeching, J.R., Marmey, P., Gavalda, M.C., Noirot, M., Haysom, H.R., et al., 1993. An assessment of genetic diversity within a collection of cassava (*Manihot esculenta* Crantz) germplasm using molecular markers. Ann. Bot. 72, 515–520.

Bellotti, A.C., Smith, L., Lapointe, S.L., 1999. Recent advances in cassava pest management. Annu. Rev. Entomol. 44, 343–370.

Berrie, L.C., Palmer, K.E., Rybicki, E.P., Rey, M.E., 2001. Molecular characterisation of a distinct South African cassava infecting Geminivirus. Arch. Virol. 143, 2253–2260.

Bi, H., Aileni, M., Zhang, P., 2010. Evaluation of cassava varieties for cassava mosaic disease resistance jointly by agro-inoculation screening and molecular markers. Afr. J. Plant Sci. 4, 330–338.

Bisaro, D.M., 2006. Silencing suppression by *Geminivirus* proteins. Virology. 344, 158–168.

Brigham, R., 1968. Natural outcrossing in dwarf-internode castor. *Ricinus communis* L. Crop Sci. 7, 353–355.

Brittaine, R., Lutaladio, N., 2010. *Jatropha*: A Smallholder Bioenergy Crop. The Potential for Pro-Poor Development. Series: Integrated Crop Management, vol. 8. FAO, Rome, Italy.

Brummer, E.C., Bouton, J.H., Kochert, G., 1995. Analysis of annual *Medicago* species using RAPD markers. Genome. 38, 362–367.

Bull, S.E., Briddon, R.W., Sserubombwe, W.S., Ngugi, K., Markham, P.G., Stanley, J., 2006. Genetic diversity and phylogeography of cassava mosaic viruses in Kenya. J. Gen. Virol. 87, 3053–3065.

Carels, N., 2013. Towards the domestication of Jatropha: The integration of science. In: Bahadur, B., Sujatha, M., Carels, N. (Eds.), Jatropha, Challenges for a New Energy Crop, vol. 2. Springer, New York, pp. 263–299.

Carter, S.E., Fresco, L.O., Jones, P.G., Fairbairn, J.N., 1995. Introduction and diffusion of cassava in Africa. Int. Inst. Trop. Agric. Res. (Guide No. 49), 45.

Ceasar, S.A., Ignacimuthu, S., 2011. Applications of biotechnology and biochemical engineering for the improvement of *Jatropha* and biodiesel: a review. Renew. Sust. Energy Rev. 15, 5176–5185.

Ceballos, H., Okogbenin, E., Pérez, J.C., López-Valle, L.A.B., Debouck, D., 2010. Cassava. In: Bradshaw, J.E. (Ed.), Root and Tuber Crops, Handbook of Plant Breeding. Springer, New York/Dordrecht, the Netherlands/Heidelberg, Germany/London, pp. 53–96.

Ceballos, H., Hershey, C., Becerra-Lopez-Lavalle, L.A., 2012. New approaches in cassava breeding. Plant Breed. Rev. 36, 427–504.

Chan, A.P., Crabtree, J., Zhao, Q., Lorenzi, H., Orvis, J., Puiu, D., et al., 2010. Draft genome sequence of the oilseed species *Ricinus communis*. Nat. Biotechnol. 28, 951–956.

Chellappan, P., Masona, M.V., Vanitharani, R., Taylor, N.J., Fauquet, C.M., 2004. Broad spectrum resistance to ssDNA viruses associated with transgene-induced gene silencing in cassava. Plant Mol. Biol. 56, 601–611.

Chellappan, P., Vanitharani, R., Fauquet, C.M., 2005. MicroRNA-binding viral protein interferes with *Arabidopsis* development. Proc. Natl. Acad. Sci. USA. 102, 10381–11036.

Colombo, C., Second, G., Valle, T.L., Charrier, A., 1998. Genetic diversity characterization of cassava cultivars (*Manihot esculenta* Crantz): I. RAPD markers. Genet. Mol. Biol. 21, 1.

da Câmara Machado, A., Frick, N., Kremen, R., Katinger, H., da Câmara Machado, M.L., 1997. Biotechnological approaches to the improvement of *Jatropha curcas*. In: Proceedings of the International Symposium on *Jatropha*. Managua, Nicaragua, p. 15.

De Oliveira, E.J., de Resende, M.D.V., da Silva Santos, V., Ferreira, C.F., Oliveira, G.A.F., da Silva, M.S., et al., 2012. Genome-wide selection in cassava. Euphytica. 187, 263–276.

Devappa, R.K., Makkar, H.P., Becker, K., 2010. Nutritional, biochemical, and pharmaceutical potential of proteins and peptides from jatropha: review. J. Agric. Food Chem. 58, 6543–6555.

Divakara, B.N., Upadhyaya, H.D., Wani, S.P., Gowda, C.L.L., 2009. Biology and genetic improvement of *Jatropha curcas* L.: a review. Appl. Energy. 13, 1–10.

Dunoyer, P., Voinnet, O., 2005. The complex interplay between plant viruses and host RNA-silencing pathways. Curr. Opin. Plant Biol. 8, 415–423.

Eckardt, N.A., 2011. Novel mechanism of viral interference of host plant suppression by BSCTV. Plant Cell. 23, 1–2.

Elias, M., Panaud, O., Robert, T., 2000. Assessment of genetic variability in a traditional cassava (*Manihot esculenta* Crantz) farming system, using AFLP markers. Heredity. 85, 219–230.

FAOSTAT, 2014. FAO Statistical Database. FAO, Rome. Available from: <http://faostat.fao.org> (accessed 14.03.14).

Fernandez-Martinez, J.M., Velasco, L., 2012. Castor. In: Gupta, S.K. (Ed.), Technological Innovations in Major World Oil Crops, vol. 1: Breeding. Springer, New York, NY, pp. 237–265.

Fondong, V.N., Pita, J.S., Rey, M.E., de Kochko, A., Beachy, R.N., Fauquet, C.M., 2000. Evidence of synergism between *African cassava mosaic virus* and a new double-recombinant *Geminivirus* infecting cassava in Cameroon. J. Gen. Virol. 81, 287–297.

Foster, J.T., Allan, G.J., Chan, A.P., Rabinowicz, P.D., Ravel, J., Jackson, P.J., et al., 2010. Single nucleotide polymorphisms for assessing genetic diversity in castor bean (*Ricinus communis*). BMC Plant Biol. 10, 13.

Gao, S., Qu, J., Chua, N.H., Ye, J., 2010. A new strain of *Indian cassava mosaic virus* causes a mosaic disease in the biodiesel crop *Jatropha curcas*. Arch. Virol. 155, 607–612.

Gibson, R.W., Otim-Nape, G.W., 1997. Factors determining recovery and reversion in mosaic-affected African cassava mosaic virus resistant cassava. Ann. Appl. Biol. 131, 259–271.

Griffiths-Jones, S., Saini, H.K., van Dongen, S., Enright, A.J., 2008. miRBase: tools for microRNA genomics. Nucleic Acids Res. 36, 154–158.

Grundhoff, A., Sullivan, C.S., 2011. Virus-encoded microRNAs. Virology. 411, 325–343.

Harrison, B.D., Zhou, X., Otim-Nape, G.W., Liu, Y., Robinson, D.J., 1997. Role of a novel type of double infection in the *Geminivirus*-induced epidemic of severe cassava mosaic in Uganda. Ann. Appl. Biol. 131, 437–448.

Hong, Y.G., Robinson, D.J., Harrison, B.D., 1993. Nucleotide sequence evidence for the occurrence of three distinct whitefly-transmitted geminiviruses in cassava. J. Gen. Virol. 74, 2437–2443.

Jansson, C., Westerbergh, A., Zhang, J., Xinwen, H., Chuanxin, S., 2009. Cassava, a potential biofuel crop in (the) People's Republic of China. Appl. Energy. 86, S95–S99.

Jarne, P., Lagoda, J.L., 1996. Microsatellites, from molecules to populations and back. Trends Ecol. Evol. 11, 424–429.

Jongschaap, R.E.E., Corré, W.J., Bindraban, P.S., Brandenburg, W.A., 2007. Claims and Facts on *Jatropha curcas* L. Global *Jatropha curcas* Evaluation, Breeding and Propagation Programme. Wageningen U.R. Plant Research International B.V, Report 158.

Kajikawa, M., Morikawa, K., Inoue, M., Widyastuti, U., Suharsono, S., Yokota, A., et al., 2012. Establishment of bispyribac selection protocols for *Agrobacterium tumefaciens*- and *Agrobacterium rhizogenes*-mediated transformation of the oil seed plant *Jatropha curcas* L. Plant Biotechnol. 29, 145–153.

Kansara, R., Jha, S., Jha Kumar, S., Mahatma, M.K., 2013. An efficient protocol for in vitro mass propagation of fusarium wilt resistant castor (*Ricinus communis* L.) Parental line skp-84 through apical meristem. Bioscan. 8, 403–408.

Kashina, B.D., Alegbejo, M.D., Banwo, O.O., Nielsen, S.L., Nicolaisen, M., 2013. Molecular identification of a new begomovirus associated with mosaic disease of *Jatropha curcas* L. in Nigeria. Arch. Virol. 158, 511–514.

Keivani, M., Ramezanpour, S.S., Soltanloo, H., Choukan, R., Naghavi, M., Ranjbar, M., 2010. Genetic diversity assessment of alfalfa (*Medicago sativa* L.) populations using AFLP markers. Aust. J. Crop. Sci. 4, 491–497.

King, A.J., Montes, L.R., Clarke, J.G., Affleck, J., Li, Y., Witsenboer, H., et al., 2013. Linkage mapping in the oilseed crop *Jatropha curcas* L. reveals a locus controlling the biosynthesis of phorbol esters which cause seed toxicity. Plant Biotechnol. J. 11, 986–996.

Kitajima, E.W., Wetter, C., Oliveira, A.R., Silva, D.M., Costa, A.S., 1965. Morphology of *Cassava common mosaic virus*. Bragantia. 24, 247–260.

Kumari, M., Grover, A., Yadav, P.V., Arif, M., Ahmed, Z., 2013. Development of EST-SSR markers through data mining and their use for genetic diversity study in Indian accessions of *Jatropha curcas* L.: a potential energy crop. Genes Genom. 35, 661–670.

Laimer, M., 2006. Virus resistance breeding in fruit trees. In: Fladung, M., Ewald, D. (Eds.), Transgenic Trees. Springer Berlin Heidelberg, pp. 181–199.

Laviola, B.G., Alves, A.A., Barros Rocha, R., Drumond, M.A., 2012. The importance of *Jatropha* for Brazil. In: Carels, N., Sujatha, M., Bahadur, B. (Eds.), *Jatropha*, Challenges for a New Energy Crop, vol. 1. Springer, New York, NY, pp. 71–94.

Lee, Y., Kim, M., Han, J., Yeom, K.H., Lee, S., Baek, S.H., et al., 2004. MicroRNA genes are transcribed by RNA polymerase II. EMBO J. 23, 4051–4060.

Legg, J.P., Thresh, J.M., 2003. Cassava virus diseases in Africa. In: Proceedings of the first international conference on plant virology in Sub-Saharan Africa, 4–8 June 2001. Ibadan, Nigeria, pp. 517–522.

Legg, J.P., Owor, B., Sseruwagi, P., Ndunguru, J., 2006. *Cassava mosaic virus* disease in East and Central Africa: epidemiology and management of a regional pandemic. Adv. Virus Res. 67, 355–418.

Li, H.Q., Sautter, C., Potrykus, I., Pounti-Kaerlas, J., 1996. Genetic transformation of cassava (*Manihot esculenta* Crantz). Nat. Biotechnol. 14, 736–740.

Li, M., Li, H., Jiang, H., Pan, X., Wu, G., 2008. Establishment of an *Agrobacterium-mediated* cotyledon disc transformation method for *Jatropha curcas*. Plant Cell Tissue Organ. Cult. 92, 173–181.

Lokko, Y., Dixon, A., Offei, S., Danquah, E., Fregene, M., 2006. Assessment of genetic diversity among African cassava *Manihot esculenta* Grantz accessions resistant to the *Cassava mosaic virus* disease using SSR markers. Genet. Resour. Crop. Evol. 53, 1441–1453.

Maghuly, F., Laimer, M., 2013. *Jatropha curcas*, a biofuel crop: functional genomics for understanding metabolic pathways and genetic improvement. Biotechnol. J. 8, 1172–1182.

Maghuly, F., Pinsker, W., Praznik, W., Fluch, S., 2006. Genetic diversity in managed subpopulations of Norway spruce [*Picea abies* (L.) Karst.]. For. Ecol. Manage. 222, 266–271.

Maghuly, F., Ramkat, R.C., Laimer, M., 2014. Virus versus host plant MicroRNAs: Who determines the outcome of the interaction? PLoS One. 9, e98263.

Makkar, H., Maes, J., De Greyt, W., Becker, K., 2009. Removal and degradation of phorbol esters during pre-treatment and transesterification of *Jatropha curcas* Oil. J. Am. Oil. Chem. Soc. 86, 173–181.

Malathi, B., Ramesh, S., Rao, K.V., Reddy, V.D., 2006. Agrobacterium-mediated genetic transformation and production of semilooper resistant transgenic castor (*Ricinus communis* L.). Euphytica. 147, 441–449.

Martinez-Herrera, J., Martinez Ayala, A.L., Makkar, H., Francis, G., Becker, K., 2010. Agroclimatic conditions, chemical and nutritional characterization of different provenances of *Jatropha Curcas* L. from Mexico. Eur. J. Sci. Res. 39, 396–407.

Maruthi, M.N., Seal, S., Colvin, J., Briddon, R.W., Bull, S.E., 2004. East african cassava mosaic zanzibar virus – a recombinant begomovirus species with a mild phenotype. Archiv. Virol. 149, 2365–2377.

Maurya, R., Gupta, A., Singh, S.K., Rai, K.M., Chandrawati, Sawant, S.V., et al., 2013. Microsatellite polymorphism in *Jatropa curcas* L. – a biodiesel plant. Ind. Crops Prod. 49, 136–142.

McKeon T.A., Chen, G.Q., 2003. Transformation of *Ricinus communis*, the castor plant. US Patent No. 6,620,986.

Monger, W.A., Seal, S., Isaac, A.M., Foster, G.D., 2001. Molecular characterization of the *Cassava brown streak virus* coat protein. Plant Pathol. 50, 527–534.

Montero-Rojas, M., Correa, A.M., Siritunga, D., 2011. Molecular differentiation and diversity of cassava (*Manihot esculenta*) taken from 162 locations across Puerto Rico and assessed with microsatellite markers. AoB Plants. plr010, 1–13.

Montes, J.M., Technow, F., Bohlinger, B., Becker, K., 2013a. Seed quality, trait associations and grouping of accessions in *Jatropha curcas* L. Ind. Crops Prod. 51, 178–185.

Montes, J.M., Technow, F., Bohlinger, B., Becker, K., 2013b. Grain quality determination by means of near infrared spectroscopy in *Jatropha curcas* L. Ind. Crops Prod. 43, 301–305.

Mukherjee, P., Varshney, A., Johnson, T.S., Jha, T.B., 2011. *Jatropha curcas*: a review on biotechnological status and challenges. Plant Biotechnol. Rep. 3, 197–215.

Müller-Starck, G., Baradat, P.H., Bergmann, F., 1992. Genetic variation within European tree species. New For. 6, 23–47.

Muthusamy, S., Kanagarajan, S., Ponnusamy, S., 2008. Efficiency of RAPD and ISSR markers system in accessing genetic variation of rice bean (*Vigna umbellata*) landraces. Electron. J. Biotechnol. 11, 1–10.

Mwine, J.T., Van Damme, P., 2011. Why do *Euphorbiaceae* tick as medicinal plants? A review of *Euphorbiaceae* family and its medicinal features. J. Med. Plant Res. 5, 652–662.

Nassar, N.M.A., Ortiz, R., 2009. Cassava genetic resources: manipulation for crop improvement. Plant Breed. Rev. 31, 247–275.

Nyaboga, E., Njiru, J., Nguu, E., Gruissem, W., Vanderschuren, H., Tripathi, L., 2013. Unlocking the potential of tropical root crop biotechnology in east Africa by establishing a genetic transformation platform for local farmer-preferred cassava cultivars. Front. Plant Sci. 4, 526.

Olsen, K., Schaal, B., 2001. Microsatellite variation in cassava (*Manihot esculenta, Euphorbiaceae*) and its wild relatives: further evidence for a southern Amazonian origin of domestication. Am. J. Bot. 88, 131–142.

Openshaw, K., 2000. A review of *Jatropha curcas*: an oil plant of unfulfilled promise. Biomass Bioenerg. 19, 1–15.

Pallas, V., Garcia, J.A., 2011. How do plant viruses induce disease? Interactions and interference with host components. J. Gen. Virol. 92, 2691–2705.

Patanun, O., Lertpanyasampatha, M., Sojikul, P., Viboonjun, U., Narangajavana, J., 2013. Computational identification of microRNAs and their targets in cassava (*Manihot esculenta* Crantz.). Mol. Biotechnol. 53, 257–269.

Patil, B.L., Fauquet, C.M., 2009. Cassava mosaic geminiviruses: actual knowledge and perspectives. Mol. Plant Pathol. 10, 685–701.

Pecina-Quintero, V., Anaya-Lopez, J.L., Zamarripa-Colmenero, A., Nunez-Colin, C.A., Montes-Garcia, N., Solís-Bonilla, J.L., et al., 2014. Genetic structure of *Jatropha curcas* L. in Mexico and probable centre of origin. Biomass Bioenerg. 60, 147–155.

Pérez-Quintero, A.L., Quintero, A., Urrego, O., Vanegas, P., López, C., 2012. Bioinformatic identification of cassava miRNAs differentially expressed in response to infection by *Xanthomonas axonopodis* pv. manihotis. BMC. Plant Biol. 12, 29.

Pfeffer, S., Zavolan, M., Grässer, F.A., Chien, M., Russo, J.J., Ju, J., et al., 2004. Identification of virus-encoded microRNAs. Science. 304, 734–736.

Pita, J.S., Fondong, V.N., Sangaré, A., Otim-Nape, G.W., Ogwal, S., Fauquet, C.M., 2001. Recombination, pseudorecombination and synergism of geminiviruses are determinant keys to the epidemic of severe cassava mosaic disease in Uganda. J. Gen. Virol. 82, 655–665.

Prins, M., Laimer, M., Noris, E., Schubert, J., Wassenegger, M., Tepfer, M., 2008. Strategies for antiviral resistance in transgenic plants. Mol. Plant Pathol. 9, 73–83.

Prochnik, S., Marri, P.R., Desany, B., Rabinowicz, P.D., Kodira, C., Mohiuddin, M., et al., 2012. The cassava genome: current progress, future directions. Trop. Plant Biol. 5, 88–94.

Puonti-Kaerlas, J., 1998. Cassava biotechnology. In: Harding, S.H., Adams, G. (Eds.), Biotechnology and Genetic Engineering Reviews, vol. 15. Nottingham University Press, Nottingham, UK, pp. 329–364.

Qiu, L., Yang, C., Tian, B., Yang, J.B., Liu, A., 2010. Exploiting EST databases for the development and characterization of EST-SSR markers in castor bean (*Ricinus communis* L.). BMC Plant Biol. 10, 278.

Qu, J., Mao, H.Z., Chen, W., Gao, S.Q., Bai, Y.N., Sun, Y.W., et al., 2012. Development of marker-free transgenic Jatropha plants with increased levels of seed oleic acid. Biotechnol. Biofuels. 5, 10.

Raj, S.K., Kumar, S., Snehi, S.K., Pathre, U., 2008. First report of cucumber mosaic virus on *Jatropha curcas* in India. Plant Dis. 92, 171.

Raj, S.K., Snehi, S.K., Gautam, K.K., Khan, M.S., 2010. First report of association of cucumber mosaic virus with blister and leaf distortion of castor bean (*Ricinus communis*) in India. Phytoparasitica. 38, 283–289.

Ramkat, R.C., 2013. Biotechnological approaches towards the genetic and phytosanitary improvement of Euphorbiaceae. (PhD thesis). Plant Biotechnology Unit, University of Natural Resources and Life Science, Vienna, Austria.

Ramkat, R.C., Calari, A., Maghuly, F., Laimer, M., 2011. Biotechnological approaches to determine the impact of viruses in the energy crop plant *Jatropha curcas*. Virol. J. 8, 386.

Reddy, K.R.K., Bahadur, B., 1989. Adventitious bud formation from leaf cultures of castor (*Ricinus communis* L.). Curr. Sci. 58, 152–154.

Reinhart, B.J., Slack, F.J., Basson, M., Pasquinelli, A.E., Bettinger, J.C., Rougvie, A.E., et al., 2000. The 21-nucleotide let-7 RNA regulates developmental timing in *Caenorhabditis elegans*. Nature. 403, 901–906.

Rivarola, M., Foster, J.T., Chan, A.P., Williams, A.L., Rice, D.W., Liu, X., et al., 2011. Castor bean organelle genome sequencing and worldwide genetic diversity analysis. PLoS One. 6, e21743.

Roger, P., Martyn, R., 1999. Annuals and Biennials. Macmillan, London, p. 106.

Rojas, M.C., Pérez, J.C., Ceballos, H., Baena, D., Morante, N., et al., 2009. Analysis of inbreeding depression in eight S1 cassava families. Crop. Sci. 49, 543–548.

Rosado, T.B., Laviola, B.G., Faria, D.A., Pappas, M.R., Bhering, L.L., Quirino, L.L., et al., 2010. Molecular markers reveal limited genetic diversity in a large germplasm collection of the biofuel crop *Jatropha curcas* L. in Brazil. Crop. Sci. 50, 2372–2382.

Rothenstein, D., Haible, D., Dasgupta, I., Dutt, N., Patil, B.L., Jeske, H., 2006. Biodiversity and recombination of cassava-infecting begomoviruses from southern India. Arch. Virol. 151, 55–69.

Roy, S., 1990. Effect of various factors on growth and productivity of *Jatropha curcas*, a hydrocarbon yielding plant. In: Grassi, G., Gosses, G., Dos Santos, G. (Eds.), Biomass for Energy and Industry. Elsevier Applied Science, London/New York, pp. 1484–1488.

Sailaja, M., Tarakeswari, M., Sujatha, M., 2008. Stable genetic transformation of castor (*Ricinus communis* L.) via particle gun-mediated gene transfer using embryo axes from mature seeds. Plant Cell Rep. 27, 1509–1519.

Sarvesh, A., Ram Rao, D.M., Reddy, T.P., 1992. Callus initiation and plantlet regeneration from epicotyl and cotyledonary explants of castor (*Ricinus communis* L.). Adv. Plant Sci. 5, 124–128.

Sato, S., Hirakawa, H., Tsuchimoto, S., Sakai, H., Shibagaki, N., 2013. Genome structure of *Jatropha curcas* L. In: Bahadur, B., Sujatha, M., Carels, N. (Eds.), Challenges for a New Energy Crop. Springer, New York, NY, pp. 577–600.

Schöpke, C., Taylor, N.J., Carcamo, R., Konan, N.K.K., Marmey, P., Henshaw, G.G., et al., 1996. Regeneration of transgenic cassava plants (*Manihot esculenta* Crantz) from microbombarded embryogenic suspension cultures. Nat. Biotechnol. 14, 731–735.

Second, G., Allem, A., Emperaire, L., Ingram, C., Colombo, C., 1997. Molecular marker (AFLP) based *Manihot* and cassava numerical taxanomy and genetic structure analysis in progress: implications for dynamic conservation and genetic mapping. Afr. J. Root Tuber Crops. 2, 140–147.

Severino, L.S., Auld, D.L., Baldanzi, M., Cândido, M.J.D., Chen, G., Crosby, W., et al., 2012. A review on the challenges for increased production of castor. Agron. J. 104, 853–880.

Soh, A.C., Wong, C.K., Ho, Y.W., Choong, C.W., 2009. Oil palm. In: Vollmann, J., Rajcan, I. (Eds.), Oil Crops, Handbook of Plant Breeding, 4. Springer, New York, NY, pp. 333–367.

Soomro, R., Memon, R.A., 2007. Establishment of callus and suspension culture in *Jatropha curcas*. Pak. J. Bot. 39, 2431–2441.

Sudhaker Johnson, T., Eswaran, N., Sujatha, M., 2011. Molecular approaches to improvement of *Jatropha curcas* Linn. as a sustainable energy crop. Plant Cell Rep. 30, 1573–1591.

Sujatha, M., Reddy, T.P., 1998. Differential cytokinin effects on the stimulation of in vitro shoot proliferation from meristematic explants of castor (*Ricinus communis* L.). Plant Cell Rep. 7, 561–566.

Sujatha, M., Reddy, T.P., 2007. Promotive effect of lysine monohydrochloride on morphogenesis in cultured seedling and mature plant tissues of castor (*Ricinus communis* L.). Ind. J. Crop Sci. 2, 11–19.

Sujatha, M., Lakshminarayana, M., Tarakeswari, M., Singh, P.K., Tuli, R., 2009. Expression of the cry1EC gene in castor (*Ricinus communis* L.) confers field resistance to tobacco caterpillar (*Spodoptera litura* Fabr) and castor semilooper (*Achoea janata* L.). Plant Cell Rep. 28, 935–946.

Sujatha, M., Sailaja, M., 2005. Stable genetic transformation of castor (*Ricinus communis* L.) via Agrobacterium tumefaciens-mediated gene transfer using embryo axes from mature seeds. Plant Cell Rep. 23, 803810.

Tagami, Y., Inaba, N., Kutsuna, N., Kurihara, Y., Watanabe, Y., 2007. Specific enrichment of miRNAs in *Arabidopsis thaliana* infected with *Tobacco mosaic virus*. DNA Res. 14, 227–233.

Tan, M., Yan, M., Wang, L., Yan, X., Fu, C., Wang, L., 2013. Replication of pistillate plants of *Ricinus communis* L. and investigation of the sex stability and genetic variation of the somaclones. Ind. Crops Prod. 50, 50–57.

Taylor, N., Chavarriaga, P., Raemakers, K., Siritunga, D., Zhang, P., 2004. Development and application of transgenic technologies in cassava. Plant Mol. Biol. 56, 671–688.

Thottappilly, G., Thresh, J.M., Calvert, L.A., Winter, S., 2003. Cassava. In: Loebenstein, G., Thottappilly, G. (Eds.), Virus and Virus-Like Diseases of Major Crops in Developing Countries. Springer Science + Business Media Dordrecht, pp. 107–165.

Thresh, J.M., Cooter, R.J.K., 2005. Strategies for controlling *Cassava mosaic virus* disease in Africa. Plant Pathol. 54, 587–614.

Tripathi, A., Mishra, D.K., Shukla, J.K., 2013. Genetic variability, heritability and genetic advance of growth and yield components of *Jatropha* (*Jatropha curcas* Linn.) genotypes. Trees. 27, 1049–1060.

Umbach, J.L., Cullen, B.R., 2009. The role of RNAi and microRNAs in animal virus replication and antiviral immunity. Genes Dev. 23, 1151–1164.

Valdes-Rodriguez, O.A., Sanchez-Sanchez, O., Perez-Vazquez, A., Caplan, J., 2013. The Mexican non-toxic *Jatropha curcas* L., food resource or biofuel? Ethnobotany Res. Appl. 11, 001–007.

Vanderschuren, H., Akbergenov, R., Pooggin, M.M., Hohn, T., Gruissem, W., Zhang, P., 2007. Transgenic cassava resistance to *African cassava mosaic virus* is enhanced by viral DNA-A bidirectional promoter-derived siRNAs. Plant Mol. Biol. 64, 549–557.

Vanderschuren, H., Moreno, I., Anjanappa, R.B., Zainuddin, I.M., Gruissem, W., 2012. Exploiting the combination of natural and genetically engineered resistance to *Cassava mosaic* and *Cassava brown streak viruses* impacting cassava production in Africa. PLoS One. 7, e45277.

Vieira, E.A., de Freitas Fialho, J., Gelape Faleiro, F., Bellon, G., da Fonseca, K.G., Carvalho, L.J.C.B., et al., 2011. Characterization of sweet cassava accessions based on molecular, quantitative and qualitative data. Crop. Breed Appl. Biotechnol. 11, 232–240.

Voinnet, O., Pinto, Y.M., Baulcombe, D.C., 1999. Suppression of gene silencing: a general strategy used by diverse DNA and RNA viruses of plants. Proc. Natl. Acad. Sci. USA. 96, 14147–14152.

Vollmann, J., Laimer, M., 2013. Novel and traditional oil crops and their biorefinery potential. In: Yang, S.-T., El Enshasy, H.A., Thongchul, N., Lo, Y.M. (Eds.), Bioprocessing Technologies in Integrated Biorefinery for Production of Biofuels, Biochemicals, and Biopolymers from Biomass. John Wiley & Sons, Inc., Hoboken, NJ, USA.

Wang, C.M., Liu, P., Sun, F., Li, L., Liu, P., Ye, J., et al., 2012. Isolation and Identification of miRNAs in *Jatropha curcas*. Int. Biol. Sci. 8, 418–429.

Wang, X., Zhang, J., Li, F., Gu, J., He, T., Zhang, X., Li, Y., 2005. MicroRNA identification based on sequence and structure alignment. Bioinformatics. 21, 3610–3614.

Welsch, R., Arango, J., Bar, C., Salazar, B., Al-Babili, S., Beltrán, J., et al., 2010. Provitamin A accumulation in cassava (*Manihot esculenta*) roots driven by a single nucleotide polymorphism in a phytoene synthase gene. Plant Cell. 22, 3348–3356.

Winter, S., Koerbler, M., Stein, B., Pietruszka, A., Paape, M., Butgereitt, A., 2010. Analysis of *Cassava brown streak viruses* reveals the presence of distinct virus species causing cassava brown streak disease in East Africa. J. Gen. Virol. 91, 1365–1372.

Xu, W., Cui, Q., Li, F., Liu, A., 2013. Transcriptome-wide identification and characterization of microRNAs from castor bean (*Ricinus communis* L.). PLoS One. 8, e69995.

Yadav, J.S., Ogwok, E., Wagaba, H., Patil, B.L., Bagewadi, B., Alicai, T., et al., 2011. RNAi-mediated resistance to cassava brown streak Uganda virus in transgenic cassava. Mol. Plant Pathol. 12, 677–687.

Ye, J., Hong, Y., Qu, J., Wang, C., 2013. Improvement of *J. curcas* oil by genetic transformation. In: Bahadur, B., Sujatha, M., Carels, N. (Eds.), Jatropha, Challenges for a New Energy Crop, vol. 2. Springer, New York, pp. 547–562.

Zainuddin, I., Schlegel, K., Gruissem, W., Vanderschuren, H., 2012. Robust transformation procedure for the production of transgenic farmer-preferred cassava cultivars. Plant Methods. 8, 24.

Zeng, C., Wang, W., Zheng, Y., Chen, X., Bo, W., Song, S., et al., 2009. Conservation and divergence of microRNAs and their functions in Euphorbiaceous plants. Nucleic Acids Res. 38, 981–995.

Zhang, P., Vanderschuren, H., Futterer, J., Gruissem, W., 2005. Resistance to cassava mosaic disease in transgenic cassava expressing antisense RNAs targeting virus replication genes. Plant Biotechnol J. 3, 385–397.

Zhou, X., Liu, Y., Calvert, L., Munoz, C., Otim-Nape, G.W., Robinson, D.J., et al., 1997. Evidence that DNA-A of a geminivirus associated with severe cassava mosaic disease in Uganda has arisen by interspecific recombination. J. Gen. Virol. 78, 2101–2111.

Zou, M., Xia, Z., Ling, P., Zhang, Y., Chen, X., Wei, Z., et al., 2011. Mining EST-derived SSR markers to assess genetic diversity in cassava (*Manihot esculenta* Crantz). Plant Mol. Biol. Rep. 29, 961–971.

Regulation framework for flowering

7

Tiziana Sgamma and Stephen Jackson
School of Life Sciences, University of Warwick, Coventry, West Midlands, UK

7.1 Introduction

Plants are sessile organisms that have developed the ability to perceive, anticipate and respond to environmental changes to maximize their ability to survive and reproduce. One of the most important aims in the life of any plant is to flower and reproduce. Plants regulate the timing of the onset of flowering to enable seed production and dispersal before environmental conditions become too adverse. For outbreeding species flowering is often synchronized with other individuals, or with insect or bird pollinators, to increase the chance of cross-pollination. Furthermore, if a plant flowers too early it might not be sufficiently well established and lack enough resources to support the production of flowers and seed. Timing is therefore the key to success and plants have developed a very elaborate network of interacting molecular pathways to control when flowering occurs. These pathways are influenced by predictable environmental cues such as diurnal and annual changes in light and temperature, and by unpredictable and internal factors such as nutrient levels, biotic and abiotic stresses, plant age and maturity. Plants modulate their development in response to these factors with short-term (e.g. diurnal) or long-term (e.g. seasonal) responses accordingly. The rate of the response is also dependent on the plant species. Generally plants are divided between annual plants that complete their development within 1 year, flowering just once at the end of their life-cycle, and biennials or perennial plants whose life cycles span two or more years, during which they can flower once (monocarpic) or several times (polycarpic).

From a farming point of view, understanding the flowering time process, anticipating and controlling it is extremely important for crop scheduling and also because flowering influences yield. As flowering time is of such commercial importance, a lot of research has been devoted to uncovering the genetic mechanisms underlying the various flowering pathways that affect flowering time. Using molecular genetic approaches in the annual model plant *Arabidopsis thaliana*, and in rice and other species, many components of these pathways have now been identified.

In this chapter, the current understanding of the molecular pathways that plants use to respond to the external environmental and internal endogenous signals to regulate flowering is summarized. The flowering repressors that act to prevent plants from flowering until the right moment, and the floral integrator genes, which activate the floral meristem identity genes are described. The main focus is on what

Applied Plant Genomics and Biotechnology.

is known about the *Arabidopsis* flowering process, as this is the best understood model and there is a lot of available information, but what happens in other plants species will also be considered.

7.2 Getting ready to flower: the juvenile to adult transition

Plants are not sensitive to inductive conditions throughout the whole course of their post-embryonic development. During this period the shoot meristem passes through three main developmental stages: the juvenile phase, the adult vegetative phase and the adult reproductive phase (Poethig, 2003) (Figure 7.1). During the juvenile phase flowering cannot be induced even when the plant is exposed to inductive conditions. It is only once the plant has reached the adult vegetative phase that reproductive competency is established and the plant can respond to florally inductive conditions. Finally, during the reproductive phase plants are capable of forming reproductive organs but they again lose the ability to respond to inductive conditions. The length of the juvenile phase varies from plant to plant and has been studied mostly in annual species where it lasts for a relatively short time. However, it can be extremely long and last for several years in perennial plants. A large number of physiological markers for juvenility, which include certain leaf characteristics and arrangement, as well as internode elongation have been identified in different species. However, most of them are not universal markers as they differ between annual and perennial plants and are affected by environmental factors (Kerstetter and Poethig, 1998; Brunner and Nilsson, 2004). The ability to flower is therefore the most robust physiological marker for marking the end of the juvenile phase. The molecular mechanisms behind the juvenile to adult transition are still not very clear. In *Arabidopsis*, several proteins including SERRATE (SE), HYPONASTIC LEAVES1 (HYL1), ARGONAUTE1 (AGO1), HASTY (HST) and SQUINT (SQN) repress this phase transition. They affect the biogenesis of microRNAs (miRNAs) from primary transcripts (pre-miRNAs), and regulate miRNA activity and accumulation

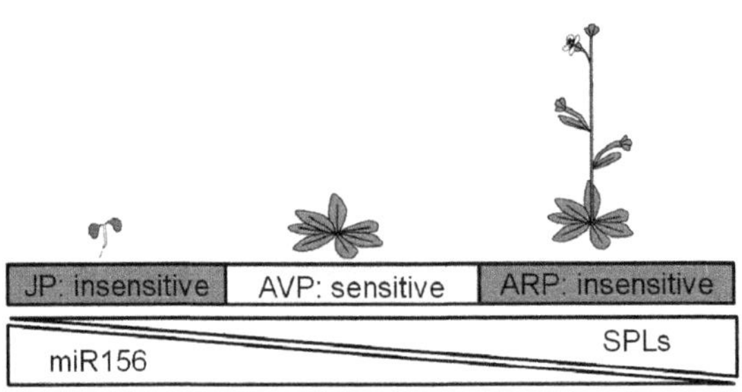

Figure 7.1 Phase changes in *Arabidopsis* plant development. The light grey bar represents the period of sensitivity to florally inductive conditions. The dark grey bars represent insensitive phases. ARP, adult reproductive phase; AVP, adult vegetative phase; JP, juvenile phase; SPL, SQUAMOSA PROMOTER BINDING PROTEIN-LIKE.

(reviewed in Rogers and Chen, 2012). miRNAs play a crucial role in the juvenile to adult transition, they are involved in pre- and post-transcriptional regulation of protein levels through the RNA interference pathway (Poethig, 2010; Yamaguchi and Abe, 2012). The main miRNA family involved in this phase transition is the *miR156* family (Yamaguchi and Abe, 2012). *miR156* levels are more abundant during the juvenile seedling stages and decrease as the plant ages; when present it inhibits SQUAMOSA PROMOTER BINDING PROTEIN-LIKE (SLP) protein production which are responsible for the juvenile to adult switch and promote flowering (as explained later in the ageing pathway) (Yamaguchi and Abe, 2012). Many other genes involved in the transition between the juvenile phase to the adult vegetative phase have been identified and characterized in *Arabidopsis* (reviewed in Huijser and Schmid, 2011). Often the environmental and internal factors that regulate the juvenile–adult transition overlap with those regulating the floral transition.

7.3 Framework controlling flowering

Once the competency to flower is acquired and the plant has completed the juvenile phase, the transition to the reproductive phase is regulated by a complex molecular network. This flowering network is usually divided into distinct interacting pathways (Photoperiodic, Vernalization, Thermosensory, Gibberellin, Sucrose, Autonomous and Ageing pathways), that converge on a very small set of genes. These genes have been referred to as floral pathway integrator genes; *APETALA1* (*AP1*), *FLOWERING LOCUS T* (*FT*), *LEAFY* (*LFY*), *TWIN SISTER OF FT* (*TSF*) and *SUPPRESSOR OF CONSTANS1* (*SOC1*), and they activate the floral meristem identity genes resulting in floral initiation (Srikanth and Schmid, 2011).

7.4 Flowering sensing the environment

Many environmental factors influence flowering time such as photoperiod and temperature. Some are predictable like the gradual change in day length and the prolonged cold temperature in winter seasons; others are less predictable like short periods of cold during spring. Plants use several different pathways to control flowering in response to their natural environment.

7.4.1 The photoperiod and light quality pathways

Except at the equator, plants can track seasonal changes by detecting the changes in day/night length within a 24-h cycle (photoperiodism). On the basis of their response to different photoperiods, plants have been classified as obligate short-day plants (SDPs) if they flower only under short days, or facultative SDPs if their flowering is accelerated by short days (SD), obligate long-day plants (LDPs) if they flower only during long days (LD), and facultative LDPs if flowering is accelerated by LD. Species that flower equally in both SD and LD photoperiods are referred to as day-neutral plants (DNPs). Some plants are not classified in any of the above categories because they respond to combinations of day lengths (Thomas and Vince-Prue, 1997). The term photoperiod was first introduced in the 1920s from Garner and Allard in a study where they proved that Maryland mammoth, a variety of tobacco, could flower only in SDs (Thomas et al., 2006). Very early experiments, giving different day length treatments to the apex or to the leaves, proved that the perception of photoperiod occurs in the leaves not in the apex, and a signal is produced in the leaves that moves to the apex. When induced leaves are grafted onto non-induced plants, flowering is induced in the receptor plants (Thomas et al., 2006). For many years since Chailakhyan proposed its existence in 1936, scientists have been looking for 'florigen', the graft-transmissible flowering signal (Thomas et al., 2006). Only in 2007 was it discovered that the FT protein could be part of the florigen signal as it could move from the leaves to the apex and induce flowering (Corbesier et al., 2007; Jaeger and Wigge, 2007; Mathieu et al., 2007). A role for *FT* mRNA, possibly acting together with FT protein, in the floral induction process has also been identified (Li et al., 2011; Jackson and Hong, 2012).

Based on the model LDP *Arabidopsis*, the perception of photoperiod is mediated by the red/far-red light-receptors phytochromes (PHYA, D and E) and the blue/UV-A light-receptors cryptochromes (CRY1 and 2) (Wang and Deng, 2002; Quail, 2002; Lin and Shalitin, 2003). Photoreceptors entrain the circadian clock (the plant's internal time-keeper) to a 24-h period (Halliday et al., 2003). A large number of genes have been classified as components of the circadian clock which function in several interacting feedback loops. As the mechanisms involved in the circadian clock are very complex and there are numerous excellent reviews (Nakamichi, 2011; Troncoso-Ponce and Mas, 2012; Carre and Veflingstad, 2013; Chow and Kay, 2013) we will not cover this in detail. It is principally composed of 3 feedback loops that operate on a cycle of about 24 h. *TIMING OF CAB EXPRESSION 1* (*TOC1*) is involved in one of these feedback loops; it down-regulates *LATE ELONGATED HYPOCOTYL* (*LHY*) and *CIRCADIAN CLOCK ASSOCIATED 1* (*CCA1*) gene expression (Carre and Veflingstad, 2013). In the morning, LHY and CCA1 proteins bind to the *TOC1* and *PSEUDO RESPONSE REGULATORS 5* (*PRR5*) gene promoters and down-regulate their expression, while activating *PRR7* and *PRR9* (Carre and Veflingstad, 2013). LHY and CCA1 protein levels fall during the day through the action of the PRRs thus allowing *TOC1* expression levels to rise (Carre and Veflingstad, 2013). Expression of the PRRs falls during the night through the action of the Evening Complex (EC) and *LHY/CCA1* transcription increase again at the following morning (Carre and Veflingstad, 2013). TOC1 is also negatively regulated at dusk by ZEITLUPE (ZTL) which marks TOC1 protein for proteasome degradation (Mas et al., 2003). A third loop involves GIGANTEA (GI) in a negative feedback loop with TOC1 (Locke et al., 2006).

As shown in Figure 7.2, whilst being a component of the clock GI also acts in an output pathway from the clock to regulate flowering. The expression of *GI* and another gene, *FLAVIN-BINDING, KELCH REPEAT F-BOX 1* (*FKF1*), are under clock control (Nakamichi, 2011). In LD, the expression of *GI* and *FKF1* peak at 10–12 h after dawn and together their proteins form a complex which repress expression of *CYCLING DOF FACTOR 1* (*CDF1*), which itself is a repressor of

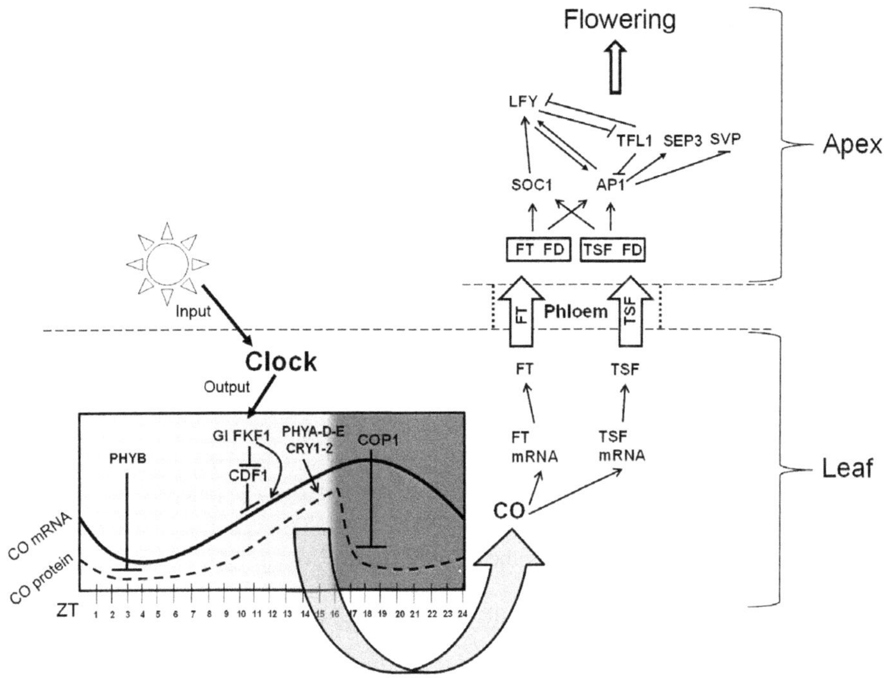

Figure 7.2 Photoperiodic pathway in *Arabidopsis thaliana*. The light and dark shading indicates day and night periods, respectively. Arrows indicate activation and T-bars show inhibition. AP1, APETALA1; CDF1, CYCLING DOF FACTOR1; CO, CONSTANS; COP1, CONSTITUTIVE PHOTOMORPHOGENIC 1; CRY1-2, CRYPTOCHROME1-2; FKF1, FLAVIN-BINDING KELCH REPEAT F-BOX1; FT, FLOWERING LOCUS T; GI, GIGANTEA; LFY, LEAFY; PHYA-E, PHYTOCHROME A-E; SEP3, SEPALLATA3; SOC1, SUPPRESSOR OF CONSTANS1, SVP, SHORT VEGETATIVE PHASE; TFL1, TERMINAL FLOWER 1; TSF, TWIN SISTER OF FT; ZT, Zeitgeber.

CONSTANS (*CO*) (Fowler et al., 1999; Sawa et al., 2007). GI and FKF1 have also been shown to promote *CO* expression by directly binding to its promoter (Mizoguchi et al., 2005; Sawa et al., 2007). At the start of the day, red light acting through PHYB stimulates the degradation of the CO protein. In LD *CO* expression rises to a peak at dusk, which leads to high levels of CO protein accumulation. CO is stabilized by blue and far-red light through PHYA, and CRY1 and CRY2; the accumulation of CO protein activates the transcription of the floral integrator gene *FT* (Valverde, 2011). The regulation of *FT* by CO probably occurs through an interaction of CO and the CCAAT-box binding protein factor with the 5′ UTR region of *FT* (Ben-Naim et al., 2006). In the dark CO protein is degraded is by the E3 ligase CONSTITUTIVE PHOTOMORPHOGENIC 1 (COP1) (Jang et al., 2008) (Figure 7.2). The accumulation of stable CO protein in the light period, is consistent with the external coincidence model (Turck et al., 2008). *CO* is expressed in the phloem companion cells of the leaf where it activates expression of *FT* and *TSF*. Studies have confirmed that it is the movement of FT protein through the phloem from the leaf to the apex that leads to flowering though the formation of a complex with FD (Wigge, 2011). Flowering occurs when the FT/FD or TSF/FD complexes activate *SOC1* and *AP1*, the flower-meristem-identity genes which activate the floral organ identity genes (Blazquez et al., 1997; Torti et al., 2012). AP1 activates *LFY* expression which in turn is also responsible for binding the *AP1* promoter and controls its expression. TERMINAL FLOWER 1 (TFL1) represses floral meristem identity genes by postponing the change from the vegetative phase to flowering by repressing *LFY* and *AP1* (Hanzawa et al., 2005). *SEPALLATA3* (*SEP3*) expression is up-regulated directly and indirectly by AP1 initiating downstream pathways involved in floral organ formation (Kaufmann et al., 2010).

The above occurs in LD in LDPs, but not in SD. As summarized in Figure 7.3, in SD the expression of *GI* peaks a few hours before *FKF1*, and thus the proteins are not able to form the complex to down-regulate *CDF1* expression, therefore *CO* remains repressed (Sawa et al., 2007). In SD, GI also regulates *FT* independently of CO through regulation of *miR172* (Jung et al., 2007). *miR172* levels increase with the age of the plant and it down-regulates the *FT* repressor TARGET OF EAT1 (TOE1), as described later (Jung et al., 2007).

In SDP the mechanism is slightly different. In rice (*Oryza sativa*) in LD the expression of the rice orthologue of *GI* follows a circadian rhythm and promotes

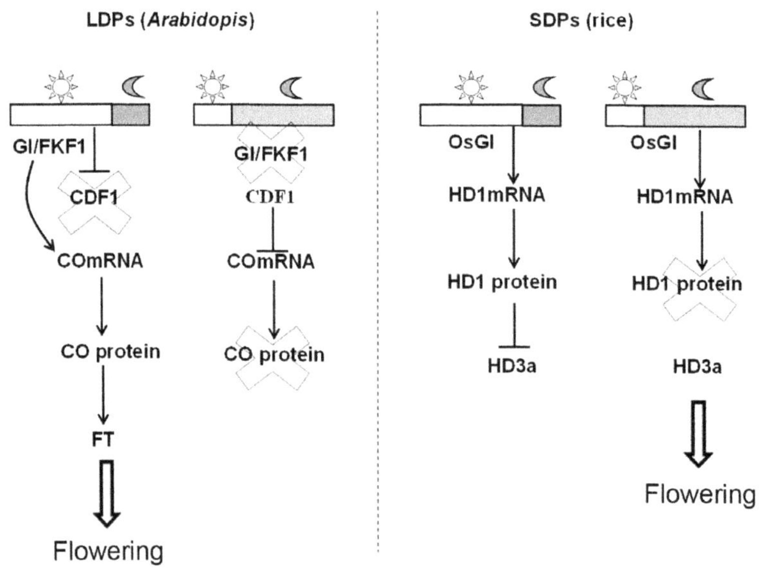

Figure 7.3 Model of the photoperiodic pathway in LDPs and SDPs. Arrows indicate activation and T-bars show inhibition. CDF1, CYCLING DOF FACTOR1; CO, CONSTANS; FKF1, FLAVIN-BINDING KELCH REPEAT F-BOX1; FT, FLOWERING LOCUS T; GI, GIGANTEA; HD1, HEADING-DATE1; HD3a, HEADING-DATE 3a; OsGI, *Oryza sativa* GI.

the expression of *HEADING-DATE1* (*HD1*), the orthologue of *CO* (Hayama et al., 2003). *HD1* expression follows a circadian rhythm similar to *CO*, peaking at dusk in LD but in rice it acts as an inhibitor of flowering, repressing the expression of *HEADING-DATE 3a* (*HD3a*), the rice orthologue of the *Arabidopsis FT* (Greenup et al., 2009). In SD, *HD1* expression peaks at night but the protein is degraded as there is no PHYB to stabilize the protein, thus it is not present to inhibit *HD3a* expression enabling HD3a protein to accumulate in phloem companion cells (Kojima et al., 2002). HD3a then moves to the apical meristem where it promotes flowering (Kojima et al., 2002).

7.4.2 Vernalization pathway

In temperate areas the rotation of the seasons allows plants to align flowering time with advantageous conditions. In 1857, Klippart proved that it is the cold temperature during the winter season that allows winter cereals to flower in spring. The word vernalization, now used for describing the phenomenon, was used for the first time by Lysenko years later in 1928. Actually the name Lysenko gave to the phenomenon was 'jarovization' after the spring cereal Jarovoe (in Russian *jar* means 'formerly fire' or 'god of spring') that flowers soon after a spring sowing. In fact the cold temperature makes a winter cereal behave as a spring one. Lysenko translated the word 'jarovization' to 'vernalization' (in latin *vernum* means 'spring'). For the first 20 years, the word vernalization was used to describe not only the capacity of flowering acquired by some plants after experienced a period of cold, but also to describe any other physiological change in plants after a period of chilling (e.g. chilling can increase tuber yield in potatoes), and even any external event that can stimulate flowering (e.g. day length, temperature rise). It was Chouard, in 1960, who explicitly defined the word vernalization as: 'the acquisition or acceleration of the ability to flower by a chilling treatment' (Chouard, 1960). Naturally, this phenomenon is only necessary in habitats where restrictive cold weather conditions are detrimental to reproductive ability. Plants therefore had adapted to ensure reproductive success by preventing flowering until after this cold period had passed. Plants that have a vernalization requirement have hastened flowering in response to a prolonged period of cold. In the dicot model plant *Arabidopsis thaliana* some ecotypes do not require vernalization, while others require different degrees of vernalization for early flowering (Chouard, 1960). During this process a range of genes show changes in their expression levels, and plants become able to respond to changing in photoperiod and eventually flower.

As shown in Figure 7.4, in *Arabidopsis* two of the key genes in the vernalization process are *FRIGIDA* (*FRI*) and *FLOWERING LOCUS* (*FLC*) (Michaels and Amasino, 1999). *FRI* is a two coiled-coil motif protein that is required for increasing *FLC* expression levels in leaf and apex (Johanson et al., 2000). FRI forms a large protein complex (FRI-C) with a collection of chromatin modification factors and general transcription factors that acts as a transcription activator complex (Choi et al., 2011). *FLC* is a MADS-box domain transcription factor that acts as a potent inhibitor of flowering by associating with another MADS-box domain protein SHORT VEGETATIVE PHASE (SVP) and repressing *FT* in leaves, and *FD* and

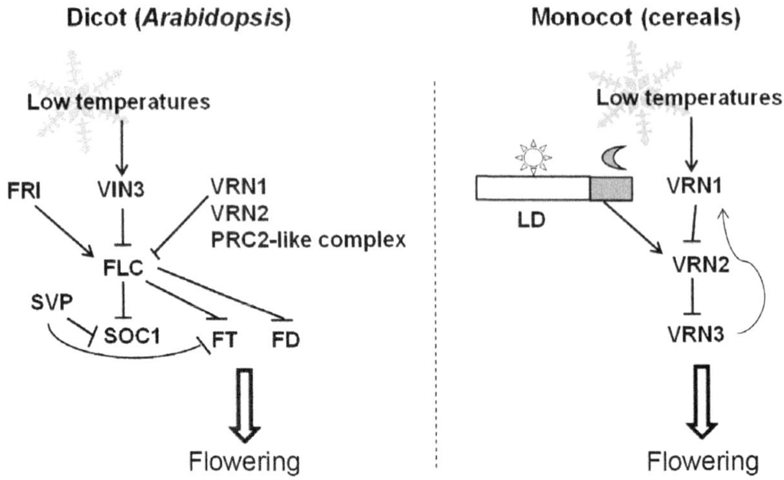

Figure 7.4 Models of the vernalization pathway in *Arabidopsis* and temperate cereals. Arrows indicate activation and T-bars show inhibition. FD, FLOWERING LOCUS D; FLC, FLOWERING LOCUS; FRI, FRIGIDA; FT, FLOWERING LOCUS T; SOC1, SUPPRESSOR OF CONSTANS1, SVP, SHORT VEGETATIVE PHASE; VIN1-3, VERNALIZATION INSENSITIVE1-3; PRC2, POLYCOMB REPRESSOR COMPLEX.

SOC1 in the apex (Kim et al., 2009). Cold treatments prevent the up-regulation of *FLC* by FRI, and enhance the expression of genes such as *VERNALIZATION INSENSITIVE 3* (*VIN3*) which represses *FLC*, and lead to histone modification of *FLC* chromatin (Kim et al., 2009; He, 2012). *FLC* repression is maintained even when the plant is returned to a higher temperature through an epigenetic mechanism involving other genes including *VRN1* and *VRN2*, and the POLYCOMB REPRESSOR COMPLEX (PRC2)-like complexes that cause histone modifications at the *FLC* locus maintaining it in an inactive chromatin state (Gendall et al., 2001; Wood et al., 2006; De Lucia et al., 2008). Repression of *FLC* is quantitative and longer cold period leads to lower *FLC* expression levels (Sheldon et al., 2000). The repression of *FLC* ends with the *meiosis*-to-*mitosis* transition so that the next generation has to experience vernalization again before flowering (Sheldon et al., 2008).

Perennial plants have longer juvenile phases compared to annual plants, and when they become competent to flower not all the meristems respond at the same time, often cycling between vegetative and reproductive phases (Albani and Coupland, 2010). A perennial relative of the model *Arabidopsis* plants, *Arabis alpine*, exhibits a decrease in the level of *PERPETUAL FLOWERING 1* (*PEP1*), the orthologue of *A. thaliana FLC*, when subject to cold treatment and a subsequent increase when the plants are exposed again to high temperature in those shoots that did not become reproductive (Wang et al., 2009; Aikawa et al., 2010). A similar process happens in perennial trees, *SVP*-like MADS-box genes (*DORMANCY ASSOCIATED MAD BOX*) control bud dormancy and are down-regulated by cold in a quantitative manner (Jimenez et al., 2010).

In the monocot cereals, *VRN1*, *2* and *3*, although they have been given the same name as in *Arabidopsis*, are not the homologues of *Arabidopsis VRN* genes and they play a different role in the vernalization process that is FLC-independent (Shitsukawa et al., 2007). As shown in Figure 7.4, *VRN1* expression levels in wheat increase in vernalized plants promoting the floral transition. *VRN2* is up-regulated by LDs and negatively regulates the flowering transition by repressing *VRN3* when the days are still long before the beginning of the winter (Shitsukawa et al., 2007; Sasani et al., 2009). *VRN3* is a homologue of the *Arabidopsis FT* and is up-regulated when the days are long, and it is responsible for up-regulating *VRN1* expression in a feedback loop through FLOWERING LOCUS D-like2 (FDL2) (Distelfeld et al., 2009). Although the vernalization response in dicots and monocots involves different genes with different functions, the post-vernalization mechanism involves epigenetic regulation both in *Arabidopsis* and temperate cereals (Gendall et al., 2001; Oliver et al., 2009).

Even in the same species the response to vernalization can involve different pathways. In *Arabidopsis* other MADS-box genes play a role in FLC-independent vernalization responses. *AGAMOUS-LIKE 24* (*AGL24*) activates the flowering transition in response to vernalization by activating *SOC1* (Michaels et al., 2003). AGL19 promotes flowering after a period of cold temperature, probably independently of *SOC1* activation (Schonrock et al., 2006). *FLC*-like genes such as *MADS AFFECTING FLOWERING 2* (*MAF2*) delay the flowering transition after a short period of cold to prevent precocious flowering when the plant experiences a short period of cold before the beginning of the winter (Ratcliffe et al., 2003).

7.4.3 Thermosensory pathway

Although there has been much research into the flowering behaviour of plants in response to vernalization, fewer studies have been conducted on how plants respond to changes in ambient temperature (i.e. between 16°C and 23°C). This aspect is acquiring increasing interest due to global climate change. The increasing temperatures experienced due to climate change have led to a precocious flowering behaviour in many plants. Many genes previously placed in the autonomous flowering pathway (described below), are now thought to be involved in an ambient temperature pathway which prevents plants from flowering according to the temperature fluctuations during each season.

As shown in Figure 7.5, in *Arabidopsis SVP* plays a role-mediating signals from *FCA* and *FVE*, which sense ambient temperature (Blazquez et al., 2003). SVP levels are higher at low temperatures (16°C) and it represses the floral transition by binding to a CArG motif present in the *FT* promoter. SVP is gradually degraded at higher temperatures thus allowing plants to flower (Lee et al., 2007, 2013). *FLOWERING LOCUS M* (*FLM*) can generate four variant proteins by temperature-dependent alternative spicing (Pose et al., 2013). The two main splice products are FLM-β and FLM-δ. FLM-β is mainly produced at low temperatures and negatively regulates flower transition by forming a complex with SVP. When the temperature rises, a decrease of the FLM−β splice form and an increase in FLM-δ production that forms an inactive complex with SVP allows flowering to occur (Pose et al., 2013).

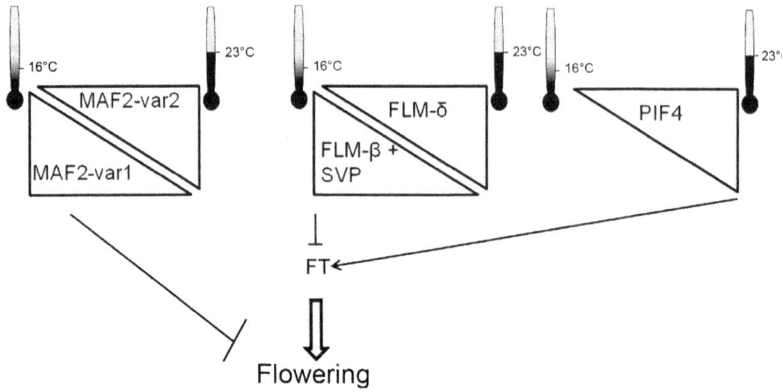

Figure 7.5 Model of the thermosensory pathway in *Arabidopsis*. Arrows indicate activation and T-bars show inhibition. FLM, FLOWERING LOCUS M; FT, FLOWERING LOCUS T; MAF2, MADS AFFECTING FLOWERING2; PIF4, PHYTOCROME INTERACTING FACTOR4; SVP, SHORT VEGETATIVE PHASE.

Another repressor that regulates flowering through a temperature-dependent alternative splicing process, is MADS AFFECTING FLOWERING 2 (MAF2) (Rosloski et al., 2013). In cold conditions the splice form MAF2- var1 (encoding the full length protein) is highly expressed and represses flowering. When the temperature rises the there is an increase in the MAF2-var2 form that generates a truncated protein, which has a limited effect on flowering (Rosloski et al., 2013). In contrast, PHYTOCROME INTERACTING FACTOR4 (PIF4) positively regulates the flowering transition when temperature rises (Kumar et al., 2012). PIF4 activates *FT* expression by binding its promoter in SD but not so much in LD as it is unstable in light and thus present at lower levels in LD (Kumar et al., 2012).

7.5 Endogenous cues regulating flowering

In addition to exogenous environmental signals, numerous endogenous cues are sensed by plants and used to regulate flowering.

7.5.1 Autonomous pathway

The autonomous pathway is not influenced by environmental signals. This pathway is important to ensure that the plant ultimately flowers, even with the lack of inductive environmental cues. Genes classified in this pathway include *LUMINIDEPENDENS* (*LD*), *FLOWERING LOCUS D* (*FLD*), *FCA*, *FY*, *FPA*, *FVE*, *FLOWERING LOCUS K* (*FLK*) and *RELATIVE OF EARLY FLOWERING6* (*REF6*) and they all target the *FLC* gene, reducing *FLC* mRNA levels via different mechanisms (Yan et al., 2010; Srikanth and Schmid, 2011). FCA, FPA and FLK interact with *FLC* mRNA while

FLD and FVE regulate *FLC* epigenetically through chromatin modification (Simpson, 2004). Some of these genes, such as *FCA* and *FY*, interact to promote *FLC* down-regulation. Interestingly, some of these genes are conserved in temperate cereals species that do not appear to have an *FLC* gene, and it has been shown that these genes play a role in other processes distinct from flowering (Simpson, 2004). How these genes are regulated remains to be investigated.

7.5.2 Gibberellic acid pathway

In 1926, Kurosawa, a Japanese scientist, noticed that rice seedlings infected by a pathogenic fungus, *Gibberella fujikuroi*, were showing atypical elongation. In 1938, the chemical responsible, gibberellic acid (GA), was isolated (Daviere and Achard, 2013). In 1957, Langridge reported a correlation of flowering time and this compound, demonstrating that the administration of exogenous GA promotes flowering. More recent studies have confirmed this theory using *Arabidopsis* mutants defective in either GA biosynthesis or signalling (Wilson et al., 1992). Gibberellins are a large family of tetracyclic diterpenoid plant hormones that regulate many developmental processes, such as seed germination, stem elongation, flowering induction and pollen maturation. In LDPs under non-inductive SDs conditions, the GA pathway plays a crucial role in determining flowering time and is essential for flower induction (Wilson et al., 1992; Mutasa-Gottgens and Hedden, 2009). Whilst several gibberellins have been isolated from plants, only some of them such as GA1, GA3, GA4 and GA7, are biologically active. The current model for the GA pathway (Figure 7.6) proposes that a bioactive GA forms a complex with GIBBERELLIN INSENSITIVE DWARF1 (GID1), a cytoplasmic/nuclear localized receptor. This complex interacts with DELLA proteins (named after a conserved protein motif starting with the amino acids D, E, L, L and A), which are negative regulators of GA signalling. DELLA proteins are then degraded through the $SCF^{GID2/SLY1}$ complex by the 26S proteasome complex (Ueguchi-Tanaka et al., 2007; Hirano et al., 2008; Achard and Genschik, 2009). In SD, GA indirectly represses *miR159* expression levels through the repression of DELLA proteins. As high levels of *miRNA159* reduce *LFY* expression GA therefore promotes expression of *LFY* and flowering (Achard et al., 2004). It has been proposed that the GA pathway has an additional role in promoting *LFY* expression through the indirect up-regulation of *SOC1* (Mutasa-Gottgens and Hedden, 2009). GA regulates *SOC1* expression by promoting expression of genes responsible of *SOC1* up-regulation such as *AGL24* and down-regulating repressors such as *SVP* (Moon et al., 2003; Li et al., 2008).

7.5.3 Sugar pathway

Soluble sugars play an essential role in plant development and have been postulated to be important for many phase changes like seed embryo development, germination and flowering time (Gibson, 2005). Early research showed that flowering in *Brassica* was enhanced by sucrose, other work in *Arabidopsis* and *Sinapis alba* showed that sucrose levels increased before the flowering transition (Corbesier et al., 1998). Observing the effect on flowering time of different *Arabidopsis* mutants, it has been proposed that

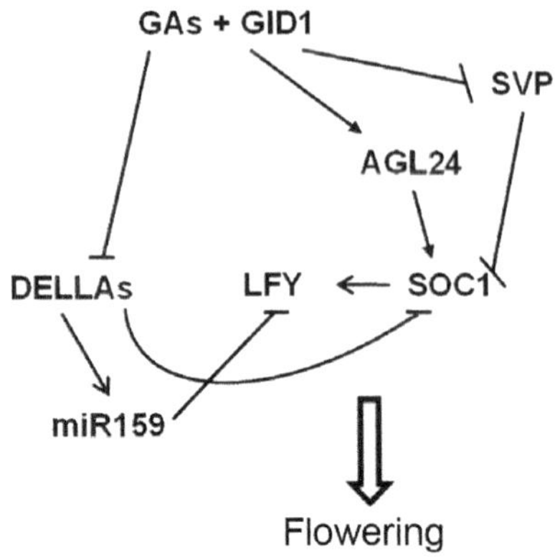

Figure 7.6 Model of the SDs gibberellins acid pathway in *Arabidopsis*. Arrows indicate activation and T-bars show inhibition. AGL24, AGAMOUS-LIKE 24; GAs, gibberellic acids; GID1, GIBBERELLIN INSENSITIVE DWARF1; LFY, LEAFY; miR159, microRNA159; SOC1, SUPPRESSOR OF CONSTANS1; SVP, SHORT VEGETATIVE PHASE.

the sucrose pathway plays a role both upstream and downstream of *FT* and *CO* (Seo et al., 2011). Before the flowering transition, sucrose accumulates in the shoot apex where flowering is promoted under SD conditions through the induction of *LFY* expression by SOC1 (Blazquez et al., 1998; Kotake et al., 2003). Although, the action of sugar in controlling flowering time has to be further investigated, it is clear that the activity of trehalose-6-phosphate (T6P) synthase is necessary for inducing flowering in *Arabidopsis* (van Dijken et al., 2004). T6P regulates sugar balance in plants, and a possible interaction with the ageing pathway influencing the production of miRNAs has been proposed (Matsoukas et al., 2012; Wahl et al., 2013).

7.5.4 Ageing pathway

In the ageing pathway (shown in Figure 7.7) the activators of flowering time belong to the SPL transcription factor gene family. SPLs are regulated by *miR156* (Yang et al., 2011). When *miRNA156* levels decrease, SPLs are free to promote flowering through both an *FT*-dependent and an *FT*-independent pathway (Fahlgren et al., 2007). SPL3, SPL4 and SPL5 directly activate the floral promoter genes *FUL*, *AP1* and *LFY*, while SPL9 promotes the transcription of the floral promoters *FRUITFUL* (*FUL*) and *SOC1* (Yamaguchi et al., 2005; Wang et al., 2009). The *FT*-dependent flowering activation involves the action of SPL9 and SPL10, which act to promote the transcription of *miR172*. Plant ageing also promotes the transcription and accumulation of *miR172* in the leaves and floral buds (Zhu and Helliwell, 2011). *miR172* targets APETALA2 (AP2)-like floral repressor genes, such as *TOE1*, *TOE2*, *SCHLAFMÜTZE* (*SMZ*) and *SCHNARCHZAPFEN* (*SNZ*), which normally repress *FT* (Yamaguchi and Abe, 2012).

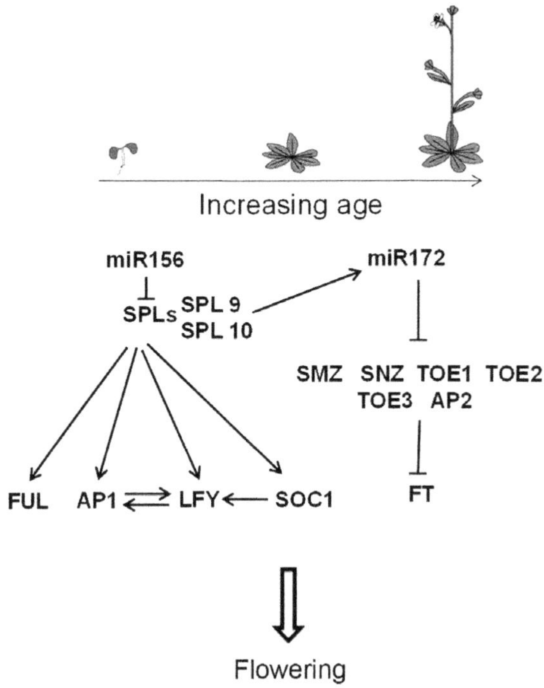

Figure 7.7 Model of the age pathway in *Arabidopsis*. Arrows indicate activation and T-bars show inhibition. AP1, APETALA1; AP2, APETALA2; FT, FLOWERING LOCUS T; FUL, FRUITFUL; LFY, LEAFY; miRNAs, microRNAs; SLP, SQUAMOSA PROMOTER BINDING PROTEIN-LIKE; SMZ, SCHLAFMÜTZE; SNZ, SCHNARCHZAPFEN; SOC1, SUPPRESSOR OF CONSTANS1; TOE1-3, TARGET OF EAT1-3.

7.6 Flowering time control and manipulation

The ability to control and manipulate plant flowering time would be a huge advantage in commercial food production, enabling better crop scheduling and reducing wastage. Furthermore, as climate change starts to influence flowering time of crops and the geographical regions where they are grown, breeders will need to produce new varieties more suited to those new climactic conditions. Understanding the molecular mechanisms controlling flowering time and the sequences of the genes involved will speed up the breeding selection process as selection can be made for specific gene combinations. Alternatively TILLING can be used to create and select new alleles of flowering time genes following mutagenesis. Whilst breeding in herbaceous plants and cereals can be done in a relatively short amount of time, tree breeding programmes can take many years. Inducing precocious flowering and shortening the juvenile phase is therefore useful in such breeding programmes. One strategy that is being developed is inducing flowering using disarmed virus vectors that express flowering time genes such as *FT* (Li et al., 2009). This approach has the advantage over normal transgenic plant approaches because it is a transient expression approach that speeds up the breeding selection process without transmitting the exogenous gene to the next generation, it thus results in non-GMO progeny.

References

Achard, P., Genschik, P., 2009. Releasing the brakes of plant growth: how GAs shutdown DELLA proteins. J. Exp. Bot. 60, 1085–1092.

Achard, P., Herr, A., Baulcombe, D.C., Harberd, N.P., 2004. Modulation of floral development by a gibberellin-regulated microRNA. Development. 131, 3357–3365.

Aikawa, S., Kobayashi, M.J., Satake, A., Shimizu, K.K., Kudoh, H., 2010. Robust control of the seasonal expression of the *Arabidopsis* FLC gene in a fluctuating environment. Proc. Natl. Acad. Sci. USA. 107, 11632–11637.

Albani, M.C., Coupland, G., 2010. Comparative analysis of flowering in annual and perennial plants. Plant Dev. 91, 323–348.

Ben-Naim, O., Eshed, R., Parnis, A., Teper-Bamnolker, P., Shalit, A., Coupland, A., et al., 2006. The CCAAT binding factor can mediate interactions between CONSTANS-like proteins and DNA. Plant J. 46, 462–476.

Blazquez, M.A., Ahn, J.H., Weigel, D., 2003. A thermosensory pathway controlling flowering time in *Arabidopsis thaliana*. Nat. Genet. 33, 168–171.

Blazquez, M.A., Green, R., Nilsson, O., Sussman, M.R., Weigel, D., 1998. Gibberellins promote flowering of *Arabidopsis* by activating the LEAFY promoter. Plant Cell. 10, 791–800.

Blazquez, M.A., Soowal, L.N., Lee, I., Weigel, D., 1997. LEAFY expression and flower initiation in *Arabidopsis*. Development. 124, 3835–3844.

Brunner, A.M., Nilsson, O., 2004. Revisiting tree maturation and floral initiation in the poplar functional genomics era. New Phytol. 164, 43–51.

Carre, I., Veflingstad, S.R., 2013. Emerging design principles in the *Arabidopsis* circadian clock. Semin. Cell. Dev. Biol. 24, 393–398.

Choi, K., Kim, J., Hwang, H.-J., Kim, S., Park, C., et al., 2011. The FRIGIDA complex activates transcription of FLC, a strong flowering repressor in *Arabidopsis*, by recruiting chromatin modification factors. Plant Cell. 23, 289–303.

Chouard, P., 1960. Vernalization and its relations to dormancy. Annu. Rev. Plant Phys. 11, 191–238.

Chow, B.Y., Kay, S.A., 2013. Global approaches for telling time: omics and the *Arabidopsis* circadian clock. Semin. Cell. Dev. Biol. 24, 383–392.

Corbesier, L., Lejeune, P., Bernier, G., 1998. The role of carbohydrates in the induction of flowering in *Arabidopsis thaliana*: comparison between the wild type and a starchless mutant. Planta. 206, 131–137.

Corbesier, L., Vincent, C., Jang, S.H., Fornara, F., Fan, Q.Z., et al., 2007. FT protein movement contributes to long-distance signaling in floral induction of *Arabidopsis*. Science. 316, 1030–1033.

Daviere, J.-M., Achard, P., 2013. Gibberellin signaling in plants. Development. 140, 1147–1151.

De Lucia, F., Crevillen, P., Jones, A.M.E., Greb, T., Dean, C., 2008. A PHD-Polycomb Repressive Complex 2 triggers the epigenetic silencing of FLC during vernalization. Proc. Natl. Acad. Sci. USA. 105, 16831–16836.

van Dijken, A.J.H., Schluepmann, H., Smeekens, S.C.M., 2004. *Arabidopsis* trehalose-6-phosphate synthase 1 is essential for normal vegetative growth and transition to flowering. Plant Physiol. 135, 969–977.

Distelfeld, A., Li, C., Dubcovsky, J., 2009. Regulation of flowering in temperate cereals. Curr. Opin. Plant Biol. 12, 178–184.

Fahlgren, N., Howell, M.D., Kasschau, K.D., Chapman, E.J., et al., 2007. High-throughput sequencing of *Arabidopsis* microRNAs: evidence for frequent birth and death of MIRNA genes. PLoS One. 2.

Fowler, S., Lee, K., Onouchi, H., Samach, A., Richardson, K., et al., 1999. GIGANTEA: a circadian clock-controlled gene that regulates photoperiodic flowering in *Arabidopsis* and encodes a protein with several possible membrane-spanning domains. EMBO J. 18, 4679–4688.

Gendall, A.R., Levy, Y.Y., Wilson, A., Dean, C., 2001. The VERNALIZATION 2 gene mediates the epigenetic regulation of vernalization in *Arabidopsis*. Cell. 107, 525–535.

Gibson, S.I., 2005. Control of plant development and gene expression by sugar signaling. Curr. Opin. Plant Biol. 8, 93–102.

Greenup, A., Peacock, W.J., Dennis, E.S., Trevaskis, B., 2009. The molecular biology of seasonal flowering-responses in *Arabidopsis* and the cereals. Ann. Bot-London. 103, 1165–1172.

Halliday, K.J., Salter, M.G., Thingnaes, E., Whitelam, G.C., 2003. Phytochrome control of flowering is temperature sensitive and correlates with expression of the floral integrator FT. Plant J. 33, 875–885.

Hanzawa, Y., Money, T., Bradley, D., 2005. A single amino acid converts a repressor to an activator of flowering. Proc. Natl. Acad. Sci. USA. 102, 7748–7753.

Hayama, R., Yokoi, S., Tamaki, S., Yano, M., Shimamoto, K., 2003. Adaptation of photoperiodic control pathways produces short-day flowering in rice. Nature. 422, 719–722.

He, Y., 2012. Chromatin regulation of flowering. Trends Plant Sci. 17, 556–562.

Hirano, K., Ueguchi-Tanaka, M., Matsuoka, M., 2008. GID1-mediated gibberellin signaling in plants. Trends Plant Sci. 13, 192–199.

Huijser, P., Schmid, M., 2011. The control of developmental phase transitions in plants. Development. 138, 4117–4129.

Jackson, S.D., Hong, Y., 2012. Systemic movement of FT mRNA and a possible role in floral induction. Frontiers Plant Sci. 3, 127.

Jaeger, K.E., Wigge, P.A., 2007. FT protein acts as a long-range signal in *Arabidopsis*. Curr. Biol. 17, 1050–1054.

Jang, S., Marchal, V., Panigrahi, K.C.S., Wenkel, S., Soppe, W., et al., 2008. *Arabidopsis* COP1 shapes the temporal pattern of CO accumulation conferring a photoperiodic flowering response. EMBO J. 27, 1277–1288.

Jimenez, S., Reighard, G.L., Bielenberg, D.G., 2010. Gene expression of DAM5 and DAM6 is suppressed by chilling temperatures and inversely correlated with bud break rate. Plant Mol. Biol. 73, 157–167.

Johanson, U., West, J., Lister, C., Michaels, S., et al., 2000. Molecular analysis of FRIGIDA, a major determinant of natural variation in *Arabidopsis* flowering time. Science. 290, 344–347.

Jung, J.-H., Seo, Y.-H., Seo, P.J., Reyes, J.L., Yun, J., et al., 2007. The GIGANTEA-regulated MicroRNA172 mediates photoperiodic flowering independent of CONSTANS in *Arabidopsis*. Plant Cell. 19, 2736–2748.

Kaufmann, K., Wellmer, F., Muino, J.M., Ferrier, T., Wuest, S.E., et al., 2010. Orchestration of floral initiation by APETALA1. Science. 328, 85–89.

Kerstetter, R.A., Poethig, R.S., 1998. The specification of leaf identity during shoot development. Annu. Rev. Cell Dev. Biol. 14, 373–398.

Kim, D.H., Doyle, M.R., Sung, S., Amasino, R.M., 2009. Vernalization: winter and the timing of flowering in plants. Annu. Rev. Cell Dev. Biol. 25, 277–299.

Kojima, S., Takahashi, Y., Kobayashi, Y., Monna, L., Sasaki, T., et al., 2002. Hd3a, a rice ortholog of the *Arabidopsis* FT gene, promotes transition to flowering downstream of Hd1 under short-day conditions. Plant Cell Physiol. 43, 1096–1105.

Kotake, T., Takada, S., Nakahigashi, K., Ohto, M., Goto, K., 2003. *Arabidopsis* TERMINAL FLOWER 2 gene encodes a heterochromatin protein 1 homolog and represses both FLOWERING LOCUS T to regulate flowering time and several floral homeotic genes. Plant Cell Physiol. 44, 555–564.

Kumar, S.V., Lucyshyn, D., Jaeger, K.E., Alos, E., et al., 2012. Transcription factor PIF4 controls the thermosensory activation of flowering. Nature. 484, 242–245.

Lee, J.H., Yoo, S.J., Park, S.H., Hwang, I., Lee, J.S., Ahn, J.H., 2007. Role of SVP in the control of flowering time by ambient temperature in *Arabidopsis*. Genes Dev. 21, 397–402.

Lee, J.H., Ryu, H.-S., Chung, K.S., Posé, D., Kim, S., Schmid, M., et al., 2013. Regulation of temperature- responsive flowering by MADS-box transcription factor repressors. Science. 342, 628–632.

Li, C., Gu, M., Shi, N., Zhang, H., Yang, X., Osman, T., et al., 2011. Mobile FT mRNA contributes to the systemic florigen signalling in floral induction. Sci. Rep. 1, 73, <http://dx.doi.org/10.1038/srep00073>.

Li, C.Y., Zhang, K., Zeng, X.W., Jackson, S., Zhou, Y., Hong, Y.G., 2009. A cis element within flowering locus T mRNA determines its mobility and facilitates trafficking of heterologous viral RNA. J. Virol. 83, 3540–3548.

Li, D., Liu, C., Shen, L., Wu, Y., Chen, H., et al., 2008. A repressor complex governs the integration of flowering signals in *Arabidopsis*. Dev. Cell. 15, 110–120.

Lin, C.T., Shalitin, D., 2003. Cryptochrome structure and signal transduction. Annu. Rev. Plant Biol. 54, 469–496.

Locke, J.C.W., Kozma-Bognar, L., Gould, P.D., Feher, B., et al., 2006. Experimental validation of a predicted feedback loop in the multi-oscillator clock of *Arabidopsis thaliana*. Mol. Syst. Biol. 2, 6.

Mas, P., Kim, W.Y., Somers, D.E., Kay, S.A., 2003. Targeted degradation of TOC1 by ZTL modulates circadian function in *Arabidopsis thaliana*. Nature. 426, 567–570.

Mathieu, J., Warthmann, N., Kuettner, F., Schmid, M., 2007. Export of FT protein from phloem companion cells is sufficient for floral induction in *Arabidopsis*. Curr. Biol. 17, 1055–1060.

Matsoukas, I.G., Massiah, A.J., Thomas, B., 2012. Florigenic and antiflorigenic signaling in plants. Plant Cell Physiol. 53, 1827–1842.

Michaels, S.D., Amasino, R.M., 1999. FLOWERING LOCUS C encodes a novel MADS domain protein that acts as a repressor of flowering. Plant Cell. 11, 949–956.

Michaels, S.D., Ditta, G., Gustafson-Brown, C., Pelaz, S., et al., 2003. AGL24 acts as a promoter of flowering in *Arabidopsis* and is positively regulated by vernalization. Plant J. 33, 867–874.

Mizoguchi, T., Wright, L., Fujiwara, S., Cremer, F., Lee, K., et al., 2005. Distinct roles of GIGANTEA in promoting flowering and regulating circadian rhythms in *Arabidopsis*. Plant Cell. 17, 2255–2270.

Moon, J., Suh, S.S., Lee, H., Choi, K.R., Hong, C.B., et al., 2003. The SOC1 MADS-box gene integrates vernalization and gibberellin signals for flowering in *Arabidopsis*. Plant J. 35, 613–623.

Mutasa-Gottgens, E., Hedden, P., 2009. Gibberellin as a factor in floral regulatory networks. J. Exp. Bot. 60, 1979–1989.

Nakamichi, N., 2011. Molecular mechanisms underlying the *Arabidopsis* circadian clock. Plant Cell Physiol. 52, 1709–1718.

Oliver, S.N., Finnegan, E.J., Dennis, E.S., Peacock, W.J., Trevaskis, B., 2009. Vernalization-induced flowering in cereals is associated with changes in histone methylation at the VERNALIZATION1 gene. Proc. Natl. Acad. Sci. USA. 106, 8386–8391.

Poethig, R.S., 2003. Phase change and the regulation of developmental timing in plants. Science. 301, 334–336.

Poethig, R.S., 2010. The past, present, and future of vegetative phase change. Plant Physiol. 154, 541–544.

Posé, D., Verhage, L., Ott, F., Yant, L., Mathieu, J., Angenent, G.C., et al., 2013. Temperature-dependent regulation of flowering by antagonistic FLM variants. Nature. 503, 414–417.

Quail, P.H., 2002. Phytochrome photosensory signalling networks. Nat. Rev. Mol. Cell Biol. 3, 85–93.

Ratcliffe, O.J., Kumimoto, R.W., Wong, B.J., Riechmann, J.L., 2003. Analysis of the *Arabidopsis* MADS AFFECTING FLOWERING gene family: MAF2 prevents vernalization by short periods of cold. Plant Cell. 15, 1159–1169.

Rogers, K., Chen, X., 2012. microRNA biogenesis and turnover in plants. Cold Spring Harb. Symp. Quant. Biol. 77, 183–194.

Rosloski, S.M., Singh, A., Jali, S.S., Balasubramanian, S., et al., 2013. Functional analysis of splice variant expression of MADS AFFECTING FLOWERING 2 of *Arabidopsis thaliana*. Plant Mol. Biol. 81, 57–69.

Sasani, S., Hemming, M.N., Oliver, S.N., Greenup, A., et al., 2009. The influence of vernalization and daylength on expression of flowering-time genes in the shoot apex and leaves of barley (*Hordeum vulgare*). J. Exp. Bot. 60, 2169–2178.

Sawa, M., Nusinow, D.A., Kay, S.A., Imaizumi, T., 2007. FKF1 and GIGANTEA complex formation is required for day-length measurement in *Arabidopsis*. Science. 318, 261–265.

Schonrock, N., Bouveret, R., Leroy, O., Borghi, L., et al., 2006. Polycomb-group proteins repress the floral activator AGL19 in the FLC-independent vernalization pathway. Genes Dev. 20, 1667–1678.

Seo, P.J., Ryu, J., Kang, S.K., Park, C.-M., 2011. Modulation of sugar metabolism by an INDETERMINATE DOMAIN transcription factor contributes to photoperiodic flowering in *Arabidopsis*. Plant J. 65, 418–429.

Sheldon, C.C., Rouse, D.T., Finnegan, E.J., Peacock, W.J., Dennis, E.S., 2000. The molecular basis of vernalization: the central role of FLOWERING LOCUS C (FLC). Proc. Natl. Acad. Sci. USA. 97, 3753–3758.

Sheldon, C.C., Hills, M.J., Lister, C., Dean, C., et al., 2008. Resetting of FLOWERING LOCUS C expression after epigenetic repression by vernalization. Proc. Natl. Acad. Sci. USA. 105, 2214–2219.

Shitsukawa, N., Ikari, C., Shimada, S., Kitagawa, S., et al., 2007. The einkorn wheat (*Triticum monococcum*) mutant, maintained vegetative phase, is caused by a deletion in the VRN1 gene. Genes Genet. Syst. 82, 167–170.

Simpson, G.G., 2004. The autonomous pathway: epigenetic and post-transcriptional gene regulation in the control of *Arabidopsis* flowering time. Curr. Opin. Plant Biol. 7, 570–574.

Srikanth, A., Schmid, M., 2011. Regulation of flowering time: all roads lead to Rome. Cell. Mol. Life Sci. 68, 2013–2037.

Thomas, B., Vince-Prue, D., 1997. Photoperiodism in Plants, second ed.
Thomas, B., Carre, I., Jackson, S., 2006. Photoperiodism and flowering. Molecular Biology and Biotechnology of Flowering, second ed., pp. 3–25.
Torti, S., Fornara, F., Vincent, C., Andres, F., et al., 2012. Analysis of the *Arabidopsis* shoot meristem transcriptome during floral transition identifies distinct regulatory patterns and a leucine-rich repeat protein that promotes flowering. Plant Cell. 24, 444–462.
Troncoso-Ponce, M.A., Mas, P., 2012. Newly described components and regulatory mechanisms of circadian clock function in *Arabidopsis thaliana*. Mol. Plant. 5, 545–553.
Turck, F., Fornara, F., Coupland, G., 2008. Regulation and identity of florigen: FLOWERING LOCUS T moves center stage. Annu. Rev. Plant Biol. 59, 573–594.
Ueguchi-Tanaka, M., Nakajima, M., Katoh, E., Ohmiya, H., et al., 2007. Molecular interactions of a soluble gibberellin receptor, GID1, with a rice DELLA protein, SLR1, and gibberellin. Plant Cell. 19, 2140–2155.
Valverde, F., 2011. CONSTANS and the evolutionary origin of photoperiodic timing of flowering. J. Exp. Bot. 62, 2453–2463.
Wahl, V., Ponnu, J., Schlereth, A., Arrivault, S., Langenecker, T., et al., 2013. Regulation of flowering by trehalose-6-phosphate signaling in *Arabidopsis thaliana*. Science. 339, 704–707.
Wang, H., Deng, X.W., 2002. Phytochrome signaling mechanism. In: Meyerowitz, E., Somerville, C. (Eds.), The *Arabidopsis* Book. American Society of Plant Biologists, Rockville, MD.
Wang, R., Farrona, S., Vincent, C., Joecker, A., Schoof, H., et al., 2009. PEP1 regulates perennial flowering in *Arabis alpina*. Nature. 459, 423–427.
Wigge, P.A., 2011. FT, a mobile developmental signal in plants. Curr. Biol. 21, R374–R378.
Wilson, R.N., Heckman, J.W., Somerville, C.R., 1992. Gibberellin is required for flowering in *Arabidopsis thaliana* under short days. Plant Physiol. 100, 403–408.
Wood, C.C., Robertson, M., Tanner, G., Peacock, W.J., et al., 2006. The *Arabidopsis thaliana* vernalization response requires a polycomb-like protein complex that also includes VERNALIZATION INSENSITIVE 3. Proc. Natl. Acad. Sci. USA. 103, 14631–14636.
Yamaguchi, A., Abe, M., 2012. Regulation of reproductive development by non-coding RNA in *Arabidopsis*: to flower or not to flower. J. Plant Res. 125, 693–704.
Yamaguchi, A., Kobayashi, Y., Goto, K., Abe, M., Araki, T., 2005. TWIN SISTER OF FT (TSF) acts as a floral pathway integrator redundantly with FT. Plant Cell Physiol. 46, 1175–1189.
Yan, Z., Liang, D., Liu, H., Zheng, G., 2010. FLC: a key regulator of flowering time in *Arabidopsis*. Russ. J. Plant Physiol. 57, 166–174.
Yang, L., Conway, S.R., Poethig, R.S., 2011. Vegetative phase change is mediated by a leaf-derived signal that represses the transcription of miR156. Development. 138, 245–249.
Zhu, Q.-H., Helliwell, C.A., 2011. Regulation of flowering time and floral patterning by miR172. J. Exp. Bot. 62, 487–495.

Epigenetic regulation during fleshy fruit development and ripening

8

Emeline Teyssier[1], *Lisa Boureauv*[2], *Weiwei Chen*[3], *Ruie Lui*[1], *Charlotte Degraeve-Guibault*[1], *Linda Stammitti*[1], *Yiguo Hong*[3] *and Philippe Gallusci*[1]
[1]Laboratory of Fruit Biology and Pathology, University of Bordeaux, INRA, Villenave D'Ornon, France; [2]Department of Pharmacology and Therapeutics, McGill University, Montreal, Quebec, Canada; [3]Research Centre for Plant RNA Signaling, College of Life and Environmental Sciences, Hangzhou Normal University, Hangzhou, People's Republic of China

8.1 Introduction

Fruit are organs unique to angiosperm plants that have gained a prominent place in human diet because they are sources of several vitamins (A, C, E and K), fibres, carbohydrates and of potent antioxidants such as lycopene (tomato) or polyphenol (grapes). Fruits are very diverse seed-containing structures that in most cases develop from ovaries after pollination has occurred. The development of fleshy fruits, which are thought to derive from a dry ancestor (for a review see Seymour et al. [2013]) is characterized by cell division and expansion events that occur following fertilization. In contrast to dry fruits there is no lignification phase, but fleshy fruits undergo a complex ripening process characterized by extensive metabolic modifications such as sugar accumulation, colour changes, and accumulation of a wide range of secondary compounds of high nutritional value (Klee and Giovannoni, 2011).

Tomato has long been the model for fleshy fruit development and ripening and an impressive array of resources has been developed in this plant (Seymour et al., 2012). These include genetic resources (introgression lines, mutant collections), tilling platforms, as well as stable and transient transformation systems, biochemical procedures for metabolomics analysis and a fully annotated genome (Consortium, 2012). Recent studies in tomato and other fruit crops (grape, strawberry and others) have now shown that the development and ripening of fleshy fruits relies on the establishment and maintenance of differential transcription patterns (Alba et al., 2005; Osorio et al., 2011; Janssen et al., 2008) and complex regulatory pathways that involve genetic and hormonal controls are operating at these developmental phases (Klee and Giovannoni, 2011). However, it appears that a full understanding of fruit development and ripening will not be achieved based only on genetic models. Epigenetic regulations are likely to be essential as well, as recently suggested by the discovery that the tomato *Cnr* (*Colour non-ripening*) mutation

Applied Plant Genomics and Biotechnology.

(characterized by abolishment of fruit ripening) was caused by the hypermethylation of part of the promoter region (Manning et al., 2006).

The concept of epigenetic refers to changes in chromatin organization, which often lead to modifications in gene expression without change in the underlying genomic DNA sequence of the organism. These modifications are, however, heritable through cell division. Epigenetic information is based on DNA methylations and histone post-translational modifications (PTMs) that determine the state of chromatin structure and regulate the transcriptionally active or inert state of DNA (Chan et al., 2005; Reyes, 2006; Li et al., 2007). These epigenetic marks are heritable through DNA replication and cell propagation (Vermaak et al., 2003) therefore determining cell lineage during development. Recent studies in animals, yeast and plants have highlighted the essential role of epigenetic control in the determination and maintenance of cell-specific gene expression programs (Lauria and Rossi, 2011; Hsieh and Fischer, 2005). Indeed, the integrative analysis of genome-wide DNA methylation pattern, in a more general sense of epigenomes, and of transcriptomic data has provided an unprecedented insight in genome functioning and organization both in animal and plant systems (Zhang, 2006; Zilberman et al., 2007; Cokus et al., 2008; Li et al., 2008; Lister et al., 2008; Wang et al., 2009). These works suggest a high number of target sequences for DNA methylation and are consistent with DNA methylation being critically important for the regulation of plant development. They also demonstrate that histone marks collaborate with DNA methylation to shape chromatin and suggest a complex interplay between gene transcription and epigenome landscapes (Roudier et al., 2011; Li et al., 2008). At present, our current understanding of the function of epigenetic marks has been mainly obtained with the model plants *Arabidopsis*, rice and maize and is by far less advanced in most crop plants. However there is increasing evidence that epigenetic regulations are essential actors that also control traits of agronomical relevance, such as plant adaptation to environmental constraints (Mirouze and Paszkowski, 2011; Sahu et al., 2013), heterosis, a phenomenon extensively used to increase agronomical production (Shen et al., 2012), flowering time (Michaels, 2009; He, 2012) or fruit quality (Seymour et al., 2008).

In this review, we will discuss recent evidence that fruit development and -ripening might also be under epigenetic control. The function of DNA methylation in fruits will be analysed along with the possible role of histone modifications. Finally, possible implications of epigenetic regulations on breeding strategies for fruit crops will be presented.

8.2 An overview of DNA methylation in plants

8.2.1 Mechanisms of DNA methylation

DNA cytosine methylation (5MeC) refers to the addition of a methyl group to the carbon 5 of cytosine. It is probably the best-characterized epigenetic mark and several recent reviews have described its function in plants and other systems

(Law and Jacobsen, 2010; He et al., 2011). It is now clearly established that 5MeC of genomic DNA is a crucial reversible epigenetic mark that impacts several biological processes. Most notably, DNA methylation is involved in protecting the genome against transposons and in controlling gene expression, therefore providing an epigenetic layer to the genetic information. In addition DNA methylation is critically important for the control of homologous recombination during meiosis (Mirouze et al., 2012). In plants, DNA methylation can occur at cytosine both in symmetrical (CG or CNG) and non-symmetrical (CNN) contexts (where N can be any nucleotide except G), and is controlled by three classes of DNA methyltransferases, namely, the DNA methyltransferase 1 (MET1), chromomethylases (CMT) and the domain rearranged methyltransferases (DRMs) (Figure 8.1) (Zhang et al., 2010; Bender, 2004; Finnegan and Kovac, 2000). DRMs control *de novo* methylation in all sequence contexts and are in charge of maintaining C methylation in the non-symmetrical CNN context after cell division has occurred. The specificity of DRMs was found to be driven by RNA-directed DNA methylation (RdDM), which is guided by small RNA (24 nt siRNA) signals produced though RNA silencing pathways (reviewed in Saze et al., 2012). Maintenance of methylation in CG and CNG symmetrical contexts occurs in a post replicative way following two different pathways. First, CG methylation is maintained by MET1, an enzyme orthologous to the DNMT1 enzyme of mammalians (Finnegan and Kovac, 2000). In *Arabidopsis*, maintenance of CG methylation in genic and repetitive sequences also requires the Variant in Methylation proteins (VIM1, 2 and 3) (Woo et al., 2008). Second, CNG methylation is maintained by the plant-specific CMTs, and requires the H3K9 methyltransferases KRYPTONITE (KYP/SUVH4), SUVH5 and SUVH6. Indeed, the *Arabidopsis* CMT3 enzyme was shown to be recruited to specific sites by binding dimethyl K9 histone H3 (H3K9Me2), the histone mark set by KYP (Du et al., 2012). Reciprocally, KYP binds CHG motives, thereby establishing a self-reinforcement loop between CHG methylation and H3K9Me2 (Johnson et al., 2007). This regulatory loop is antagonized by Increase in Bonzai methylation (IBM1) that contains a jmjC domain. The IBM1 protein erases H3K9Me2 and prevents non-CpG methylation in genes (Saze et al., 2008; Miura et al., 2009). Both CG and CNG methylation requires the activity of the chromatin remodeler Decrease in DNA Methylation 1 (DDM1). Mutations in the *DDM1* gene lead a progressive loss of methylation in CG and CNG context both in repetitive (Vongs et al., 1993) and unique sequences after repeated self-fertilisation (Saze and Kakutani, 2007). It should also be noted and DDM1 has recently been shown to facilitate CHH methylation by the chromothylase CMT2 at heterochromatin loci, independently from the RdDM that requires DRD1 (Decrease in RNA-Dependant DNA methylation 1) to methylate transposons in euchromatic regions (Zemach et al., 2013; Cell 153: 195_203). Finally, plant DNA methylation can be actively reversed by DNA Glycosylase-Lyase (DNA-GL) together with 3′ phosphatase. DNA-GLs are essential enzymes involved in parental imprinting and protection of the genome against extensive methylation, (Zhu, 2009; Martínez-Macías María et al., 2012). DNA-GLs seem to be plant specific and removal of DNA methylation is thought to occur following different mechanisms in mammalian cells (Zhang and Zhu, 2012).

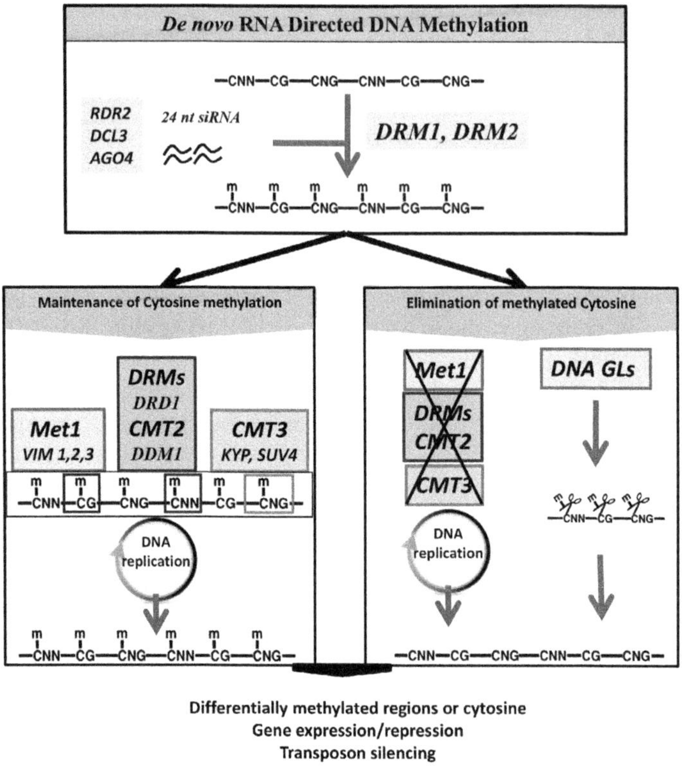

Figure 8.1 Overview of DNA methylation controls in plants. Summary of enzymes and processes involved in the regulation of DNA methylation of plant genomic DNA as determined in the *Arabidopsis* model. *De novo* DNA methylation is set up by DRM1 and DRM2, which are targeted to specific genomic loci by small interference RNA (24 nt siRNA). The RNA-directed DNA methylation (RdDM) requires several additional proteins involved in the production of these small RNAs, a subset of which is indicated. This process is described in detail in Law and Jacobsen (2010). Briefly, RNA-Dependant RNA Polymerase 2 (RDRP2) is thought to produce dsRNA that are most likely the substrate of Dicer Like 3 (DCL3), which produces 24 nt siRNA. The Argonaute protein AGO4 binding to the 24 nt siRNAs is required for RdDM. MET1 and CMT3 are then necessary for maintenance of methylation respectively in the CG and CNG context, and requires Cedrease in DNA Methylation 1 (DDM1) for their activity. CNN methylation in heterochromatic regions is performed by CMT2, which is targeted to DNA through its interaction with DDM1, whereas CNN methylation of TE in euchromatic regions is done by DRM2, which requires Defective in RNA-Directed DNA methylation (DRD1). Methylation removal is thought to occur by simple dilution in all contexts when maintenance of methylation is not active, a process which therefore on active replication. Active demethylation occurs independently of replication and involves DNA Glycosylase-lyases that can remove methylated cytosine independently of replication. Active DNA methylation can target specific loci or affect all genome. The dynamic of DNA methylation determines gene transcriptional states and plays important functions during plant development and adaptation to environmental constraints.

8.2.2 Targets and distribution of DNA methylation in plants

The integrative analysis of genome-wide DNA methylation distribution and of transcriptomic data has now revealed that DNA methylation targets several loci in plant genomes and is likely to play critical functions in regulating gene expression throughout development. Clearly, the highest methylation levels are found at tandem and dispersed repeats. Transposons are homogeneously and heavily methylated in all sequence contexts (CG, CNG, CNN) (Zhang, 2006; Zilberman et al., 2007; Cokus et al., 2008; Li et al., 2008; Lister et al., 2008; Wang et al., 2009). However, many *Arabidopsis* genes are also methylated even though the distribution of DNA methylation within genes is clearly distinct from transposons and suggest complex effects on gene expression that depend both on the location of DNA methylation in genes and on the methylation context (Zhang, 2006; Cokus et al., 2008; Stroud et al., 2013). Hence, DNA methylation tends to be depleted at gene ends. However, 5% of the genes were methylated in their promoter region and more than 30% within the transcribed region, a phenomenon also called gene body methylation (gbM) (Zhang, 2006). Highly methylated promoters correspond to genes with low expression levels or which are expressed in a tissue-specific manner. In contrast gbM, which occurs only in CG context, was found in *Arabidopsis* genes that tend to be expressed constitutively at rather high levels (Zhang, 2006). gbM was recently shown to be maintained in different plants between ortholog genes suggesting an evolutionary conserved function for this type of methylation (Takuno and Gaut, 2013; Aceituno et al., 2008). Very recently, the methylome description of 86 *Arabidopsis* mutants affected by DNA methylation has revealed part of the complex regulatory pathways that control the methylation processes in this plant (Stroud et al., 2013).

In addition to *Arabidopsis*, the description of the genome-wide distribution of 5MeC has also been initiated in rice and maize. Although the distribution of methylation is essentially conserved between these three species, a few distinctive features can be listed: (i) contrary to *Arabidopsis*, DNA methylation peaks at the ATG in rice and maize genes rather than in the body of genes (Wang et al., 2009); (ii) the CHH methylation level is 2 to 3 times higher in maize than in *Arabidopsis* and rice and (iii) maize genes tend to be poorly methylated and are separated by long stretches of heavily methylated intergenic regions. This global methylation landscape contrast with the mosaic distribution of methylation found in *Arabidopsis*, and may reflect differences between large plant genomes enriched in transposons and small genomes that do not contain many of them.

8.2.3 DNA methylation in fruit development

8.2.3.1 DNA methylation patterns are highly dynamic during fruit development

DNA methylation has long been thought to be a stable epigenetic mark. However, recent works have now shown dynamic changes in DNA methylation level and pattern that are either controlled by developmental factors (for review see Law and

Jacobsen, 2010) or in response to environmental cues (for reviews see Chinnusamy and Zhu, 2009; Sahu et al., 2013). As far as fruit are concerned, analysis of the dynamic of DNA methylation is still at its infancy and was limited to the tomato fruits. Indeed, first evidence of genomic DNA methylation changes was obtained in this plant more than two decades ago following the demonstration of variations of CG content between plant tissues and stable inheritance of allele-specific methylation states between parents and offspring (Messeguer et al., 1991). Another pioneering work also showed that demethylation occurred at fruit ripening-specific promoters at the onset of fruit ripening (Hadfield et al., 1993). More recently, the genome-wide description of DNA methylation demonstrated dynamic changes in 5MeC distribution during fruit development (Zhong et al., 2013). Results indicate that fruit pericarp DNA loses methylation during development (Teyssier et al., 2008; Zhong et al., 2013), and this appears to be a tissue-specific event not observed in the locular tissue (Teyssier et al., 2008). Interestingly many additional differences clearly allow distinguishing pericarp from locular tissue: (i) in young fruits, the global methylation level is higher in the pericarp than in the locular tissue and becomes similar only during ripening; (ii) a decrease in the total content in 5mMeC of genomic DNA during ripening and (iii) a transient increase in CNG methylation at tandem and dispersed repeats during fruit growth are only observed in the pericarp. This indicates that fruit tissues might be subjected to different epigenetic controls as it is also suggested by the differential expression of the *DNMT* genes during the development of fruit tissues (Figure 8.2) (Teyssier et al., 2008).

In addition, description of the tomato fruit methylome revealed unusual features of the distribution of methylation in this plant. Most notably, in contrast with the low CNN methylation level found in rice and *Arabidopsis* (1.5% and 2.2% respectively, Feng et al., 2010), and to a lower extend in maize (5.4%), (Gent et al., 2013), the CNN methylation content is surprisingly high in tomato (8.8% in leaves and 13.5% in fruits). Interestingly, in tomato CNN methylation is enriched at the promoter region of the most highly expressed genes (Zhong et al., 2013), as was also observed in maize but not in *Arabidopsis* (Gent et al., 2013). As mentioned above, in euchromatic regions CNN methylation is a non-symmetrical type of methylation which depends on RdDM, a mechanism that set up methylation in all sequence contexts and is guided to specific sequences by 24 nt siRNAs (Law and Jacobsen, 2010), (Figure 8.1). Thus, in maize, CNN methylation in promoters is associated with enrichment in 24 nt siRNAs and *de novo* methylation of nearby transposons, supporting the hypothesis that CNN methylation may be recruited to silence near gene transposons (Gent et al., 2013). In tomato, 24 nt siRNAs that are the most abundant class of small RNAs in fruits are also enriched in the 5′ part of the most highly expressed tomato genes. This is consistent with active *de novo* DNA methylation targeting tomato promoter region presumably in link with gene expression regulation by small RNAs at the transcriptional level in this organ (Mohorianu et al., 2011). High CNN methylation levels might reflect an increased proportion of *de novo* methylation over maintenance methylation in tomato fruits. In line with these observations, *DRM* genes present stage and tissue preferential expression in tomato fruits, consistent with *de novo* methylation being critically important for proper fruit development (Teyssier et al., 2008).

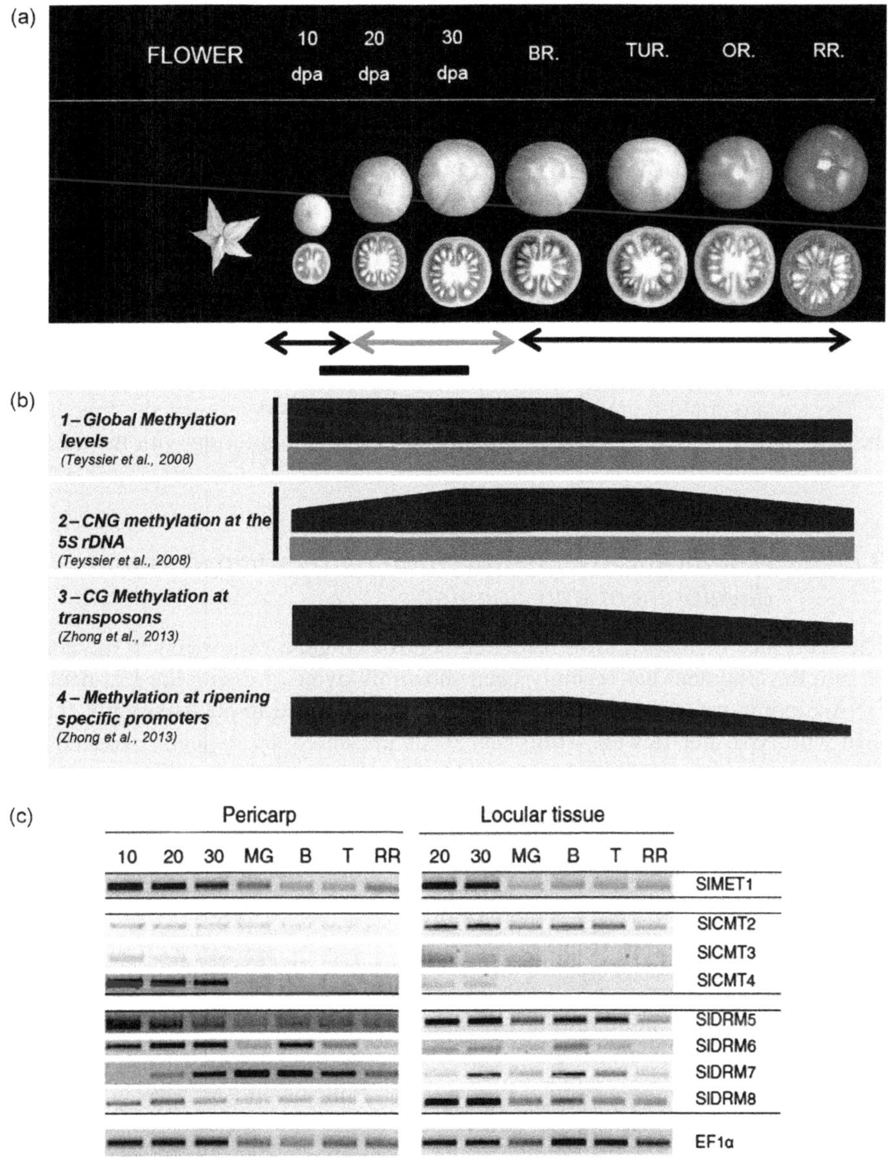

Figure 8.2 Variations of DNA methylation in tomato fruits. (a) Fruit development is initiated following flower pollination and occurs in three main phases; the cell division phase (black arrow) followed by the cell elongation phase (green arrow) and ripening (red arrow). Black bars indicate the period of maximum endoreduplication activity. dpa, days post anthesis; Br., Breaker; Tur., Turning; Or., Orange and RR., Red Ripe stages. (b) Description of major 5 MeC content and profile variations in tomato fruits. 1 – Total percentage of 5 MeC (upper panel) in pericarp (grey area) is higher in pericarp of developing fruits (30%) than in the locular tissues (20%). After the breaker stage, the 5 MeC content of the pericarp genomic DNA drops down to a level similar to the methylation level of the locular tissue genomic DNA. No decrease is observed in the locular tissue (Teyssier et al., 2008).

◀ 2 – CNG methylation was shown to increase sharply at the 5S rDNA and GYPSY transposons in the pericarp during fruit development, but not in the locular tissue (Teyssier et al., 2008). 3 – Transposon methylation was reported to decrease in the CG context specifically in the pericarp of developing tomato fruit (Zhong et al., 2013). 4 – Methylation in all sequence context decreases at RIN-binding sites during fruit ripening (Zhong et al., 2013). (c) Expression analysis of Tomato DNA methyltransferases genes in pericarp and locular tissues of tomato fruits at various developmental stages. Developmetal stages are indicated in days post pollination (dap). Ripening stages are as in (a). Met1: DNA methyltransferase 1, CMT: chromomethylase; DRM: Domain Rearranged Methyltransferase. *Elongation Factor 1* (AF1) is used as a control gene.
Source: (a) Adapted from Teyssier et al. (2008); (b) Adapted from Teyssier et al. (2008).

The tomato could therefore provide an opportunity to determine the function of the near gene CNN type of methylation by specifically interfering with the mechanisms that regulate this type of methylation.

8.2.3.2 Modifications of methylation pattern impact tomato fruit development and ripening

The relevance of methylation mediated control on gene expression in the context of fruit development has recently been shown by overexpressing the UV-damaged DNA-Binding protein 1 (DDB1) in tomato. This resulted in plants slightly smaller than wild type, but flowers, fruits and seeds presented dramatically reduced size. The size reduction of fruits has been correlated with an increased expression of genes encoding negative regulators of cell division and reduced methylation of the promoter region of one of these genes. This result suggested a direct link between fruit size and epigenetic control of cell division, in this case mediated by DDB1 overexpression (Liu et al., 2012). The *Colourless non ripening* (*Cnr*) epiallele also provided strong evidence of a direct link between epigenetic control and fruit ripening. The tomato *Cnr* mutation is caused by the hypermethylation of part of the promoter region of the *CNR* gene, which explained the repression of this gene expression and inhibition of ripening (Manning et al., 2006). More recently, the treatment of young tomato fruits with 5 azacytidine (5 AzaC), a general inhibitor of DNA methylation, resulted in premature fruit ripening sectors, further illustrating that DNA methylation is critically important for this developmental process (Zhong et al., 2013). Indeed, the *CNR* promoter was hypomethylated in premature red sectors but not in adjacent non-ripening sectors, consistent with a direct link between *CNR* hypomethylation and ripening induction. It is noteworthy that 5 AzaC was injected in fruits during the growth phase when endoreduplication was still active (Teyssier et al., 2008) and may therefore have resulted in a more general decrease in DNA methylation levels and distribution. Hence, in these fruits premature ripening might be due to a more global decrease in genomic DNA methylation than *CNR* hypomethylation. Several observations would support this latter possibility: (i) there is a general tendency that methylation level at 5′ end of genes decreases during tomato fruit development; (ii) most known fruit ripening genes are demethylated during fruit ripening, among which those that were expressed in premature red sectors induced by 5 AzaC treatment; (iii) RIN (Ripening Inhibitor)-binding sites at 292 genes with known function during ripening are progressively demethylated during fruit ripening and their methylation states is negatively correlated with expression of the corresponding genes; (iv) there is a 30% decrease in global DNA methylation during fruit ripening, which is unlikely to be solely explained by gene demethylation and (v) a decrease in CNG methylation was also observed at tandem repeats in ripe fruits (Teyssier et al., 2008; Zhong et al., 2013). Interestingly, genomic DNA remained methylated in *rin* and *cnr* mutants both impaired in ripening. As both the *RIN* and *CNR* genes tend to be demethylated during ripening, this suggests a regulatory loop between these two transcription factors, and mechanisms controlling methylation that could contribute to control both the timing and kinetic of ripening (Manning et al., 2006, Zhong et al., 2013). Altogether these results strongly suggest a methylation control over fruit development and ripening that might not be limited to the methylation state of the *CNR* gene, but also involves multiple other target sequences.

8.2.3.3 DNA methylation might also control fruit quality in other plants

Tomato is used as model for fleshy fruit development. Yet tomato fruits are climacteric and necessitate ethylene at the onset of fruit ripening (Klee and Giovannoni, 2011; Cara and Giovannoni, 2008). An important question would therefore be whether the putative function of DNA methylation for fruit development and ripening is specific to the tomato, or more globally applies to other climacteric and non-climacteric fruits. Unfortunately, very few studies have analysed the relevance of DNA methylation as a regulatory mechanism coordinating fruit developmental processes in other plants. In this context, it has been recently demonstrated that the silencing of the pear *MYB* gene, *PcMYB10*, which led to the blocking of anthocyanin accumulation in the skin of pear fruits, is due to the hypermethylation of its promoter region. This provided a direct evidence that DNA methylation might also be a regulator of fruit development and ripening in other fruits (Wang et al., 2013). Interestingly, this effect was due to an increased level of methylation at CNN motives suggesting an important regulatory function role for this type of methylation in pear fruit, in line with the observation that CNN methylation is enriched in the promoter region of fruit tomato genes (Zhong et al., 2013).

So far, other studies addressing the question of DNA methylation function in fruits have mainly been correlative in nature. Indeed, Banana (*Musa* spp.) plants with lower methylation levels also present altered fruit development and ripening kinetic, although it is unclear whether this is a direct effect due to improper methylation of genes involved in these developmental processes (Msogoya et al., 2011). Similarly, the mantled phenotype of the palm oil tree which leads to the formation of abnormal carpels and fruits has been tentatively associated with improper methylation of target genes following somatic embryogenesis, but clear evidence of a direct causal effect is still missing (Jaligot et al., 2000, 2011; Rival et al., 2008). Further research is necessary to determine the functional role of DNA methylation in other fruit crops.

8.3 Histone marks are likely to play fundamental role in fruit development

8.3.1 Overview

In addition to DNA methylation, histone PTMs can influence chromatin structure and gene expression. They impact chromatin organization in different ways, including nucleosome net charge modification (Strahl and Allis, 2000). However, combination of marks can also be used as signals for proteins that modify chromatin organization and gene expression (Kouzarides, 2007; Berr et al., 2011). Indeed, histone PTMs, which affect mainly but not exclusively histone tails, are very diverse and include the phosphorylation, methylation, acetylation or ubiquitination of various amino acids. Levels and distributions of histone PTMs on chromatin are extremely dynamic depending on the activity of a plethora of enzymes (reviewed in Kouzarides, 2007). In *Arabidopsis* combinations of histone PTMs have been shown to define four different chromatin states preferentially associated either with active genes, repressed genes, silent repetitive sequences or intergenic regions (Roudier et al., 2011). This is similar to the five major chromatin states identified in *Drosophila* following the genome-wide mapping of 53 chromatin proteins (Filion et al., 2010). Hence, it is now considered that combination of epigenetic marks ultimately specifies chromatin states. In addition, each individual histone mark has been tentatively associated to specific states of the chromatin. For example, histone acetylation appears to be associated with gene expression, whereas histone methylation may lead to activation or repression of genes depending on the residue modified. Hence dimethylation of Lysine 9 or trimethylation of Lysine 27 of Histone H3 (H3K9me2 and H3K27me3) are associated with repressed genes, whereas H3K4me3, H3K9me3 and H3K36me3 with active genes (for reviews see Li et al., 2007, Pfluger and Wagner, 2007; Jenuwein and Allis, 2001; Kim et al., 2007; Santos-Rosa et al., 2002; and Zhang et al., 2007). Finally, PTMs collaborate with DNA methylation to remodel chromatin establishing regulatory loops that allows to reinforce specific chromatin states as demonstrated for CHG methylation and H3K9me2 (see part I-1 in Du et al., 2012).

Several recent works have now demonstrated that the dynamic regulation of histone marks is essential for plant development to proceed (Berr et al., 2011; Ahmad et al., 2010). Among the plethora of proteins that affect plant development through histone PTMs, the polycomb group (PcG) proteins play a pivotal role by controlling phase transitions during development and cell fate determination as well as cellular differentiation (Bemer and Grossniklaus, 2012; Holec and Berger, 2012). PcG proteins were initially identified in *Drosophila* where they repress the expression of homeotic genes and other genes (Orlando and Paro, 1995; Grimaud et al., 2006). Since then, three different polycomb-repressive complexes (PRCs) have been identified, called PRC1, PRC2 and PhoRC (for a review see Kohler and Villar, 2008). PcG activity is counteracted by proteins that are collectively referred to as *trithorax* group proteins (Köhler and Hennig, 2010; Bemer and Grossniklaus, 2012). The best-studied PcG proteins in plants are the protagonist of the PRC2 complex, which is responsible for the reversible histone H3 trimethylation on lysine K27 (Bemer and Grossniklaus, 2012; Butenko and Ohad, 2011).

An increasing amount of data also indicates that histone deacetylases (HDACs) and histone acetyltransferase (HATs) play key regulatory roles in plant growth and development, in addition to their function in plant stress response (Chen and Tian, 2007; Hollender and Liu, 2008; Servet et al., 2010; Kim, 2013; Ma, 2013). Histone acetylation homoeostasis is regulated by the antagonist actions of HATs and deacetylases which are encoded by complex multigenic families in plants (Chen and Tian, 2007).

8.3.2 Genes involved in histone PTMs are differentially regulated during fruit development

Currently, most studies aiming at deciphering the role of histone PTMs in fruits have focused on the identification and expression analysis of genes potentially involved in histone PTMs. Although correlative, these studies indicate that in all species analysed, including apple (Janssen et al., 2008), grapevine (Aquea et al., 2010, 2011; Almada et al., 2011) and tomato (Cigliano et al., 2013), several genes encoding histone-modifying enzymes are specifically or preferentially expressed in fruits and therefore may have been recruited for regulating this developmental programme. Two HDAC genes are specifically expressed during early apple fruit development (Janssen et al., 2008). Similarly, genes encoding HAT and HDAC in *Vitis vinifera* and in tomato also present fruit-specific or preferential expression (Aquea et al., 2010; Cigliano et al., 2013). In addition several tomato histone methyltransferase (HMT) genes are preferentially expressed at early stages of fruit development during the cell division phase (Cigliano et al., 2013), as was previously demonstrated for the *SlEZ2* gene, which encodes an HMT of the Enhancer of zeste group (How Kit et al., 2010). These results suggest that histone methylation is very active during the cell division phase of fruit development suggesting an early programming of chromatin structure at this stage.

Of course, members of *HMT* and *HDAC* gene families may also play critical roles at later fruit developmental stages. For example, some tomato *HMT* and *HDAC* genes are expressed during the fruit growth phase, or in a wave-like manner or even later during the ripening phases. Among those, *SlSDG33* gene (SDG standing for SET-DOMAIN group) is closely related to the gene encoding the *Arabidopsis* HMT *AtSDG8* (Cigliano et al., 2013) that was recently shown to control the carotenoid biosynthetic pathway in this plant (Cazzonelli et al., 2009). The increased expression of *SlDG33* during fruit ripening makes this gene an interesting candidate that could modulate lycopene accumulation in tomato fruit. In a similar fashion, *SlSDG44*, which is particularly highly expressed at the mature green stage, could be involved in chromatin remodelling during fruit ripening. Indeed this gene is related to *ATX1*, which encodes a TRITHORAX protein. These proteins are known to be positive regular of gene expression by antagonizing the function of the PcG proteins (Schuettengruber et al., 2007). Whether, as its *Arabidopsis* counterpart that has been shown to regulate the expression of the xyloglucanase gene *XTH33*, (Ndamukong, 2009), *SlDG44* also controls the expression of gene-encoding cell wall-modifying proteins during tomato fruit ripening remains to be determined (Cigliano et al., 2013). In a more general way, the function of most epigenetic regulator involved in PTMs will necessitate to be determined to decipher their real function in fruits.

8.3.3 DET1 and PRC2s are involved in the control of tomato fruit development

A role for epigenetic regulations involving histones during fruit development was first suggested by the analysis of tomato plants silenced for the *SlDET1* gene. These transgenic plants produced fruits characterized by an increase in pigments content (Davuluri et al., 2004, 2005). Although DET1 function is not yet entirely deciphered, this negative regulator of photomorphogenesis is thought to act through chromatin regulation (Lau and Deng, 2012), since both in tomato and *Arabidopsis* DET1 was shown to interact with non-acetylated tails of histone H2B (Benvenuto et al., 2002).

Additional evidence for a direct involvement of histone PTMs in the regulation of fruit development and ripening comes from the characterization of transgenic tomato plants affected in the expression of different *PcG* genes encoding protagonists of tomato PRC2s. PRC2 complexes are composed of four different core proteins, which were initialy identified in *Drosophila* and more recently in humans and plants. These proteins are named according to their *Drosophila* counterparts: Enhancer of Zeste [E(Z)], Extra sex combs (Esc), Suppressor of Zeste12 [Su(z) 12] and p55. When properly assembled in a protein complex with Esc and p55, E(Z) catalyses the trimethylation of H3K27. The PRC2 complexes have been best studied in *Arabidopsis* leading to the characterization of three PRC2s that differ in their composition in E(Z) and Su(z)12 paralogs (Holec and Berger, 2012; Makarevich et al., 2008; Hennig and Derkacheva, 2009; Bemer and Grossniklaus, 2012). *Arabidopsis* PcG proteins forms a family of eight homologues of PRC2 components: MEDEA (MEA), CURLY LEAF (CLF) and SWINGER (SWN) are the homologues of E(z); EMBRYONIC FLOWER 2 (EMF2), FERTILIZATION-INDEPENDENT SEED 2 (FIS2) and VERNALIZATION 2 (VRN2) the homologues of Su(z)12 and FERTILIZATION INDEPENDENT ENDOSPERM (FIE) and MULTISUPPRESSOR OF IRA 1 (MSI 1) are the single homologues of Esc and p55, respectively. Specific *Arabidopsis* PRC2s control particular aspects of plant development. For example, the VRN2 containing PRC2 is required for vernalization-induced flowering, whereas MEDEA- and FIS2-containing complexes control seed development (Hennig and Derkacheva, 2009; Köhler and Hennig, 2010). In tomato, the composition of the PcG genes family is different, with only one *Esc* and one *Su(z)12* gene and three *EZ* genes (*SlEZ1*, *SlEZ2* and *SlEZ3*). *SlEZ2* and *SlEZ3* are orthologous to the *Arabidopsis CLF* gene, whereas *SlEZ1 is* orthologous to the *Arabidopsis SWN* gene (How Kit et al., 2010). The functional analysis of SlEZ1 (How Kit et al., 2010) and SlEZ2 (Boureau et al., submitted) provides evidence for a role of polycomb in fruit development and ripening. Hence, tomato fruit plants with reduced levels of *SlEZ1* gene expression present a moderate increase in carpel number suggesting a direct role of SlEZ1 on carpel initiation (How Kit et al., 2010). A similar function might be fulfilled by the SlEZ2 proteins as fruits formed on *SlEZ2* RNAi plants are characterized by the formation of additional carpels. These fruits also have shiny skins, are softer than WT and showed modified shapes and colour (Boureau et al., submitted). In line with these studies, *FIE* co-suppressed tomato lines were also characterized by strong phenotypes such as flower abnormalities, parthenocarpic fruit formation and fruit shape that resembles to some extent those of *SlEZ1* RNAi lines (How Kit et al., 2010). Indeed *FIE* is encoded by a unique gene in tomato and is therefore expected to be present in all PRC2 complexes. Knock down of *FIE* may therefore have a stronger effect on polycomb regulated processes than the repression of single *EZ* gene.

At the time being, the molecular bases of PcG role on fruit phenotypic modifications are still unknown, although these results clearly demonstrate a role for PRC2s in fruit development. However, *AGAMOUS*, *SHOOT MERISTEM LESS* and SHATTERPROOF are primary targets of PRC2 in *Arabidopsis* (Schubert et al., 2006). The tomato orthologues of these genes, respectively *TAG1*, *LeT6* and *TAGL1*, that have been shown to play central roles in tomato fruit development and ripening (Pan et al., 2010) may also be targeted by PRC2s in tomato leaves (Boureau et al., submitted) further supporting the idea of a central role of PcG proteins in the control of fruit development.

8.4 Concluding remarks

There is a growing body of evidence that epigenetic regulators are key actors of fruit development and ripening. In a more general way, several recent studies demonstrate that epigenome polymorphisms – DNA methylation and histone marks – have a profound effect on genome functioning (Lauria and Rossi, 2011), and may contribute to phenotypic diversity (Zhang et al., 2008), control complex traits (Johannes et al., 2009) and eventually be inherited (Schmitz et al., 2011). Furthermore, the generation of epigenetic recombinant inbred lines by crossing *Arabidopsis* plants with hypomethylated genomes and WT plants with normal methylation levels has demonstrated that mosaic epigenomes could be associated with quantitative traits affecting plant development and flowering and stably inherited over several generations (Johannes et al., 2009; Johannes and Colomé-Tatché, 2011).

This indicates the heritable nature of some epigenetic variations that can be artificially generated or may occur naturally.

Indeed, mechanisms that govern the generation of epigenetic diversity are not yet well understood; neither are the mechanisms underlying the heredity of such variations. Of course, environmental factors, such as non-biotic stresses may trigger epigenetic variations, (Feil and Fraga, 2012). In addition, it was recently demonstrated that stable transgressive phenotypes observed in the progeny of crosses between cultivated tomato and wild relatives are associated with transgressive small RNA that may eventually lead to new methylation imprint. This suggests that unpredicted epigenctic variations which may impact plant phenotypes in the following generations could be generated in breeding strategies. Notwithstanding only a limited number of spontaneous epigenetic variants, referred to as 'epimutants' have been described so far (Manning et al., 2006; Cubas et al., 1999; Das and Messing, 1994). Obviously, the contribution of epigenomic diversity to the phenotypic diversity of fruit crops, and more of plants in general is nowadays largely underestimated. There is no doubt that the development of high throughput sequencing technologies will allow the analysis of epigenome diversity in fruit crops, providing novel information concerning the extend and function of this layer of regulation on traits of agronomical relevance including fruit quality. Taking into account such epigenetic variations in breeding is clearly of critical importance and may therefore lead to new improvements of crop species (Springer, 2013).

References

Aceituno, F.F., Moseyko, N., Rhee, S.Y., Gutiérrez, R.A., 2008. The rules of gene expression in plants: organ identity and gene body methylation are key factors for regulation of gene expression in *Arabidopsis thaliana*. BMC Genomics. 9, 348.

Ahmad, A., Zhang, Y., Cao, X.-F., 2010. Decoding the epigenetic language of plant development. Mol. Plant. 3, 719–728.

Alba, R., Payton, P., Fei, Z., McQuinn, R., Debbie, P., Martin, G.B., et al., 2005. Transcriptome and selected metabolite analyses reveal multiple points of ethylene control during tomato fruit development. Plant Cell. 17, 2954–2965.

Almada, R., Cabrera, N., Casaretto, J., Peña-Cortés, H., Ruiz-Lara, S., González Villanueva, E., 2011. Epigenetic repressor-like genes are differentially regulated during grapevine (Vitis vinifera L.) development. Plant Cell Reps. 30, 1959–1968.

Aquea, F., Timmermann, T., Arce-Johnson, P., 2010. Analysis of histone acetyltransferase and deacetylase families of Vitis vinifera. Plant Physiol. Biochem. 48, 194–199.

Aquea, F., Vega, A., Timmermann, T., Poupin, M., Arce-Johnson, P., 2011. Genome-wide analysis of the SET DOMAIN GROUP family in Grapevine. Plant Cell Reps. 30, 1087–1097.

Bemer, M., Grossniklaus, U., 2012. Dynamic regulation of Polycomb group activity during plant development. Curr. Opin. Plant Biol. 15, 523–529.

Bender, J., 2004. DNA methylation and epigenetics. Ann. Rev. Plant Biol. 55, 41–68.

Benvenuto, G., Formiggini, F., Laflamme, P., Malakhov, M., Bowler, C., 2002. The photomorphogenesis regulator DET1 binds the amino-terminal tail of histone H2B in a nucleosome context. Curr. Biol. 12, 1529–1534.

Berr, A., Shafiq, S., Shen, W.-H., 2011. Histone modifications in transcriptional activation during plant development. Biochim. Biophys. Acta. 1809, 567–576.

Butenko, Y., Ohad, N., 2011. Polycomb-group mediated epigenetic mechanisms through plant evolution. Biochim. Biophys. Acta. 1809, 395–406.

Cara, B., Giovannoni, J.J., 2008. Molecular biology of ethylene during tomato fruit development and maturation. Plant Sci. 175, 106–113.

Cazzonelli, C.I., Cuttriss, A.J., Cossetto, S.B., Pye, W., Crisp, P., Whelan, J., et al., 2009. Regulation of carotenoid composition and shoot branching in arabidopsis by a chromatin modifying histone methyltransferase, SDG8. Plant Cell Online. 21, 39–53.

Chan, S.W., Henderson, I.R., Jacobsen, S.E., 2005. Gardening the genome: DNA methylation in *Arabidopsis thaliana*. Nat. Rev. Genet. 6, 351–360.

Chen, Z.J., Tian, L., 2007. Roles of dynamic and reversible histone acetylation in plant development and polyploidy. Biochim. Biophys. Acta. 1769, 295–307.

Chinnusamy, V., Zhu, J.-K., 2009. Epigenetic regulation of stress responses in plants. Curr. Opin. Plant Biol. 12, 133–139.

Cigliano, R.A., Sanseverino, W., Cremona, G., Ercolano, M., Conicella, C., Consiglio, F., 2013. Genome-wide analysis of histone modifiers in tomato: gaining an insight into their developmental roles. BMC Genomics. 14, 57.

Cokus, S.J., Feng, S., Zhang, X., Chen, Z., Merriman, B., Haudenschild, C.D., et al., 2008. Shotgun bisulphite sequencing of the *Arabidopsis* genome reveals DNA methylation patterning. Nature. 452, 215–219.

Consortium, T.T.G., 2012. The tomato genome sequence provides insights into fleshy fruit evolution. Nature. 485, 635–641.

Cubas, P., Vincent, C., Coen, E., 1999. An epigenetic mutation responsible for natural variation in floral symmetry. Nature. 401, 157–161.

Das, O.P., Messing, J., 1994. Variegated phenotype and developmental methylation changes of a maize allele originating from epimutation. Genetics. 136, 1121–1141.

Davuluri, G.R., van Tuinen, A., Mustilli, A.C., Manfredonia, A., Newman, R., Burgess, D., et al., 2004. Manipulation of DET1 expression in tomato results in photomorphogenic phenotypes caused by post-transcriptional gene silencing. Plant J. 40, 344–354.

Davuluri, G.R., van Tuinen, A., Fraser, P.D., Manfredonia, A., Newman, R., Burgess, D., et al., 2005. Fruit-specific RNAi-mediated suppression of DET1 enhances carotenoid and flavonoid content in tomatoes. Nat. Biotechnol. 23, 890–895.

Du, J., Zhong, X., Bernatavichute Yana, V.B., Stroud, H., Feng, S., Caro, E., et al., 2012. Dual binding of chromomethylase domains to H3K9me2-containing nucleosomes directs DNA methylation in plants. Cell. 151, 167–180.

Feil, R., Fraga, M.F., 2012. Epigenetics and the environment: emerging patterns and implications. Nat. Rev. Genet. 13, 97–109.

Feng, S., Cokus, S.J., Zhang, X., Chen, P.-Y., Bostick, M., Goll, M.G., et al., 2010. Conservation and divergence of methylation patterning in plants and animals. Proc. Natl. Acad. Sci. USA. 107, 8689–8694.

Filion, G.J., van Bemmel, J.G., Braunschweig, U., Talhout, W., Kind, J., Ward, L.D., et al., 2010. Systematic protein location mapping reveals five principal chromatin types in drosophila cells. Cell. 143, 212–224.

Finnegan, E.J., Kovac, K.A., 2000. Plant DNA methyltransferases. Plant Mol. Biol. 43, 189–201.

Gent, J.I., Ellis, N.A., Guo, L., Harkess, A.E., Yao, Y., Zhang, X., et al., 2013. CHH islands: de novo DNA methylation in near-gene chromatin regulation in maize. Genome Res. 23, 628–637.

Grimaud, C., Negre, N., Cavalli, G., 2006. From genetics to epigenetics: the tale of Polycomb group and trithorax group genes. Chromosome Res. 14, 363–375.

Hadfield, K.A., Dandekar, A.M., Romani, R.J., 1993. Demethylation of ripening specific genes in tomato fruit. Plant Sci. 92, 13–18.

He, X.-J., Chen, T., Zhu, J.-K., 2011. Regulation and function of DNA methylation in plants and animals. Cell Res. 21, 442–465.

He, Y., 2012. Chromatin regulation of flowering. Trends Plant Sci. 17, 556–562.

Hennig, L., Derkacheva, M., 2009. Diversity of Polycomb group complexes in plants: same rules, different players? Trends Genet. 25, 414–423.

Holec, S., Berger, F., 2012. Polycomb group complexes mediate developmental transitions in plants. Plant Physiol. 158, 35–43.

Hollender, C., Liu, Z., 2008. Histone deacetylase genes in *Arabidopsis* development. J. Integr. Plant Biol. 50, 875–885.

How Kit, A., Boureau, L., Stammitti-Bert, L., Rolin, D., Teyssier, E., Gallusci, P., 2010. Functional analysis of SlEZ1 a tomato enhancer of zeste (E(z)) gene demonstrates a role in flower development. Plant Mol. Biol. 74, 201–213.

Hsieh, T.-F., Fischer, R.L., 2005. Biology of chromatin dynamics. Annu. Rev. Plant Biol. 56, 327–351.

Jaligot, E., Rival, A., Beulé, T., Dussert, S., Verdeil, J.L., 2000. Somaclonal variation in oil palm (*Elaeis guineensis* Jacq.): the DNA methylation hypothesis. Plant Cell Rep. 19, 684–690.

Jaligot, E., Adler, S., Debladis, É., Beulé, T., Richaud, F., Ilbert, P., et al., 2011. Epigenetic imbalance and the floral developmental abnormality of the *in vitro*-regenerated oil palm *Elaeis guineensis*. Ann. Bot. 108, 1453–1462.

Janssen, B., Thodey, K., Schaffer, R., Alba, R., Balakrishnan, L., Bishop, R., et al., 2008. Global gene expression analysis of apple fruit development from the floral bud to ripe fruit. BMC Plant Biol. 8, 16.

Jenuwein, T., Allis, C.D., 2001. Translating the histone code. Science. 293, 1074–1080.

Johannes, F., Colomé-Tatché, M., 2011. Quantitative epigenetics through epigenomic perturbation of isogenic lines. Genetics. 188, 215–227.

Johannes, F., Porcher, E., Teixeira, F.K., Saliba-Colombani, V., Simon, M., Agier, N., et al., 2009. Assessing the impact of transgenerational epigenetic variation on complex traits. PLoS Genet. 5, e1000530.

Johnson, L.M., Bostick, M., Zhang, X., Kraft, E., Henderson, I., Callis, J., et al., 2007. The SRA methyl-cytosine-binding domain links DNA and histone methylation. Curr. Biol. 17, 379–384.

Kim, A., Kiefer, C.M., Dean, A., 2007. Distinctive signatures of histone methylation in transcribed coding and noncoding human Î²-globin sequences. Mol. Cell. Biol. 27, 1271–1279.

Kim, W., Latrasse, D., Servet, C., Zhou, D.-X., 2013. Arabidopsis histone deacetylase HDA9 regulates flowering time through repression of AGL19. Biochem. Biophys. Res. Commun. 432, 394–398.

Klee, H.J., Giovannoni, J.J., 2011. Genetics and control of tomato fruit ripening and quality attributes. Annu. Rev. Genet. 45, 41–59.

Köhler, C., Hennig, L., 2010. Regulation of cell identity by plant Polycomb and trithorax group proteins. Curr. Opin. Genet. Dev. 20, 541–547.

Kohler, C., Villar, C.B.R., 2008. Programming of gene expression by Polycomb group proteins. Trends Cell Biol. 18, 236–243.

Kouzarides, T., 2007. Chromatin modifications and their function. Cell. 128, 693–705.

Lau, O.S., Deng, X.W., 2012. The photomorphogenic repressors COP1 and DET1: 20 years later. Trends Plant Sci. 17, 584–593.

Lauria, M., Rossi, V., 2011. Epigenetic control of gene regulation in plants. Biochim. Biophys. Acta. 1809, 369–378.

Law, J.A., Jacobsen, S.E., 2010. Establishing, maintaining and modifying DNA methylation patterns in plants and animals. Nat. Rev. Genet. 11, 204–220.

Li, B., Carey, M., Workman, J.L., 2007. The role of chromatin during transcription. Cell. 128, 707–719.

Li, X., Wang, X., He, K., Ma, Y., Su, N., He, H., et al., 2008. High-resolution mapping of epigenetic modifications of the rice genome uncovers interplay between DNA methylation, histone methylation, and gene expression. Plant Cell. 20, 259–276.

Lister, R., O'Malley, R.C., Tonti-Filippini, J., Gregory, B.D., Berry, C.C., Millar, A.H., et al., 2008. Highly integrated single-base resolution maps of the epigenome in *Arabidopsis*. Cell. 133, 523–536.

Liu, J., Tang, X., Gao, L., Gao, Y., Li, Y., Huang, S., et al., 2012. A role of tomato UV-damaged DNA binding protein 1 (DDB1) in organ size control via an epigenetic manner. PLoS One. 7, e42621.

Ma, X., Lv, S., Zhang, C., Yang, C., 2013. Histone deacetylases and their functions in plants. Plant Cell Reps. 32, 465–478.

Makarevich, G., Villar, C.B.R., Erilova, A., Kohler, C., 2008. Mechanism of PHERES1 imprinting in *Arabidopsis*. J. Cell Sci. 121, 906–912.

Manning, K., Tör, M., Poole, M., Hong, Y., Thompson, A.J., King, G.J., et al., 2006. A naturally occuring epigenetic mutation in a gene encoding an SBP-box transcription factor inhibits tomato fruit ripening. Nat. Genet. 38, 948–952.

Martínez-Macías María, I., Qian, W., Miki, D., Pontes, O., Liu, Y., Tang, K., et al., 2012. A DNA 32 phosphatase functions in active DNA demethylation in *Arabidopsis*. Mol. Cell. 45, 357–370.

Messeguer, R., Ganal, M., Steffens, J., Tanksley, S.D., 1991. Characterization of the level, target sites and inheritance of cytosine methylation in tomato nuclear DNA. Plant Mol. Biol. 16, 753–770.

Michaels, S.D., 2009. Flowering time regulation produces much fruit. Curr. Opin. Plant Biol. 12, 75–80.

Mirouze, M., Paszkowski, J., 2011. Epigenetic contribution to stress adaptation in plants. Curr. Opin. Plant Biol. 14, 267–274.

Mirouze, M., Lieberman-Lazarovich, M., Aversano, R., Bucher, E., Nicolet, J., Reinders, J., et al., 2012. Loss of DNA methylation affects the recombination landscape in *Arabidopsis*. Proc. Natl. Acad. Sci. USA. 109, 5880–5885.

Miura, A., Nakamura, M., Inagaki, S., Kobayashi, A., Saze, H., Kakutani, T., 2009. An *Arabidopsis* jmjC domain protein protects transcribed genes from DNA methylation at CHG sites. EMBO J. 28, 1078–1086.

Mohorianu, I., Schwach, F., Jing, R., Lopez-Gomollon, S., Moxon, S., Szittya, G., et al., 2011. Profiling of short RNAs during fleshy fruit development reveals stage-specific sRNAome expression patterns. Plant J. 67, 232–246.

Msogoya, T.J., Grout, B.W., Roberts, A., 2011. Reduction in genome size and DNA methylation alters plant and fruit development in tissue culture induced off-type banana (*Musa* spp.). J. Anim. Plant Sci. 2011, 1450–1456.

Ndamukong, I., Chetram, A., Saleh, A., Avramova, Z., 2009. Wall-modifying genes regulated by the Arabidopsis homolog of trithorax, ATX1: repression of the XTH33 gene as a test case. lant. J. 58, 541–553.

Orlando, V., Paro, R., 1995. Chromatin multiprotein complexes involved in the maintenance of transcription patterns. Curr. Opin. Genet. Dev. 5, 174–179.

Osorio, S., Alba, R., Damasceno, C.M.B., Lopez-Casado, G., Lohse, M., Zanor, M.I., et al., 2011. Systems biology of tomato fruit development: combined transcript, protein, and metabolite analysis of tomato transcription factor (nor, rin) and ethylene receptor (Nr) mutants reveals novel regulatory interactions. Plant Physiol. 157, 405–425.

Pan, I.L., McQuinn, R., Giovannoni, J.J., Irish, V.F., 2010. Functional diversification of AGAMOUS lineage genes in regulating tomato flower and fruit development. J. Exp. Bot. 61, 1795–1806.

Pfluger, J., Wagner, D., 2007. Histone modifications and dynamic regulation of genome accessibility in plants. Curr. Opin. Plant Biol. 10, 645–652.

Reyes, J.C., 2006. Chromatin modifiers that control plant development. Curr. Opin. Plant Biol. 9, 21–27.

Rival, A., Jaligot, E., Beulé, T., Finnegan, E.J., 2008. Isolation and expression analysis of genes encoding MET, CMT, and DRM methyltransferases in oil palm (*Elaeis guineensis* Jacq.) in relation to the 'mantled' somaclonal variation. J. Exp. Bot. 59, 3271–3281.

Roudier, F., Ahmed, I., Berard, C., Sarazin, A., Mary-Huard, T., Cortijo, S., et al., 2011. Integrative epigenomic mapping defines four main chromatin states in *Arabidopsis*. EMBO J. 30, 1928–1938.

Sahu, P., Pandey, G., Sharma, N., Puranik, S., Muthamilarasan, M., Prasad, M., 2013. Epigenetic mechanisms of plant stress responses and adaptation. Plant Cell Rep. 32, 1151–1159.

Santos-Rosa, H., Schneider, R., Bannister, A.J., Sherriff, J., Bernstein, B.E., Emre, N.C.T., et al., 2002. Active genes are tri-methylated at K4 of histone H3. Nature. 419, 407–411.

Saze, H., Kakutani, T., 2007. Heritable epigenetic mutation of a transposon-flanked *Arabidopsis* gene due to lack of the chromatin-remodeling factor DDM1. EMBO J. 26, 3641–3652.

Saze, H., Shiraishi, A., Miura, A., Kakutani, T., 2008. Control of genic DNA methylation by a jmjC domain-containing protein in *Arabidopsis thaliana*. Science. 319, 462–465.

Saze, H., Tsugane, K., Kanno, T., Nishimura, T., 2012. DNA methylation in plants: relationship to small RNAs and histone modifications, and functions in transposon inactivation. Plant Cell Physiol. 53, 766–784.

Schmitz, R.J., Schultz, M.D., Lewsey, M.G., O'Malley, R.C., Urich, M.A., Libiger, O., et al., 2011. Transgenerational epigenetic instability is a source of novel methylation variants. Science. 334, 369–373.

Schubert, D., Primavesi, L., Bishopp, A., Roberts, G., Doonan, J., Jenuwein, T., et al., 2006. Silencing by plant Polycomb-group genes requires dispersed trimethylation of histone H3 at lysine 27. EMBO J. 25, 4638–4649.

Schuettengruber, B., Chourrout, D., Vervoort, M., Leblanc, B., Cavalli, G., 2007. Genome regulation by polycomb and trithorax proteins. Cell. 128, 735–745.

Servet, C., Conde e Silva, N., Zhou, D.-X., 2010. Histone acetyltransferase AtGCN5/HAG1 is a versatile regulator of developmental and inducible gene expression in *Arabidopsis*. Mol. Plant. 3, 670–677.

Seymour, G., Poole, M., Manning, K., King, G.J., 2008. Genetics and epigenetics of fruit development and ripening. Curr. Opin. Plant Biol. 11, 58–63.

Seymour, G.B., Chapman, N.H., Chew, B.L., Rose, J.K.C., 2012. Regulation of ripening and opportunities for control in tomato and other fruits. Plant Biotechnol. J. 11, 269–278.

Seymour, G.B., Østergaard, L., Chapman, N.H., Knapp, S., Martin, C., 2013. Fruit development and ripening. Ann. Rev. Plant Biol. 64, 219–241.

Shen, H., He, H., Li, J., Chen, W., Wang, X., Guo, L., et al., 2012. Genome-wide analysis of DNA methylation and gene expression changes in two *Arabidopsis* ecotypes and their reciprocal hybrids. Plant Cell Online. 24, 875–892.

Springer, N.M., 2013. Epigenetics and crop improvement. Trends Genet. 29, 241–247.

Strahl, B.D., Allis, C.D., 2000. The language of covalent histone modifications. Nature. 403, 41–45.

Stroud, H., Greenberg Maxim, V.C., Feng, S., Bernatavichute Yana, V., Jacobsen Steven, E., 2013. Comprehensive analysis of silencing mutants reveals complex regulation of the *Arabidopsis* methylome. Cell. 152, 352–364.

Takuno, S., Gaut, B.S., 2013. Gene body methylation is conserved between plant orthologs and is of evolutionary consequence. Proc. Natl. Acad. Sci. USA. 110, 1797–1802.

Teyssier, E., Bernacchia, G., Maury, S., How Kit, A., Stammitti-Bert, L., Rolin, D., et al., 2008. Tissue dependent variations of DNA methylation and endoreduplication levels during tomato fruit development and ripening. Planta. 228, 391–399.

Vermaak, D., Ahmad, K., Henikoff, S., 2003. Maintenance of chromatin states: an open-and-shut case. Curr. Opin. Cell Biol. 15, 266–274.

Vongs, A., Kakutani, T., Martienssen, R., Richards, E.J., 1993. *Arabidopsis thaliana* DNA methylation mutants. Science. 260, 1926–1928.

Wang, X., Elling, A.A., Li, X., Li, N., Peng, Z., He, G., et al., 2009. Genome-wide and organ-specific landscapes of epigenetic modifications and their relationships to mRNA and small RNA transcriptomes in maize. Plant Cell. 21, 1053–1069.

Wang, Z., Meng, D., Wang, A., Li, T., Jiang, S., Cong, P., et al., 2013. The methylation of the PcMYB10 promoter is associated with green-skinned sport in max red bartlett pear. Plant Physiol. 162, 885–896.

Woo, H.R., Dittmer, T.A., Richards, E.J., 2008. Three SRA-domain methylcytosine-binding proteins cooperate to maintain global CpG methylation and epigenetic silencing in *Arabidopsis*. PLoS Genet. 4, e1000156.

Zemach, A., Kim, M.Y., Hsieh, P.-H., Coleman-Derr, D., Eshed-Williams, L., Thao, K., et al., 2013. The arabidopsis nucleosome remodeler DDM1 allows DNA methyltransferases to access H1-containing heterochromatin. Cell. 153, 193–205.

Zhang, H., Zhu, J.K., 2012. Active DNA demethylation in plants and animals. Cold Spring Harb. Symp. Quant. Biol. 77, 161–173.

Zhang, K., Sridhar, V.V., Zhu, J., Kapoor, A., Zhu, J.-K., 2007. Distinctive core histone post-translational modification patterns in *Arabidopsis thaliana*. PLoS One. 2, e1210.

Zhang, M., Kimatu, J.N., Xu, K., Liu, B., 2010. DNA cytosine methylation in plant development. J. Genet. Genomics. 37, 1–12.

Zhang, X., 2006. Genome-wide high-resolution mapping and functional analysis of DNA methylation in *Arabidopsis*. Cell. 126, 1189–1201.

Zhang, X., Shiu, S.-H., Cal, A., Borevitz, J.O., 2008. Global analysis of genetic, epigenetic and transcriptional polymorphisms in *Arabidopsis thaliana* using whole genome tiling arrays. PLoS Genet. 4, e1000032.

Zhong, S., Fei, Z., Chen, Y.-R., Zheng, Y., Huang, M., Vrebalov, J., et al., 2013. Single-base resolution methylomes of tomato fruit development reveal epigenome modifications associated with ripening. Nat. Biotech. 31, 154–159.

Zhu, J.-K., 2009. Active DNA demethylation mediated by DNA glycosylases. Annu. Rev. Genet. 43, 143–166.

Zilberman, D., Gehring, M., Tran, R.K., Ballinger, T., Henikoff, S., 2007. Genome-wide analysis of *Arabidopsis thaliana* DNA methylation uncovers an interdependence between methylation and transcription. Nat. Genet. 39, 61–69.

Tomato fruit quality improvement facing the functional genomics revolution

9

Dominique Rolin[1], Emeline Teyssier[1], Yiguo Hong[2] and Philippe Gallusci[1]
[1]Laboratory of Fruit Biology and Pathology, University of Bordeaux, INRA, Villenave D'Ornon, France; [2]Research Centre for Plant RNA Signaling, College of Life and Environmental Sciences, Hangzhou Normal University, Hangzhou, China

9.1 Introduction

Fruits derive from specific tissues of the flower, the ovaries and in some cases from accessory tissues. Fruits are the means by which angiosperm plants disseminate seeds. As far as human are concerned, fruits constitute a commercially important and nutritionally indispensable food commodity. Indeed, eating fruit provides health benefits through an important source of health-related substances, among which are minerals, antioxidants and vitamins. A diet enriched in vegetables and fruits is thought to reduce risk for heart disease including heart attack and stroke (Threapleton et al., 2013), protect against certain types of cancers (Albuquerque et al., 2014), reduce the risk of obesity and type 2 diabetes (Boeing et al., 2012), as well as kidney stones and helps to decrease bone loss (Baia Lda et al., 2012).

Many fleshy fruits are major food crops of great economic importance. However, strategies for fruit production are facing two contradictory pressures operating on fruit production strategy. On the one hand, the drive to improve diets in the developing world has had a profound impact on global production of fruits and vegetables, which has risen by nearly one-third over the past decade (Hood et al., 2012). For example, China has significantly increased its production capacity for these typically high-value, high-return crops to keep up with growing domestic consumption and capitalize on export opportunities to the developed world (Huang, 2011). This strategy drives many large global buyers to demand for more standardization of the production and strong and robust supply chains worldwide. On the other hand, sustainable agriculture has progressed from a focus primarily on a low input, organic farming approach with a strong emphasis on small and local fruit farms. These farmers are exploring and testing again the existing available biodiversity for local production (Wagner, 1999). These two opposite fruit production strategies will most likely impact on the way research has to improve fruit in the future.

In agriculture, the improvement of crop species has been a fundamental human pursuit since cultivation began. For fruits, improvement is linked with the concept of 'consumer's satisfaction', often improperly named 'quality'. At the same time,

Applied Plant Genomics and Biotechnology.

fruit quality may be regarded as a very complex trait where implicit needs (essential and evident such as the improvement of pest resistance or safety) and explicit needs (declared needs conforming to user requests such as nutritional and organoleptic traits) have to be satisfied. The challenges that face modern plant breeders are to develop higher yielding, nutritious, healthy and environmentally friendly varieties that improve our quality of life without harnessing additional natural habitats to agricultural production.

In the last 50 years, the process of fruit development has been the object of many studies aimed to investigate how biochemical composition from primary to secondary metabolites is modified during growth and how genetic factors control fruit growth and maturation. The majority of these investigations have been carried out on tomato (*Solanum lycopersicum* L.), which is considered the model plant for fleshy fruit development. Tomato has long served as a model system for plant genetics, development, physiology, pathology and fruit ripening resulting in the accumulation of substantial information regarding the biology of this economically important organism. The genome of tomato has been sequenced by an international consortium of 10 countries (Sato et al., 2012); tilling mutant collections are available (Okabe et al., 2011). With the development of new genomic tools such as transcriptomics, proteomics and metabolomics, the strategies to study fruit development are changing and open up new perspectives and opportunities.

The objective of this short review is to describe the state of the art in the studies using functional genomics tools to improve the quality of tomato fruits. We propose to revisit the concept of fleshy fruit quality, and to discuss the impact of recent progress in functional genomics on crop selection for fruit quality improvement.

9.2 What is meant by fruit quality?

For consumers, the quality of fruits relies on a combination of attributes that can be related to their biochemical characteristics (composition affecting appearance such as texture, consistency, colour, smell, taste and safety), the ease of consuming the fruit (post-harvest life and convenience) and the satisfaction or pleasure generated by fruit consumption. However, from the 1950s to about 1970, the main focus of fruit breeding was to improve production rates and profitability in order to fulfil farmers' and food retail requirements. In line with these objectives, efforts have also been made in pest protection in order to minimize losses.

From 1970, the highly perishable nature of fleshy fruits became an important limiting factor for fruit distribution and post-harvest shelf life has become a major challenge for crop improvement in fruit species. Indeed, fruit ripening is a developmental process unique to fleshy fruits characterized by the activation of metabolic pathways leading to the accumulation of primary and secondary products that determine fruit quality, but also impact fruit shelf life (Bapat et al., 2010; Klee and Giovannoni, 2011; Seymour et al., 2013; Osorio et al., 2013). Ripening was shown to be controlled by a complex network of developmental and hormonal signalling but also by a variety of environmental cues. Importantly, fleshy fruits are divided into two main categories, climacteric and non-climacteric, depending on whether or not their ripening is controlled by the phytohormone ethylene. Substantial insights have now been gained into the mechanistic basis of ethylene biosynthesis, perception and signalling and the identity of master regulators of ripening such as ripening-inhibitor (RIN), non-ripening (NOR) and colorless non-ripening (CNR) genes that operate upstream of, or in concert with, the regulatory pathway mediated by this plant hormone. These results provided guides to breeders for new varieties with reduced perishability and greater transportability. Since the 1990s an amazing variety of food products is available in grocery stores. In the United States, on average 280 different fresh fruit and vegetable items were stocked year-round in the 1990s, with 310 different items in summer months. Many of these commodities travel long distance across country or even internationally, between production and consumption regions (Lohr and Hanson, 1995).

Since about 1990, environmental concerns have intensified, methods of sustainable agriculture are favoured, and organic or semi-organic production is required to produce fruit perceived to be safe and nutritious by the public. Today, consumers are becoming increasingly demanding with regard to the external appearance, taste, nutritional and organoleptic characteristics of fruits. In addition to nutritional quality, sensory quality (i.e. visual aspect, firmness and taste) is of utmost importance for fruit consumption. Although visual appearance is a critical factor driving initial consumer choice, in subsequent purchases, gustative quality becomes the most influential factor. To satisfy consumer expectations, fruit breeders are now pursuing sensory quality as one of their major breeding objectives, although the complex nature of many of the sensory traits and the lack of efficient selection criteria make it a difficult task.

To facilitate local, national and international trade and given the complexities and subtleties mentioned, there is a need for normalization and quality certification between the different actors (breeders, producer, commercial distribution and consumers). The actual scheme of quality certification, used by consumers and other actors, requires a system accrediting its conformance to the applicable documents (codes of practice, protocols, etc.). Efforts are underway to implement the accreditation of quality certification and food safety systems for global market as well as for the products of organic farming (Chyau, 2009). Independently of this task, the remaining question is how to improve the fruit quality in this competitive and complex network of actors.

9.3 Genomics-assisted breeding for improving fruit quality

The tomato plant was introduced in Europe during the fifteenth century by the Spanish, but tomato domestication as an edible vegetable started during the fourteenth century. Initial tomato breeding was performed by farmers who used open pollination strategies. Development of new cultivars happened by natural outcrossing, recombination of pre-existing genetic variation or spontaneous mutation (Bauchet and Causse, 2012) leading to the modification of a wide range of morphological and physiological traits compared to wild tomato ancestors.

During the second half of the twentieth century, breeding objectives have evolved over time with cultivar release and improvement of growing systems. Four main objectives were however maintained: adaptability to the environment, fruit yield, fruit quality and resistance to diseases and pests. (Bauchet and Causse, 2012; Bai and Lindhout, 2007; Causse et al., 2007; Foolad, 2007). Of course fruit production has always been the major focus of breeders. Most efforts were devoted to improve yield and shelf life and more recently taste and nutritional value. Recently, numerous efforts have been made to support traditional breeding to achieve tomato fruit with better quality, at least in several large projects pushed by public interest (in Solanaceae, for instance, see: SOL <http://solgenomics.net/solanaceae-project> and EUSOL <http://www.eu-sol.net/science>).

Crop improvement and efficient breeding strategies rely on the availability of genetic diversity and the heritability of the interest traits between wild and closely related species (Hajjar and Hodgkin, 2007). In cultivated tomato, genetic variability is limited, largely because of population bottlenecks due to the forms of founder events like natural and artificial selections that occurred during domestication and breeding of modern cultivars. Nowadays, huge efforts are underway worldwide to preserve more of 80,000 tomato and related wild species accessions in seed banks. The main collections are the Tomato Genetic Resource Center in California (USA, TGRC, <http://tgrc.ucdavis.edu>), the USDA collection (USA, <www.ars.usda.gov>), the World Vegetable Center in Taiwan (<www.avrdc.org>) and other collections in Europe. These resources, along with new breeding tools such as molecular markers and quantitative trait locus (QTL) analysis have provided enormous amounts of genetic variability from cultivated and wild species stored in international germplasm banks, and have allowed breeders to develop innovative strategies to efficiently incorporate new genetic variability into modern varieties. A non-exhaustive list of agronomic traits of interest available from wild tomato species is presented in Bergougnoux (2014, Table 6). Closely out-group related species such as *L. chilense*, *L. cheesmanii* and *L. hirsutum* have been used to improve drought tolerance, soluble solids and salt- and cold tolerance, respectively, by introgressing 'wild traits' in the cultivated tomato (Hobson and Grierson, 1993). More recently the same objective has been pursued with *L. pennellii* for sugar content (Fridman et al., 2000) and yield and fitness (Semel et al., 2006).

However, many agronomic traits including yield, biotic and abiotic resistance and biochemical composition result from the segregation of numerous interacting QTLs, the expression of which is under the control of environmental factors. To make the analysis of these complex traits easier, introgression lines (ILs) have been initiated years ago (Eshed and Zamir, 1995). Many attempts have been done using wild species related to *S. lycopersicum*, including *S. cheesmaniae* (Robert et al., 2001), *S. lycopersicoides* (Canady et al., 2005), *S. sitiens* (Pertuze et al., 2003); *S. peruviavum* (Yates et al., 2004); *S. chmielewskii* (Frary et al., 2003) and *S. habrochaites*, *S. neorickii* and *S. pimpinelifollium* (Fridman et al., 2004).

Many success stories have been related in the literature where yield was increased (Gur and Zamir, 2004) or biotic and abiotic stress resistance traits have been introduced in *S. lycopersicum* from wild species (Rose et al., 2005). The two emblematic stories concern the *Brix 9-2-5* QTL and the *fw2.2* QTL controlling tomato fruit size (Frary et al., 2000). *Brix 9-2-5* QTL corresponds to fruit-specific invertase that was mapped in five different tomato species (Fridman et al., 2004). The authors demonstrated that polymorphism is localized to an amino acid in close proximity to the fructosyl-binding site of the invertase crystal and that a single base change can affect both the enzyme kinetics and the sink strength of the invertase. Many studies have shown that *fw2.2* is a major fruit weight QTL in tomato and is likely due to a single gene, whose function remains unknown (Frary et al., 2000; Xu et al., 2013).

Databases resulting from multiple independent QTL experiments and integrating genotypic and phenotypic data have been created in many fruit crops (see Figure 1 in Orzaez et al., 2010). These molecular maps based on crosses between cultivated and various wild tomato species are very useful to finding candidate genomic regions involved in agronomic traits of interest. But all this information has not yet been incorporated into breeding programs. This is primarily due to the low resolution of the QTL positions, which usually expands large chromosome regions that may contain several hundred of genes, making their marker-assisted introgressions less efficient and involving many more genes apart from the one responsible for the trait.

It is a long and complex path from the detection of QTL through high density QTL map to the introduction of relevant alleles into crops by seamless marker-free genetic engineering. The first step is to transform the broad candidate region of the genome in candidate gene that can be tested through transient assays using VIGS technology. The precision of the QTL map is limited (depending on population size, heritability of the trait, number of QTLs, accuracy of the phenotyping, etc.). In most cases QTL intervals correspond to large genomic regions that cover 10–20 cM. Such large genomic regions may include hundreds to thousands of genes (Wurschum, 2012), making it virtually impossible to identify those of interest. Of course, fine mapping can be used to narrow down QTLs more accurately, i.e. to a region of less than 1 cM, which contains only tens of gene, although this step can be time consuming.

9.4 Future potential of tomato breeding using omics approaches

However, for classical plant breeding, the characterization of genes underlying traits of interest is not an absolute requirement, whereas it becomes crucial when genetic engineering is the strategy selected for crop improvement. Noteworthy, a large number of ESTs are now available in many crop species that can be used as resources for gene finding and expression studies (Labate et al., 2009; Frary et al., 2005). For tomato ESTs, several repositories are available worldwide, including the Dana-Farber Cancer Institute (DFCI) Tomato Gene Index, The SOL Genomics

Network (SGN) (Mueller et al., 2005), MiBASE (Aoki et al., 2010), PlantGDB (Lushbough et al., 2011) and TomatEST (D'Agostino et al., 2007), with more than 300,000 tomato ESTs reported in these databases. In 2003, the International Solanaceae Project (SOL) launched the tomato sequencing, which was published in 2012 (Sato et al., 2012). A high-quality tomato genome sequence has now been published (Sato et al., 2012), which offers new opportunities for breeders willing to extend the use of omics strategies in breeding scheme. The development of functional genomics tools, including next-generation sequencing, proteomics and metabolomics, is a direct consequence of recent and forthcoming publication of genome sequences, leading to new interesting perspectives in genomic research. Understanding the nature (mainly the function) of a gene underlying an agronomic trait of interest will be crucial for breeders who want to explore genetically engineered approach. Breeders and experts in functional genomics and in fruit biotechnology can work together on the production of successive waves of next-generation fruit crops to improve fruit organoleptic and nutritional quality and therefore generate wider public acceptance.

Improvement of fruit quality (metabolic content, taste, aroma, flavour, size, morphology, etc.) using GM-based approaches is however greatly limited by the difficulty to identify relevant candidate genes controlling traits that are interesting for breeders and consumers habits. Thankfully, functional genomics tools such as transcriptomics, proteomics and metabolomics constitute a trilogy in the post-genomics era to elucidate key steps in cellular events and especially in gene function (Kusano and Fukushima, 2013).

Over the last 15 years, serious developments in our ability to experimentally determine the content and function of genomes have taken place. Transcriptome analysis such as microarrays and now next-generation sequencing have produced a mature science whereby gene networks and cascades have been able to provide mechanistic insight (Usadel and Fernie, 2013; Koenig et al., 2013). The upgrade of the technology to sensitive quantitative PCR have created the opportunity to follow transcription factors and lowly expressed genes. And more recently, the next generation of sequencing named RNAseq has greatly improved the coverage of transcript profiling and provides tools for understanding plant transcriptomes at a global scale.

Metabolomics is aimed at identifying and quantifying all possible metabolites of a biological sample, thereby defining the steady-state levels of the intermediates of metabolic networks that constitute the metabolic phenotype. The science of metabolomics is being extensively used to gain insight into the biochemical composition of the tomato fruit. Indeed tomato can be considered at the same time as a fruit model for flesh fruit genomics and a product of considerable commercial importance, conferring to the tomato metabolome an important value (de Vos et al., 2011; Stewart et al., 2011). Metabolomics technology (GC-MS, LC-MS and NMR) has been used to assess metabolite changes in a diverse range of commercial cultivars of *S. lycopersicum* and wild species, as well as in various mutants and transgenic tomato plants (Le Gall et al., 2003). The use of metabolomics technologies has boosted our current understanding of important aspects of fruit development and ripening process (de Vos et al., 2011; Zanor et al., 2009; Schauer et al., 2006; Eckardt, 2008).

The research of volatile compounds for taste (Tikunov et al., 2005, 2010), of polyphenolic components with potential health-promoting anti-oxidant effects (Butelli et al., 2008; Bovy et al., 2007) and of carotenoids has been the focus of major research efforts (Fraser et al., 2007a,b). There has also been some focus on the influence of biotic or abiotic stress on the metabolome (Sanchez Perez et al., 2009). In contrast, the effects of cultivation practices on fruit metabolic content (Deborde et al., 2009) has received little attention, although their study may also provide diagnostic tools and relevant knowledge on the mechanisms of plant response to environmental modifications.

Of course, improving taste and quality, including potentially health-beneficial metabolites (e.g. antioxidants) as well as stress tolerance, requires at the same time the consideration of the biochemical composition of tomato plants, therefore metabolomics as well as transcriptomics, and eventually the integration of these complementary approaches. Indeed, integration of these two omics strategies has now been used to investigate the metabolic networks of tomato (Barone et al., 2009). The development of high-throughput data collection techniques helps to determine how and when the molecules interact with each other. Profiling large populations to define novel metabolic QTL (mQTL) have opened a new way for breeding strategies. Combining metabolomics, transcriptomics analysis and extensive phenotyping of large and genetically diverse populations with an integrated bioinformatics platform have facilitated the identification of novel mQTL and the underlying genetics of the trait of interest. In 2002, Causse et al. have found that few chromosome regions control sensory and biochemical traits. A broad profiling of tomato volatiles, which are the determinant flavour molecules in fruit, in a population of 74 *S. lycopersicum* × *S. pennellii* ILs yielded more than 100 QTL (Tieman et al., 2006). By using a combination of metabolic profiling and flux analysis alongside reverse genetic studies, the authors were able to confirm the metabolic pathway of the 2-phenylethanol and 2-phenylacetaldehyde, the two main aromatic volatiles in tomato. Similarly, GC-MS profiling approaches to analyse primary metabolites on the same *S. pennellii* ILs allowed to identify 889 mQTL governing the accumulation of 74 metabolites that were obtained from 2 independent experiments across 2 years (Schauer et al., 2006). Compounds include sugars, organic acids and amino acids as well as intermediate metabolites and vitamins. In many lines the metabolite content was increased along with yield. To integrate metabolic and phenotypic traits of the ILs, correlation network analysis was conducted; the network was visualized by graph representation in order to show the patterns of intra- and internode connections (Guimera and Nunes Amaral, 2005). The results showed that more than 50% of the metabolites were associated with a specific morphological phenotype. Supplemental introgression breeding in tomato was generated to identify important 'hubs', which are highly connected genes and metabolites in coexpression modules, (heterozygous hybrids between ILs and M82 [ILHs] and homozygous ILs). The results show that most of the 332 putative mQTL from the wild species are linked with metabolite content when they are compared to *S. lycopersicum* ILs.

9.5 Fruit modelling to establish fleshy fruit traits of interest

Understanding the effects of climate and management techniques as well as genotype on fruit quality are now the subject of intensive research efforts explored by using mathematical models as useful frameworks. Early process-based fruit models have focused on carbon relationships leading to predictions of fruit dry mass changes during growth, see Heuvelink and Bertin (1994) for tomato model. Considering size and composition of fleshy fruit evolution relies on several processes at the cellular, organ and plant levels with a complex interplay between genotype and environment, fruit modelling has mainly focused on molecular and physical processes occurring during the successive phases of fruit development (Martre et al., 2011).

Initial models were therefore rather simple taking into account only elementary biophysical processes such as carbon assimilation and allocation with a limited number of environmental signals and by assuming that fruit pericarp behaves as a single cell. Some models have dealt with nitrogen content, representing nitrogen and carbon dynamics on a similar conceptual basis of sink-driven assimilation and allocation using priority rules (Wermelinger et al., 1991). Others models have taken water and carbon fluxes into account (Lescourret et al., 2001). More recently, several modelling attempts have focused on fruit metabolism describing biochemical synthesis and degradation reactions that determine sugar (Genard et al., 2003) and citric acid accumulation (Lobit et al., 2003).

Nowadays, fruit models which quantify how a fruit responds to genetic and environmental factors are based on a dynamic mathematical simulation of the biophysical and physiological processes involved. Because the size and composition of fruit is determined through successive developmental phases, current modelling approaches consider the biochemical mechanisms and environmental conditions that occur during these successive phases, namely cell division, cell expansion and ripening phases. A phenomenological simulation model of cell division has also been developed to describe cell dynamics during the early growth of tomato fruit under non-limiting conditions (Bertin et al., 2003). But this model has limited value in analysing genotype × environment interactions. More recently, a model coupling cell division and endoreduplication was proposed, describing the phenomenological development of mitotic cycles, the transition from mitosis to endoreduplication and further endoreduplication cycles (Bertin et al., 2007).

Modelling fruit composition requires however a more mechanistic approach. This was developed for the construction of a peach fruit model named SUGAR in which the fruit is considered as a single metabolic compartment and respiration and sugar metabolisms are described in biochemical terms (Genard et al., 2003). More recently, a modified version of SUGAR-simulating developmental, growth and temperature-related variations in the relative rates of different sugar transformations has been adapted to tomato (Prudent et al., 2011) and grape (Dai et al., 2009). This model allowed identifying three factors (sucrose supply, metabolic activity and dilution) that contribute to the genetic variability of total sugar concentration in tomato fruit (Prudent et al., 2011).

An alternative possibility for modelling fruit composition is to develop detailed kinetic models of primary metabolism (glycolysis, oxidative pentose phosphate, sucrose metabolism). Despite the considerable advances made in the topological and stoichiometry analysis of metabolic pathways, the development of kinetic models of metabolic networks is hampered by the numerous parameters required to describe the enzyme kinetics rate laws, which are often difficult to determine experimentally (Rontein et al., 2002). This approach have been developed for crops such as barley (Grafahrend-Belau et al., 2009), rapeseed (Schwender et al., 2003) and sugarcane (Uys et al., 2007) and is under progress in our laboratory for tomato fruit.

So far, fruit modelling is still at its infancy, even though some models may give predictions for quality traits as a function of the environment or management practices. In general only few genotypic parameters have been taken in account. There are two main reasons for this; first of all, some genetic information on the traits of interest and underlying biochemical processes are missing in the models, and secondly models are still unable to simulate the complexity of fruit development and functioning. Modelling needs to integrate more information from genetics and functional genomics including gene function and regulation, to newly consider and strengthen the physiological assumptions and equations used in their models and reduce uncertainty related to differences in cultivar responses to environmental variations. This requires close collaborations between geneticists, physiologists and modellers.

9.6 Concluding remarks

In this short review, we have highlighted the huge effort provided by geneticists to preserve and make that genetic variability was accessible for tomato breeding programme and all key breeding methods that have been developed to efficiently incorporate new genetic variability in modern tomato varieties. In parallel we have presented the current status of omics and specially metabolomics in the biochemical characterization of tomato and the first efforts made to modelling fruit development and function. However, this potential for the improvement of varieties has only been used for a small number of traits, such as increased productivity, fruit size, shelf-live, biotic and abiotic stress tolerance. Organoleptic properties have not only been neglected as an objective, but the increase in productivity has been accompanied by a decline.

There are still a number of challenges that require attention, such as identifying genes and allelic forms associated with traits of interest and improving the precision and stability of the transferred DNA. Although a large amount of metabolomics data have already been published on *S. lycopersicum*, its application to widely divergent genetic populations of tomato has to be extended. These efforts should include assessments of the relative contribution of genotype and environment on metabolite composition, analyses of metabolite heritability and the integration of metabolite data with morphological phenotyping. The application of post-genomics tools should accelerate the selection process. Indeed the combined use of metabolomics associated with computational and statistical approaches with an emphasis on network-based inference, combined with genome sequencing, high-throughput reverse genetics extensive fruit modelling will probably considerably shorten the time required for the production of fruit elite lines. Many scientists have noticed already that high-throughput experimental validation platforms for systematic gene function elucidation in plants are needed. The other solution that is in early phase of deployment is the directed genome engineering for genome optimization (D'Halluin and Ruiter, 2013). The development of site-specific nucleases for precise genome modification has expanded the repertoire of tools for the development and optimization of traits, already including mutation breeding, molecular breeding and transgenesis. Through directed genome engineering technology, the huge amount of information provided by genomics and systems biology should now more effectively be used for the creation of plants with improved or new traits, and for the dissection of gene functions. For this reason, we strongly believe that metabolomics as well as other omics assisted by fruit modelling can be applied to fleshy fruit species breeding in a similar manner to that which has already proven successful in breeding programs in others crops (Joshi et al., 2014; Swamy and Kumar, 2013)

References

Albuquerque, R.C., Baltar, V.T., Marchioni, D.M., 2014. Breast cancer and dietary patterns: a systematic review. Nutr. Rev. 72 (1), 1–17.

Aoki, K., Yano, K., Suzuki, A., Kawamura, S., Sakurai, N., Suda, K., et al., 2010. Large-scale analysis of full-length cDNAs from the tomato (*Solanum lycopersicum*) cultivar Micro-Tom, a reference system for the Solanaceae genomics. BMC Genomics. 11, 210.

Bai, Y.L., Lindhout, P., 2007. Domestication and breeding of tomatoes: what have we gained and what can we gain in the future? Ann. Bot. (Lond). 100 (5), 1085–1094.

Baia Lda, C., Baxmann, A.C., Moreira, S.R., Holmes, R.P., Heilberg, I.P., 2012. Noncitrus alkaline fruit: a dietary alternative for the treatment of hypocitraturic stone formers. J. Endourol. 26 (9), 1221–1226.

Bapat, V.A., Trivedi, P.K., Ghosh, A., Sane, V.A., Ganapathi, T.R., Nath, P., 2010. Ripening of fleshy fruit: molecular insight and the role of ethylene. Biotechnol. Adv. 28 (1), 94–107.

Barone, A., Di Matteo, A., Carputo, D., Frusciante, L., 2009. High-throughput genomics enhances tomato breeding efficiency. Curr. Genomics. 10 (1), 1–9.

Bauchet, G., Causse, M., 2012. Genetic diversity in tomato (*Solanum lycopersicum*) and its wild relatives. In: Caliskan, M. (Ed.), Genetic Diversity in Plants. InTech, ISBN: 978-953-51-0185-7, http://dx.doi.org/10.5772/33073. Available from: <http://www.intechopen.com/books/genetic-diversity-in-plants/genetic-diversity-in-tomato-solanum-lycopersicum-and-its-wild-relatives>.

Bergougnoux, V., 2014. The history of tomato: from domestication to biopharming. Biotechnol. Adv. 32 (1), 170–189

Bertin, N., Genard, M., Fishman, S., 2003. A model for an early stage of tomato fruit development: cell multiplication and cessation of the cell proliferative activity. Ann. Bot. 92 (1), 65–72.

Bertin, N., Lecomte, A., Brunel, B., Fishman, S., Genard, M., 2007. A model describing cell polyploidization in tissues of growing fruit as related to cessation of cell proliferation. J. Exp. Bot. 58 (7), 1903–1913.

Boeing, H., Bechthold, A., Bub, A., Ellinger, S., Haller, D., Kroke, A., et al., 2012. Critical review: vegetables and fruit in the prevention of chronic diseases. Eur. J. Nutr. 51 (6), 637–663.

Bovy, A., Schijlen, E., Hall, R.D., 2007. Metabolic engineering of flavonoids in tomato (*Solanum lycopersicum*): the potential for metabolomics. Metabolomics. 3, 399–412.

Butelli, E., Titta, L., Giorgio, M., Mock, H.P., Matros, A., Peterek, S., et al., 2008. Enrichment of tomato fruit with health-promoting anthocyanins by expression of select transcription factors. Nat. Biotechnol. 26 (11), 1301–1308.

Canady, M.A., Meglic, V., Chetelat, R.T., 2005. A library of *Solanum lycopersicoides* introgression lines in cultivated tomato. Genome. 48 (4), 685–697.

Causse, M., Chaib, J., Lecomte, L., Buret, M., Hospital, F., 2007. Both additivity and epistasis control the genetic variation for fruit quality traits in tomato. Theor. Appl. Genet. 115 (3), 429–442.

Causse, M., Saliba-Colombani, V., Lecomte, L., Duffe, P., Rousselle, P., Buret, M., 2002. QTL analysis of fruit quality in fresh market tomato: a few chromosome regions control the variation of sensory and instrumental traits. J. Exp. Bot. 53 (377), 2089–2098.

Chyau, J., 2009. Casting a global safety net–a framework for food safety in the age of globalization. Food Drug Law J. 64 (2), 313–334.

D'Agostino, N., Aversano, M., Frusciante, L., Chiusano, M.L., 2007. TomatEST database: *in silico* exploitation of EST data to explore expression patterns in tomato species. Nucleic Acids Res. 35 (Database issue), D901–D905.

D'Halluin, K., Ruiter, R., 2013. Directed genome engineering for genome optimization. Int. J. Dev. Biol. 57 (6–8), 621–627.

Dai, Z.W., Vivin, P., Robert, T., Milin, S., Li, S.H., Génard, M., 2009. Model-based analysis of sugar accumulation in response to source–sink ratio and water supply in grape (*Vitis vinifera*) berries. Funct. Plant Biol. 36, 527–540.

de Vos, R.C.H., Hall, R.D., Moing, A., 2011. Metabolomics of a model fruit: tomato. In: Hall, R.D. (Ed.), Annual Plant Reviews, vol. 43. Wiley-Blackwell, Oxford, UK.

Deborde, C., Maucourt, M., Baldet, P., Bernillon, S., Biais, B., Talon, G., et al., 2009. Proton NMR quantitative profiling for quality assessment of greenhouse-grown tomato fruit. Metabolomics. 5 (2), 183–198.

Eckardt, N.A., 2008. Heritability of the tomato fruit metabolome. Plant Cell. 20 (3), 501.

Eshed, Y., Zamir, D., 1995. An introgression line population of *Lycopersicon pennellii* in the cultivated tomato enables the identification and fine mapping of yield-associated QTL. Genetics. 141, 1147–1162.

Foolad, M.R., 2007. Genome mapping and molecular breeding of tomato. Int. J. Plant Genomics. 2007, 64358.

Frary, A., Doganlar, S., Frampton, A., Fulton, T., Uhlig, J., Yates, H., et al., 2003. Fine mapping of quantitative trait loci for improved fruit characteristics from *Lycopersicon chmielewskii* chromosome 1. Genome. 46 (2), 235–243.

Frary, A., Nesbitt, T.C., Grandillo, S., Knaap, E., Cong, B., Liu, J., et al., 2000. fw2.2: a quantitative trait locus key to the evolution of tomato fruit size. Science. 289 (5476), 85–88.

Frary, A., Xu, Y., Liu, J., Mitchell, S., Tedeschi, E., Tanksley, S., 2005. Development of a set of PCR-based anchor markers encompassing the tomato genome and evaluation of their usefulness for genetics and breeding experiments. Theor. Appl. Genet. 111 (2), 291–312.

Fraser, P.D., Enfissi, E.M., Goodfellow, M., Eguchi, T., Bramley, P.M., 2007a. Metabolite profiling of plant carotenoids using the matrix-assisted laser desorption ionization time-of-flight mass spectrometry. Plant J. 49 (3), 552–564.

Fraser, P.D., Enfissi, E.M., Halket, J.M., Truesdale, M.R., Yu, D., Gerrish, C., et al., 2007b. Manipulation of phytoene levels in tomato fruit: effects on isoprenoids, plastids, and intermediary metabolism. Plant Cell. 19 (10), 3194–3211.

Fridman, E., Pleban, T., Zamir, D., 2000. A recombination hotspot delimits a wild-species quantitative trait locus for tomato sugar content to 484 bp within an invertase gene. Proc. Natl. Acad. Sci. USA. 97 (9), 4718–4723.

Fridman, E., Carrari, F., Liu, Y.S., Fernie, A.R., Zamir, D., 2004. Zooming in on a quantitative trait for tomato yield using interspecific introgressions. Science. 305 (5691), 1786–1789.

Genard, M., Lescourret, F., Gomez, L., Habib, R., 2003. Changes in fruit sugar concentrations in response to assimilate supply, metabolism and dilution: a modeling approach applied to peach fruit (*Prunus persica*). Tree Physiol. 23 (6), 373–385.

Grafahrend-Belau, E., Schreiber, F., Koschutzki, D., Junker, B.H., 2009. Flux balance analysis of barley seeds: a computational approach to study systemic properties of central metabolism. Plant Physiol. 149 (1), 585–598.

Guimera, R., Nunes Amaral, L.A., 2005. Functional cartography of complex metabolic networks. Nature. 433 (7028), 895–900.

Gur, A., Zamir, D., 2004. Unused natural variation can lift yield barriers in plant breeding. PLoS Biol. 2 (10), e245.

Hajjar, R., Hodgkin, T., 2007. The use of wild relatives in crop improvement: a survey of developments over the past 20 years. Euphytica. 156, 1–13.

Heuvelink, E., Bertin, N., 1994. Dry-matter partitioning in a tomato crop: comparison of two simulation models. J. Hortic. Sci. 69, 885–903.

Hobson, G.E., Grierson, D., 1993. Tomato. In: Seymour, G.B., Taylor, J., Tucker, G. (Eds.), *Biochemistry of Fruit Ripening*. Springer, Chapman and Hall, London, UK, pp. 405–442.

Hood, C., Martinez-Donate, A., Meinen, A., 2012. Promoting healthy food consumption: a review of state-level policies to improve access to fruits and vegetables. WMJ. 111 (6), 283–288.

Huang, P.C., 2011. China's new-age small farms and their vertical integration: agribusiness or co-ops? Mod. China. 37 (2), 107–134.

Joshi, T., Fitzpatrick, M.R., Chen, S., Liu, Y., Zhang, H., Endacott, R.Z., et al., 2014. Soybean knowledge base (SoyKB): a web resource for integration of soybean translational genomics and molecular breeding. Nucleic Acids Res. 42 (1), D1245–D1252.

Klee, H.J., Giovannoni, J.J., 2011. Genetics and control of tomato fruit ripening and quality attributes. Annu. Rev. Genet. 45, 41–59.

Koenig, D., Jimenez-Gomez, J.M., Kimura, S., Fulop, D., Chitwood, D.H., Headland, L.R., et al., 2013. Comparative transcriptomics reveals patterns of selection in domesticated and wild tomato. Proc. Natl. Acad. Sci. USA. 110 (28), E2655–E2662.

Kusano, M., Fukushima, A., 2013. Current challenges and future potential of tomato breeding using omics approaches. Breed Sci. 63 (1), 31–41.

Labate, J.A., Robertson, L.D., Wu, F., Tanksley, S.D., Baldo, A.M., 2009. EST, COSII, and arbitrary gene markers give similar estimates of nucleotide diversity in cultivated tomato (*Solanum lycopersicum* L.). Theor. Appl. Genet. 118 (5), 1005–1014.

Le Gall, G., Colquhoun, I.J., Davis, A.L., Collins, G.J., Verhoeyen, M.E., 2003. Metabolite profiling of tomato (*Lycopersicon esculentum*) using ^{1}H NMR spectroscopy as a tool to detect potential unintended effects following a genetic modification. J. Agric. Food Chem. 51 (9), 2447–2456.
Lescourret, F., Genard, M., Habib, R., Fishman, S., 2001. Variation in surface conductance to water vapor diffusion in peach fruit and its effects on fruit growth assessed by a simulation model. Tree Physiol. 21 (11), 735–741.
Lobit, P., Genard, M., Wu, B.H., Soing, P., Habib, R., 2003. Modelling citrate metabolism in fruits: responses to growth and temperature. J. Exp. Bot. 54 (392), 2489–2501.
Lohr, L., Hanson, S.D., 1995. Diversity of sources for fresh produce: implications for local markets. J. Agric. Appl. Econ. 27 (2), 510–521.
Lushbough, C.M., Jennewein, D.M., Brendel, V.P., 2011. The BioExtract server: a web-based bioinformatic workflow platform. Nucleic Acids Res. 39 (Web server issue), W528–W532.
Martre, P., Bertin, N., Salon, C., Genard, M., 2011. Modelling the size and composition of fruit, grain and seed by process-based simulation models. New Phytol. 191 (3), 601–618.
Mueller, L.A., Solow, T.H., Taylor, N., Skwarecki, B., Buels, R., Binns, J., et al., 2005. The SOL Genomics Network: a comparative resource for Solanaceae biology and beyond. Plant Physiol. 138 (3), 1310–1317.
Okabe, Y., Asamizu, E., Saito, T., Matsukura, C., Ariizumi, T., Bres, C., et al., 2011. Tomato TILLING technology: development of a reverse genetics tool for the efficient isolation of mutants from Micro-Tom mutant libraries. Plant Cell Physiol. 52 (11), 1994–2005.
Orzaez, D., Monforte, A.J., Granell, A., 2010. Using genetic variability available in the breeder's pool to engineer fruit quality. GM Crops. 1 (3), 120–127.
Osorio, S., Scossa, F., Fernie, A.R., 2013. Molecular regulation of fruit ripening. Front Plant Sci. 4, 198.
Pertuze, R.A., Ji, Y., Chetelat, R.T., 2003. Transmission and recombination of homologous *Solanum sitiens* chromosomes in tomato. Theor. Appl. Genet. 107 (8), 1391–1401.
Prudent, M., Lecomte, A., Bouchet, J.P., Bertin, N., Causse, M., Genard, M., 2011. Combining ecophysiological modelling and quantitative trait locus analysis to identify key elementary processes underlying tomato fruit sugar concentration. J. Exp. Bot. 62 (3), 907–919.
Robert, V.J.M., West, M.A.L., Inai, S., Caines, A., Amtzen, L., Smith, J.K., et al., 2001. Marker-assisted introgression of blackmold resistance QTL alleles from wild *Lycopersicon cheesmannii* to cultivated tomato (*L. esculentum*) and evaluation of QTL phenotypic effects. Mol. Breed. 8, 217–233.
Rose, L.E., Langley, C.H., Bernal, A.J., Michelmore, R.W., 2005. Natural variation in the Pto pathogen resistance gene within species of wild tomato (Lycopersicon). I. Functional analysis of Pto alleles. Genetics. 171 (1), 345–357.
Sanchez Perez, I., Culzoni, M.J., Siano, G.G., Gil Garcia, M.D., Goicoechea, H.C., Martinez Galera, M., 2009. Detection of unintended stress effects based on a metabonomic study in tomato fruits after treatment with carbofuran pesticide. Capabilities of MCR-ALS applied to LC-MS three-way data arrays. Anal. Chem. 81 (20), 8335–8346.
Sato, S., Tabata, S., Hirakawa, H., Asamizu, E., Shirasawa, K., et al., 2012. The tomato genome sequence provides insights into fleshy fruit evolution. Nature 485 (7400), 635–641.
Schauer, N., Semel, Y., Roessner, U., Gur, A., Balbo, I., Carrari, F., et al., 2006. Comprehensive metabolic profiling and phenotyping of interspecific introgression lines for tomato improvement. Nat. Biotechnol. 24 (4), 447–454.

Schwender, J., Ohlrogge, J.B., Shachar-Hill, Y., 2003. A flux model of glycolysis and the oxidative pentosephosphate pathway in developing *Brassica napus* embryos. J. Biol. Chem. 278 (32), 29442–29453.

Semel, Y., Nissenbaum, J., Menda, N., Zinder, M., Krieger, U., Issman, N., et al., 2006. Overdominant quantitative trait loci for yield and fitness in tomato. Proc. Natl. Acad. Sci. USA. 103 (35), 12981–12986.

Seymour, G.B., Ostergaard, L., Chapman, N.H., Knapp, S., Martin, C., 2013. Fruit development and ripening. Annu. Rev. Plant Biol. 64, 219–241.

Stewart, D., Shepherd, L.V.T., Hall, R.D., Fraser, P.D., 2011. Crops and tasty, nutritious food – how can metabolomics help? In: Hall, R.D. (Ed.), Annual Plant Reviews, vol. 43. Wiley-Blackwell, Oxford, UK.

Swamy, B.P., Kumar, A., 2013. Genomics-based precision breeding approaches to improve drought tolerance in rice. Biotechnol. Adv. 31 (8), 1308–1318.

Threapleton, D.E., Greenwood, D.C., Evans, C.E., Cleghorn, C.L., Nykjaer, C., Woodhead, C., et al., 2013. Dietary fibre intake and risk of cardiovascular disease: systematic review and meta-analysis. BMJ. 347, f6879.

Tieman, D.M., Zeigler, M., Schmelz, E.A., Taylor, M.G., Bliss, P., Kirst, M., et al., 2006. Identification of loci affecting flavour volatile emissions in tomato fruits. J. Exp. Bot. 57 (4), 887–896.

Tikunov, Y., Lommen, A., de Vos, C.H., Verhoeven, H.A., Bino, R.J., Hall, R.D., et al., 2005. A novel approach for nontargeted data analysis for metabolomics. Large-scale profiling of tomato fruit volatiles. Plant Physiol. 139 (3), 1125–1137.

Tikunov, Y.M., de Vos, R.C., Gonzalez Paramas, A.M., Hall, R.D., Bovy, A.G., 2010. A role for differential glycoconjugation in the emission of phenylpropanoid volatiles from tomato fruit discovered using a metabolic data fusion approach. Plant Physiol. 152 (1), 55–70.

Usadel, B., Fernie, A.R., 2013. The plant transcriptome-from integrating observations to models. Front Plant Sci. 4, 48.

Uys, L., Botha, F.C., Hofmeyr, J.H., Rohwer, J.M., 2007. Kinetic model of sucrose accumulation in maturing sugarcane culm tissue. Phytochemistry. 68 (16–18), 2375–2392.

Wagner, W.C., 1999. Sustainable agriculture: how to sustain a production system in a changing environment. Int. J. Parasitol. 29 (1), 1–5.

Wermelinger, B., Baumgärtner, J., Gutierrez, A.P., 1991. A demographic model of assimilation and allocation or carbon and nitrogen in grapevines. Ecol. Model. 53, 1–26.

Wurschum, T., 2012. Mapping QTL for agronomic traits in breeding populations. Theor. Appl. Genet. 125 (2), 201–210.

Xu, J., Xiong, W., Cao, B., Yan, T., Luo, T., Fan, T., et al., 2013. Molecular characterization and functional analysis of "fruit-weight 2.2-like" gene family in rice. Planta. 238 (4), 643–655.

Yates, H.E., Frary, A., Doganlar, S., Frampton, A., Eannetta, N.T., Uhlig, J., et al., 2004. Comparative fine mapping of fruit quality QTLs on chromosome 4 introgressions derived from two wild tomato species. Euphytica. 135 (3), 283–296.

Zanor, M.I., Rambla, J.L., Chaib, J., Steppa, A., Medina, A., Granell, A., et al., 2009. Metabolic characterization of loci affecting sensory attributes in tomato allows an assessment of the influence of the levels of primary metabolites and volatile organic contents. J. Exp. Bot. 60 (7), 2139–2154.

Rice genomics and biotechnology

10

Dawei Xue[1,3], *Hua Jiang*[2] *and Qian Qian*[1,3]
[1]College of Life and Environmental Sciences, Hangzhou Normal University, Hangzhou, China; [2]Institute of Plant Protection and Microbiology, Zhejiang Academy of Agricultural Science, Hangzhou, China;
[3]State Key Laboratory of Rice Biology, China National Rice Research Institute, Hangzhou, China

10.1 Golden age of genomics, biological engineering and paddy rice

The emergence of new technologies has led to rapid developments in rice productivity and provided a strong driving force directed towards biological research. Such a boost in technology has also led to the golden years of bioengineering and genomics research (Gartland et al., 2013). These fields benefit from the consolidated foundation laid by numerous studies on paddy rice, and various studies range from karyotype analysis to linkage group naming (Kadam and Ramiah, 1943), as well as from mutant separation, to genetic analysis, to mutant gene naming (Nagao and Takahashi, 1963). All of these studies lay the basis for the golden age of biological research on paddy rice. Full-length cDNA library construction (Rice Full-Length c et al., 2003), numerous mutant library initiative systems (Krishnan et al., 2009; Chang et al., 2012), continuous map-based cloning and reverse-genetics research (Hiei et al., 1994; Song et al., 1995; Terada et al., 2000) and publication of the entire genome sequences of japonica rice Nipponbare and indica rice 9311 (Goff et al., 2002; Yu et al., 2002) have all helped propel paddy rice into the limelight of genomic and biological research (International Rice Genome Sequencing project, 2005). Genbank searches using '*Oryza sativa*' as a species keyword yield 247,494 items in sequence. A search in PubMed using 'rice' and 'gene' as keywords yields 14,693 articles, including 14,395 papers published after 1990 (97.97%) and 12,355 articles published after 2000 (84.09%) (NCBI, 2013-11-19). Although imprecise, the reported data adequately reflect the rate at which rice research has been conducted. At present, the number of genes used in basic paddy rice research is difficult to determine from statistics alone. The estimated number exceeds 1,000 but remains considerably smaller than the number of paddy rice genes (37,544). What is certain is that these studies have involved rice growth and development, as well as all aspects of rice interaction with the environment (Jiang et al., 2007; Jiang et al., 2012). The results all contribute to a clearer understanding of paddy rice.

Research on rice genomics and advances in technology are inseparable. Prior to the publication of the rice genome sequence, related research on rice genomics,

Applied Plant Genomics and Biotechnology.

similar to many other species, was scarce and progressed slowly. With the publication of the genomic sequence of rice and related genomics research, identification of gene families, analysis of gene expression (Yang et al., 2013) and analysis of comparative genomics have become the main foci of rice genome studies. The continuous development of gene chip technology has led to the maturity of commercial chip technology. The use of different mutant gene chips to analyse the transcriptome expression of mutants under different environments and stress conditions has gradually become a vital component of rice genomics research. The rise of next-generation sequencing technology, which is known for its high throughput, has also significantly changed the face of rice genomics research, particularly in genome-wide association analysis (Han and Huang, 2013; Huang et al., 2010, 2012), rice origin analysis via resequencing (X. Huang et al., 2013), gene chip comparison and high-accuracy transcriptome analysis and small RNA analysis (Watanabe et al., 2013), among others. Under the impetus of new technology, rice genomics research has gradually transformed from impending obsolescence into an indispensable part of biological research.

As research becomes increasingly intensive, studies that rely on rice mutants are no longer limited to mutations involving single mutant genes (Li et al., 2003). Instead, studies tend to involve the same type of systematic research on mutant phenotypes (Wang and Li, 2011). Dwarf multibranches have become increasingly important in studies of this type. Dwarf multibranched mutants contain at least two plant height and tiller phenotypes; the resulting mutations are classified based on the mutant genes, which are categorized as plant height-associated hormones (Ueguchi-Tanaka et al., 2000) or tiller-associated genes (Challis et al., 2013). The rice brittle culm mutant, which is characterized by its high breakability, is an important product of this type of research (Wu et al., 2012). In addition, all special mutant types are formed from leaf-colour and floral organ mutations. Although the mutant gene that causes the same phenotype mutation is different in each case, the research material can serve as a basis for a more complete understanding of related molecular mechanisms.

The reverse-genetics approach is used in preliminary research to determine gene functions. This technique is also used in intensive studies of genes with the same function or of a family of genes with similarities or differences that are repressed in the process of evolution. Studies on the WRKY family involve the most extensive reverse genetic research in existence (Nuruzzaman et al., 2013). Members of the WRKY family are essential transcription factors that play vital roles in various biological processes (Berri et al., 2009). These types of research supplement the studies on similar phenotypic mutants. However, the safety of using transgenesis over reverse genetics remains under debate because mutants of certain rice plant phenotypes are more likely to be prioritized for use in inbreeding.

A large number of correlation studies on the epigenetics of paddy rice (Zhao and Zhou, 2012; Macovei et al., 2012) and on small RNA and proteomics (L. Liu et al., 2013) have also been reported. These studies provide a clearer understanding of rice gene regulation and may serve as a basis for the systematic understanding of rice biology.

10.2 Research on the important agronomic traits in rice biology

According to the demand for rice resulting from global population growth, the total output of paddy rice should maintain a 1.2% average annual growth rate from 2001 to 2030 to meet the needs of population growth for rice (FAOSTAT, http://faostat.fao.org/faostat) (Van Nguyen and Ferrero, 2006). Limitations in water resources and arable land have caused the rice farming areas of rice-producing countries to exhibit a declining trend. Therefore, increases in rice yield mainly rely on only the increase in per unit area yield. Modern scientific practice has proven that breeding high-yielding rice is crucial to increase total rice yields. This breakthrough achievement depends on findings regarding the specific germplasm (gene) and its effective use in rice production. The discovery and application of the half dwarfing gene (*sd1*) facilitated the first green revolution in rice production (Khush, 2001). In 1970, Chinese researchers discovered the wild abortive cytoplasmic male sterility gene and the relevant restoring gene. This discovery paved the way for the invention of the 'three-line' hybrid rice, which facilitated a second breakthrough in rice yields and contributed significantly to solving the hunger problem stemming from limited paddy rice areas in the world (Xing and Zhang, 2010). With the continuing increase in gravity of the food problem, facilitating a third breakthrough in rice yield has become the focal point of not only rice researchers but also of national governments worldwide. Super rice breeding by using a combination of ideal plant type and indica–japonica heterosis is undoubtedly a feasible option (Xue and Qian, 2007). Current research on rice molecular biology and functional genomics is tightly focused on the major needs of 'yield', and progress has been made in improving plant height, tiller, ear type, grain weight, ideal plant type and other important agronomic traits. These traits of paddy rice will continue to be the main focus of rice research.

Plant height is one of the most important traits affecting rice yield. Dwarf rice is controlled by dwarf-type genes or half-dwarf genes. The single recessive dwarf gene possessed by Aizizhan, Dee-geo-woo-gen, Aijiaonante and its derived varieties is *sd1*. This gene encodes an oxidase in the synthetic route of gibberellin (GA20–oxidase) (Sasaki et al., 2002). This gene not only restricts the height of the stalk but also increases the tillers and enhances the erection of the blade. Recent research has shown that *sd1* is significantly influenced by artificial selection in the domestication of japonica rice (Asano et al., 2011).

Rice tillering is another important agronomic trait of rice yield. Reasonable control of the occurrence of rice tillers to reduce ineffective tillers as much as possible is beneficial to rice yield (Wang and Li, 2011). The *MOC1* gene is the first gene related to tiller. This gene is a member of the GRAS transcription factor family and closely related to the LAS gene of *Arabidopsis thaliana* and the *ls* gene of tomato. It is also necessary for the growth of the axillary meristem (Li et al., 2003). The loss of *MOC1* function hinders the plant lateral bud primordium from growing normally such that the tiller cannot form. By contrast, the excessive expression of

MOC1 facilitates the formation of a large number of transgenic plant tillers. The *MOC1* gene is likely to be a key aspect in controlling rice tillers. This gene not only affects the tiller of the vegetative organ stem; its mutation also significantly reduces the number of panicle branches. By cloning the *MOC1* gene, researchers have acquired an in-depth understanding of the mechanism of rice tiller regulatory molecules. Studies in this area mark important progress in plant morphogenesis, particularly in the research on collateral formation.

Grain number per ear is a factor directly affecting plant yield. Increasing the grain number per ear is important in cultivating high-yielding varieties. Most high-yielding varieties have high grain numbers per ear. Several genes (DEP1 and DEP2) considerably affecting grain number per ear have been cloned. Map-based cloning shows the presence of a cytokinin oxidase in Gn1 encoding OsCKX2 (Ashikari et al., 2005). In addition, down-regulation of *Gn1* expression leads to the accumulation of cytokinin in the inflorescence meristem, where an increase in the quantity of glumous flowers, namely, grain number per ear, eventually increases the yield. Different genes are commonly used to increase the yield of indica varieties. The pear type is mainly decided by the form, quantity and length of the primary and secondary branches. Ear shape is generally divided into three types, namely, erect, half erect and curved. Erect panicles are an important characteristic of an ideal type of rice plant in North China. DEP1 is the key multiple-effect gene controlling rice yield. This gene has been successfully isolated from the Northeast China super rice variety Shen Nong 265 (Huang et al., 2009b). The dominant gene on the DEP1 locus is caused by gain-of-function mutation. The DEP1 gene of this mutation can promote cell division, reduce spike neck length, densify rice ears and increase branch numbers to increase the yield of paddy rice by 15%–20%. Researchers have also found that high-yielding rice varieties (large-erect and half-erect ear types) in Northeast China and in the middle and lower reaches of the Yangtze River region feature DEP1 gene mutations. The DEP1 gene has a key function in increasing rice yields in China (Huang et al., 2009b). DEP2 is another rice erect dense panicle gene located at chromosome 7. This gene encodes an unknown protein and mainly affects plant development by promoting rapid elongation of the primary and secondary branches. DEP2 mutations promote cell proliferation, resulting in disordered spike differentiation and the dense panicle phenotype (Li et al., 2010). The erect-type dense panicle gene DEP3 located in chromosome 6 was recently isolated. Compared with the wild type phenotype, the DEP3 mutant vascular bundle has a smaller size and coarser stem. DEP3 encodes a protein that contains patatin phospholipase A2 of the superfamily structural domain. Current research shows that DEP3 may have an important function in vascular bundle formation (Qiao et al., 2011).

Grain size is an important factor affecting cereal grain yield. Grain shape is an important agronomic trait directly related to yield. Clarification of the genetic and developmental mechanism of rice grain shape and its application to breeding is important in increasing the per unit area yield of rice (R. Huang et al., 2013). To date, the main genes related to cloned grain shape include GS3, GW2, GW5, GS5, GW8, etc. GS3 is the major gene affecting the quantitative trait locus (QTL) of the first clone that controls the rice grain length and weight; it is also the major gene affecting the QTL of the clone that controls the rice grain width and grain fullness (Fan et al., 2006). Results show that nearly all of the robust japonica varieties with complete GS3 protein are expressed as medium grain shapes and that the GS3 protein of the long-grained indica rice type has no function. The grain shape of rice varieties can be effectively changed through mutation and replacement of this gene (Mao et al., 2010).

GW2 is the major gene that controls grain width and weight. Song et al. (2007) found that GW2 encodes a Ring E3 ubiquitin ligase located in the cytoplasm. The substrate of GW2 anchors to the proteasome to function in degradation. GW2 also negatively regulates cell division. The deficiency of GW2 function disrupts the transfer of ubiquitin to the target protein and the substrate that should be degraded cannot be specifically identified. In addition, the division of the spikelet hull is activated, which increases the width of spikelet hull. Conversely, the grain milk filling rate is indirectly improved, and the endosperm size is increased. Finally, the width of the grain and the grain weight are increased.

GW5 encodes a 144 nuclear localization composed of amino acid, and the protein contains a nuclear localization signal and a rich arginine area. The allele gene from Asominori significantly increases the width and weight of the grain (Weng et al., 2008). The grain shape genes of the cloned GS3, GW2 and GW5 are negatively correlated with grain shape. GS5 positively regulates the quantitative characteristics of grain width, seed setting rate and thousand seed weight. Up-regulated GS5 expression may be involved in promoting cell cycle, which enhances the transverse division of rice glume shell cells, increases the width of glume shells, speeds up the growth of grain filling and endosperm, and eventually increases the seed yield, seed size and grain weight (Li et al., 2011).

Researchers have recently discovered that the GW8 gene, which can affect the quality and yield of rice, encodes the SBP type transcription factor OsSPL16. This gene also controls grain size, particle type and rice quality (Wang et al., 2012). High expression of this gene can promote cell division and increase grain size, filling speed and thousand seed weight, which will eventually increase rice yield. This study also found that most high-yielding rice varieties in China contain the GW8 gene.

Rice yield, which is determined by tiller number, number of grains per ear, grain weight, grain filling rate, plant type and other agronomic traits, is a complex quantitative trait resulting from the coordinative control of multiple genes and the environment. To increase the yield potential of high-yield varieties, rice-breeding scientists have proposed the concept of a new plant type. The new, ideal plant type is characterized by decreased but valid tiller, a large number of ear grains, a thick and solid stalk and a developed root system. Theoretical analysis has shown that under equator drought conditions, the yields of rice varieties with characteristics of ideal plant types can be increased by 25% compared with those of the current variety (Khush, 2001). In 2010, Chinese scientists isolated and cloned the main-effect quantitative trait genes of an ideal rice plant type by using rice material 'little tiller', which has ideal plant type characteristics Ideal Plant Architecture 1 (IPA1)

(Jiao et al., 2010). IPA1 encodes the transcription factor OsSPL14, which contains the SBP structural domain, and positions it in the nucleus by transcriptional activation. IPA1 contains the miR156 target site in the body; it can be regulated and controlled by transcription cutting and translation inhibition. RNA sequencing and further analysis show that IPA1 can regulate the rice tiller through DEP1, regulate rice plant height and ear length through DEP1 and transfer mutation of the IPA1 gene into the rice variety Xiushui 11 through backcross transformation. Analysis of the near-isogenic line of the backcross descendant revealed that compared with the biological parent of Xiushui 11, the strain containing the mutation IPA1 gene has typical characteristics of the ideal plant type, and yield increases more than usual in a field plot experiment. Therefore, mutations of this gene reduce the tiller number of paddy rice, increase the thickness of the stem and remarkably increase the number of grains per ear and the thousand seed weight; this variety has the characteristics of the ideal plant type, is the plant model for the improvement of rice cultivars, is a powerful tool to increase rice yields and has great potential application in rice breeding (Jiao et al., 2010). A research group led by Japanese scientists cloned this gene using the high-yield rice variety ST-12 and introduced the allelic gene to low-yield Nipponbare paddy rice; here, the rice yield increased by 40% (Miura et al., 2010).

10.3 Emergence of new technologies will extend the golden age of paddy rice

Before the rice genome sequence was published, research on rice-related gene functions was slow. In the post-genome era, however, biological research on paddy rice quickly expanded (Sasaki et al., 2005). Thus, technological progress promotes rice research. Second-generation sequencing technology has matured. Soon after entering the genome era, the individual genome era was ushered in.

The most important result brought about by high-throughput sequencing is a large number of molecular markers (Subbaiyan et al., 2012; Arai-Kichise et al., 2011; Huang et al., 2009a). More molecular markers mean more detailed maps and more convenient positioning. In the early stages of high-throughput sequencing, sequencing techniques had limited coverage because of price limitations (Huang et al., 2010). However, technological progress has dramatically reduced the price of whole genome sequencing. Related research on the whole genome level, such as correlation analysis (Huang et al., 2012) and group origin research (Huang et al., 2012), can be easily performed. The function of molecular markers remains important, but, in contrast to traditional PCR, high-throughput sequencing can simultaneously analyse tens of thousands of molecular markers (Han and Huang, 2013). In addition, comparison of different species is not limited to only some sections of the genome but extends to the scope of the whole genome (Bennetzen and Ma, 2003).

High-throughput sequencing has not only changed the research methods of genomics and population genetics but also enabled relevant mutations to be positioned and separated efficiently by next-generation sequencing (Abe et al., 2012; Kawahara et al., 2013). High-throughput sequencing is a cutting-edge tool for quantitative trait gene (QTG) cloning, which is dependent on the mutant gene map of a large-scale segregation population as the clone complex trait. Moreover, the use of high-throughput sequencing for several applications, such as crop improvement (Berkman et al., 2012) and accelerated domestication (Henry, 2012), also reflects remarkable advantages. The development of new sequencing technologies will drive molecular marker-assisted breeding or molecular design breeding in the laboratory and increased production may be expected.

If high-throughput sequencing brings innovations to research methods on rice genomes, subsequent technologies must then promote new directions with great potential in the rice bioengineering field. Zinc-finger nucleases (ZFN), transcription activator-like effector nucleases (TALEN) and clustered regulatory interspaced short palindromic repeat (CRISPR) are popular genome-editing methods used in biological engineering (Gaj et al., 2013). These three technologies can process a particular area of the genome to facilitate gene function research or gene therapy. Related reports have been made on the applications of TALEN (Li et al., 2012) and CRISPR technologies (Miao et al., 2013) in paddy rice. Current studies in the field of animals and microbes indicate that CRISPR technology may have favourable prospects (Pennisi, 2013). Research on CRISPR technology in paddy rice shows great potential (Miao et al., 2013; Jiang et al., 2013; Xie and Yang, 2013). These technologies will bring new features to rice functional genome research.

The emergence of new technologies significantly affects rice genomics and biotechnology research, thereby extending the golden age of rice genomics and biotechnology.

10.4 Rice requires further research development to remain a monocotyledon model plant

Rice yield research has long been the focus of scientific attention, but important agronomic traits related to yield determine the dominant direction of rice research. Research of rice responses under biotic and abiotic stress (Chen and Ronald, 2011; Helliwell and Yang, 2013; Mizoi and Yamaguchi-Shinozaki, 2013), plant hormones (Tong and Chu, 2012), (Ueguchi-Tanaka and Matsuoka, 2010), flowering and development (Tsuji et al., 2011; Zhang et al., 2011; Yoshida and Nagato, 2011), nutrition utilization (Makino, 2011) and other aspects has yielded significant achievements. The genome sequences of other crops, such as corn (Schnable et al., 2009) and barley (International Barley Genome Sequencing et al., 2012), have been successfully published, which means that rice biology research has gradually lost its advantage of possessing known genome sequences. New model plants, such as *Brachypodium distachyon* (Brkljacic et al., 2011) and maize (Schnable et al., 2009), have also emerged, thereby challenging the model plant status of rice. Rice can retain its status of being a monocotyledonous model in the field of plant research but further research development on this plant is required (Flavell, 2009).

Research on various aspects of rice biology has been performed using methods of forward and reverse genetics, but research dealing with complex properties, such as yield, tiller, plant height, rice quality and resistance require more attention (Hao and Lin, 2010). These complex properties can ultimately affect rice yield. Given the importance of rice yield, research on these complex properties provides a reference for further studies on these properties in other plants and helps clarify the complex control mechanism of yield, which may become a topic for more distinctive research on rice.

Plants are rooted in soil and cannot move to avoid adversity, unlike animals. Therefore, plants consume large amounts of energy to build complete stress-coping mechanisms and resist negative environmental effects. Biological stress is an important category of adversity stress. Complex interactions can arise between two different biological systems. The interaction between paddy rice and biological organisms causing threats can be a good platform for future research. *Magnaporthe oryzae*, which causes rice blast, is included in current fungal research (W. Liu et al., 2013) because its genome sequence is known (Dean et al., 2005). In-depth interaction research on this pathogen is also a large component of paddy rice biological research.

Heterosis is a popular topic in crop research, but its complexity results in many difficulties. Paddy rice is a good model for research on heterosis (Zhang, 2007). With high-throughput sequencing and new biological engineering technology, research on heterosis can yield great progress. Moreover, research of reproductive isolation in paddy rice (Ouyang and Zhang, 2013) shows certain advantages.

The Chinese government has approved the production of transgenic rice (Lu, 2010), but public concerns on the safety of transgenic rice have not ceased. Strategies for the effective evaluation of transgenic rice are limited, and a large number of transgenic studies are currently in the laboratory stage (Gaudin et al., 2013). Research on transgenic rice safety is necessary, and numerous practice explorations are also required to highlight the benefits of transgenic technology. Golden rice (Beyer, 2010) is a benchmark, but it is only the beginning.

References

Abe, A., Kosugi, S., Yoshida, K., Natsume, S., Takagi, H., Kanzaki, H., et al., 2012. Genome sequencing reveals agronomically important loci in rice using MutMap. Nat. Biotechnol. 30 (2), 174–178.

Arai-Kichise, Y., Shiwa, Y., Nagasaki, H., Ebana, K., Yoshikawa, H., Yano, M., et al., 2011. Discovery of genome-wide DNA polymorphisms in a landrace cultivar of Japonica rice by whole-genome sequencing. Plant Cell Physiol. 52 (2), 274–282.

Asano, K., Yamasaki, M., Takuno, S., Miura, K., Katagiri, S., Ito, T., et al., 2011. Artificial selection for a green revolution gene during japonica rice domestication. Proc. Natl. Acad. Sci. USA. 108 (27), 11034–11039.

Ashikari, M., Sakakibara, H., Lin, S., Yamamoto, T., Takashi, T., Nishimura, A., et al., 2005. Cytokinin oxidase regulates rice grain production. Science. 309 (5735), 741–745.

Bennetzen, J.L., Ma, J., 2003. The genetic colinearity of rice and other cereals on the basis of genomic sequence analysis. Curr. Opin. Plant Biol. 6 (2), 128–133.

Berkman, P.J., Lai, K., Lorenc, M.T., Edwards, D., 2012. Next-generation sequencing applications for wheat crop improvement. Am. J. Bot. 99 (2), 365–371.

Berri, S., Abbruscato, P., Faivre-Rampant, O., Brasileiro, A.C., Fumasoni, I., Satoh, K., et al., 2009. Characterization of WRKY co-regulatory networks in rice and Arabidopsis. BMC Plant Biol. 9, 120.

Beyer, P., 2010. Golden Rice and 'Golden' crops for human nutrition. New Biotechnol. 27 (5), 478–481.

Brkljacic, J., Grotewold, E., Scholl, R., Mockler, T., Garvin, D.F., Vain, P., et al., 2011. Brachypodium as a model for the grasses: today and the future. Plant Physiol. 157 (1), 3–13.

Challis, R.J., Hepworth, J., Mouchel, C., Waites, R., Leyser, O., 2013. A role for more axillary growth1 (MAX1) in evolutionary diversity in strigolactone signaling upstream of MAX2. Plant Physiol. 161 (4), 1885–1902.

Chang, Y., Long, T., Wu, C., 2012. Effort and contribution of T-DNA insertion mutant library for rice functional genomics research in China: review and perspective. J. Integr. Plant Biol. 54 (12), 953–966.

Chen, X., Ronald, P.C., 2011. Innate immunity in rice. Trends Plant Sci. 16 (8), 451–459.

Dean, R.A., Talbot, N.J., Ebbole, D.J., Farman, M.L., Mitchell, T.K., Orbach, M.J., et al., 2005. The genome sequence of the rice blast fungus *Magnaporthe grisea*. Nature. 434 (7036), 980–986.

Fan, C., Xing, Y., Mao, H., Lu, T., Han, B., Xu, C., et al., 2006. GS3, a major QTL for grain length and weight and minor QTL for grain width and thickness in rice, encodes a putative transmembrane protein. Theor. Appl. Genet. 112 (6), 1164–1171.

Flavell, R., 2009. Role of model plant species. Methods Mol. Biol. 513, 1–18.

Gaj, T., Gersbach, C.A., Barbas III, C.F., 2013. ZFN, TALEN, and CRISPR/Cas-based methods for genome engineering. Trends Biotechnol. 31 (7), 397–405.

Gartland, K.M., Bruschi, F., Dundar, M., Gahan, P.B., Viola Magni, M., Akbarova, Y., 2013. Progress towards the 'Golden Age' of biotechnology. Curr. Opin. Biotechnol. 24 (Suppl. 1), S6–S13.

Gaudin, A.C., Henry, A., Sparks, A.H., Slamet-Loedin, I.H., 2013. Taking transgenic rice drought screening to the field. J. Exp. Bot. 64 (1), 109–117.

Goff, S.A., Ricke, D., Lan, T.H., Presting, G., Wang, R., Dunn, M., et al., 2002. A draft sequence of the rice genome (*Oryza sativa* L. ssp. japonica). Science. 296 (5565), 92–100.

Han, B., Huang, X., 2013. Sequencing-based genome-wide association study in rice. Curr. Opin. Plant Biol. 16 (2), 133–138.

Hao, W., Lin, H.X., 2010. Toward understanding genetic mechanisms of complex traits in rice. J. Genet. Genomics. 37 (10), 653–666.

Helliwell, E.E., Yang, Y., 2013. Molecular strategies to improve rice disease resistance. Methods Mol. Biol. 956, 285–309.

Henry, R.J., 2012. Next-generation sequencing for understanding and accelerating crop domestication. Brief. Funct. Genomics. 11 (1), 51–56.

Hiei, Y., Ohta, S., Komari, T., Kumashiro, T., 1994. Efficient transformation of rice (*Oryza sativa* L.) mediated by Agrobacterium and sequence analysis of the boundaries of the T-DNA. Plant J. 6 (2), 271–282.

Huang, R., Jiang, L., Zheng, J., Wang, T., Wang, H., Huang, Y., et al., 2013. Genetic bases of rice grain shape: so many genes, so little known. Trends Plant Sci. 18 (4), 218–226.

Huang, X., Feng, Q., Qian, Q., Zhao, Q., Wang, L., Wang, A., et al., 2009a. High-throughput genotyping by whole-genome resequencing. Genome Res. 19 (6), 1068–1076.

Huang, X., Qian, Q., Liu, Z., Sun, H., He, S., Luo, D., et al., 2009b. Natural variation at the DEP1 locus enhances grain yield in rice. Nat. Genet. 41 (4), 494–497.

Huang, X., Wei, X., Sang, T., Zhao, Q., Feng, Q., Zhao, Y., et al., 2010. Genome-wide association studies of 14 agronomic traits in rice landraces. Nat. Genet. 42 (11), 961–967.

Huang, X., Kurata, N., Wei, X., Wang, Z.X., Wang, A., Zhao, Q., et al., 2012. A map of rice genome variation reveals the origin of cultivated rice. Nature. 490 (7421), 497–501.

Huang, X., Lu, T., Han, B., 2013. Resequencing rice genomes: an emerging new era of rice genomics. Trends Genet. 29 (4), 225–232.

International Barley Genome Sequencing C, Mayer, K.F., Waugh, R., Brown, J.W., Schulman, A., Langridge, P., et al., 2012. A physical, genetic and functional sequence assembly of the barley genome. Nature. 491 (7426), 711–716.

International Rice Genome Sequencing Project, 2005. The map-based sequence of the rice genome. Nature. 436 (7052), 793–800.

Jiang, H., Guo, L.B., Qian, Q., 2007. Recent progress on rice genetics in China. J. Integr. Plant Biol. 49 (6), 776–790.

Jiang, W., Zhou, H., Bi, H., Fromm, M., Yang, B., Weeks, D.P., 2013. Demonstration of CRISPR/Cas9/sgRNA-mediated targeted gene modification in Arabidopsis, tobacco, sorghum and rice. Nucleic Acids Res. 41 (20), e188.

Jiang, Y., Cai, Z., Xie, W., Long, T., Yu, H., Zhang, Q., 2012. Rice functional genomics research: progress and implications for crop genetic improvement. Biotechnol. Adv. 30 (5), 1059–1070.

Jiao, Y., Wang, Y., Xue, D., Wang, J., Yan, M., Liu, G., et al., 2010. Regulation of OsSPL14 by OsmiR156 defines ideal plant architecture in rice. Nat. Genet. 42 (6), 541–544.

Kadam, G.S., Ramiah, K., 1943. Symbolization of genes in rice. Indian J. Genet. Plant Breed. 3 (1), 7–27.

Kawahara, Y., de la Bastide, M., Hamilton, J.P., Kanamori, H., McCombie, W.R., Ouyang, S., et al., 2013. Improvement of the *Oryza sativa* Nipponbare reference genome using next generation sequence and optical map data. Rice. 6 (1), 1–10.

Khush, G.S., 2001. Green revolution: the way forward. Nat. Rev. Genet. 2 (10), 815–822.

Krishnan, A., Guiderdoni, E., An, G., Hsing, Y.I., Han, C.D., Lee, M.C., et al., 2009. Mutant resources in rice for functional genomics of the grasses. Plant Physiol. 149 (1), 165–170.

Li, F., Liu, W., Tang, J., Chen, J., Tong, H., Hu, B., et al., 2010. Rice DENSE AND ERECT PANICLE 2 is essential for determining panicle outgrowth and elongation. Cell Res. 20 (7), 838–849.

Li, T., Liu, B., Spalding, M.H., Weeks, D.P., Yang, B., 2012. High-efficiency TALEN-based gene editing produces disease-resistant rice. Nat. Biotechnol. 30 (5), 390–392.

Li, X., Qian, Q., Fu, Z., Wang, Y., Xiong, G., Zeng, D., et al., 2003. Control of tillering in rice. Nature. 422 (6932), 618–621.

Li, Y., Fan, C., Xing, Y., Jiang, Y., Luo, L., Sun, L., et al., 2011. Natural variation in GS5 plays an important role in regulating grain size and yield in rice. Nat. Genet. 43 (12), 1266–1269.

Liu, L., Waters, D.L., Rose, T.J., Bao, J., King, G.J., 2013. Phospholipids in rice: significance in grain quality and health benefits: a review. Food Chem. 139 (1–4), 1133–1145.

Liu, W., Liu, J., Ning, Y., Ding, B., Wang, X., Wang, Z., et al., 2013. Recent progress in understanding PAMP- and effector-triggered immunity against the rice blast fungus Magnaporthe *oryzae*. Mol. Plant. 6 (3), 605–620.

Lu, C., 2010. The first approved transgenic rice in China. GM Crops. 1 (3), 113–115.

Macovei, A., Gill, S.S., Tuteja, N., 2012. microRNAs as promising tools for improving stress tolerance in rice. Plant Signal Behav. 7 (10), 1296–1301.

Makino, A., 2011. Photosynthesis, grain yield, and nitrogen utilization in rice and wheat. Plant Physiol. 155 (1), 125–129.

Mao, H., Sun, S., Yao, J., Wang, C., Yu, S., Xu, C., et al., 2010. Linking differential domain functions of the GS3 protein to natural variation of grain size in rice. Proc. Natl. Acad. Sci. USA. 107 (45), 19579–19584.

Miao, J., Guo, D., Zhang, J., Huang, Q., Qin, G., Zhang, X., et al., 2013. Targeted mutagenesis in rice using CRISPR-Cas system. Cell Res. 23 (10), 1233–1236.

Miura, K., Ikeda, M., Matsubara, A., Song, X.J., Ito, M., Asano, K., et al., 2010. OsSPL14 promotes panicle branching and higher grain productivity in rice. Nat. Genet. 42 (6), 545–549.

Mizoi, J., Yamaguchi-Shinozaki, K., 2013. Molecular approaches to improve rice abiotic stress tolerance. Methods Mol. Biol. 956, 269–283.

Nagao, S., Takahashi, M., 1963. Genetical studies on rice plant. XXVII. Trial construction of twelve linkage groups of Japanese rice. J. Fac. Agric. Hokkaido Univ. 53 (1), 72–130.

Nuruzzaman, M., Sharoni, A.M., Satoh, K., Kumar, A., Leung, H., Kikuchi, S., 2013. Comparative transcriptome profiles of the WRKY gene family under control, hormone-treated, and drought conditions in near-isogenic rice lines reveal differential, tissue specific gene activation. J. Plant Physiol. 171 (1), 2–13.

Ouyang, Y., Zhang, Q., 2013. Understanding reproductive isolation based on the rice model. Annu. Rev. Plant Biol. 64, 111–135.

Pennisi, E., 2013. The CRISPR craze. Science. 341 (6148), 833–836.

Qiao, Y., Piao, R., Shi, J., Lee, S.I., Jiang, W., Kim, B.K., et al., 2011. Fine mapping and candidate gene analysis of dense and erect panicle 3, DEP3, which confers high grain yield in rice (*Oryza sativa* L.). Theor. Appl. Genet. 122 (7), 1439–1449.

Rice Full-Length cDNA C, National Institute of Agrobiological Sciences Rice Full-Length c DNAPT, Kikuchi, S., Satoh, K., Nagata, T., Kawagashira, N., et al., 2003. Collection, mapping, and annotation of over 28,000 cDNA clones from japonica rice. Science. 301 (5631), 376–379.

Sasaki, A., Ashikari, M., Ueguchi-Tanaka, M., Itoh, H., Nishimura, A., Swapan, D., et al., 2002. Green revolution: a mutant gibberellin-synthesis gene in rice. Nature. 416 (6882), 701–702.

Sasaki, T., Matsumoto, T., Antonio, B.A., Nagamura, Y., 2005. From mapping to sequencing, post-sequencing and beyond. Plant Cell Physiol. 46 (1), 3–13.

Schnable, P.S., Ware, D., Fulton, R.S., Stein, J.C., Wei, F., Pasternak, S., et al., 2009. The B73 maize genome: complexity, diversity, and dynamics. Science. 326 (5956), 1112–1115.

Song, W.Y., Wang, G.L., Chen, L.L., Kim, H.S., Pi, L.Y., Holsten, T., et al., 1995. A receptor kinase-like protein encoded by the rice disease resistance gene, Xa21. Science. 270 (5243), 1804–1806.

Song, X.J., Huang, W., Shi, M., Zhu, M.Z., Lin, H.X., 2007. A QTL for rice grain width and weight encodes a previously unknown RING-type E3 ubiquitin ligase. Nat. Genet. 39 (5), 623–630.

Subbaiyan, G.K., Waters, D.L., Katiyar, S.K., Sadananda, A.R., Vaddadi, S., Henry, R.J., 2012. Genome-wide DNA polymorphisms in elite indica rice inbreds discovered by whole-genome sequencing. Plant Biotechnol. J. 10 (6), 623–634.

Terada, R., Nakajima, M., Isshiki, M., Okagaki, R.J., Wessler, S.R., Shimamoto, K., 2000. Antisense waxy genes with highly active promoters effectively suppress waxy gene expression in transgenic rice. Plant Cell Physiol. 41 (7), 881–888.

Tong, H., Chu, C., 2012. Brassinosteroid signaling and application in rice. J. Genet. Genomics. 39 (1), 3–9.

Tsuji, H., Taoka, K., Shimamoto, K., 2011. Regulation of flowering in rice: two florigen genes, a complex gene network, and natural variation. Curr. Opin. Plant Biol. 14 (1), 45–52.

Ueguchi-Tanaka, M., Matsuoka, M., 2010. The perception of gibberellins: clues from receptor structure. Curr. Opin. Plant Biol. 13 (5), 503–508.

Ueguchi-Tanaka, M., Fujisawa, Y., Kobayashi, M., Ashikari, M., Iwasaki, Y., Kitano, H., et al., 2000. Rice dwarf mutant d1, which is defective in the alpha subunit of the heterotrimeric G protein, affects gibberellin signal transduction. Proc. Natl. Acad. Sci. USA. 97 (21), 11638–11643.

Van Nguyen, N., Ferrero, A., 2006. Meeting the challenges of global rice production. Paddy Water Environ. 4 (1), 1–9.

Wang, S., Wu, K., Yuan, Q., Liu, X., Liu, Z., Lin, X., et al., 2012. Control of grain size, shape and quality by OsSPL16 in rice. Nat. Genet. 44 (8), 950–954.

Wang, Y., Li, J., 2011. Branching in rice. Curr. Opin. Plant Biol. 14 (1), 94–99.

Watanabe, K.A., Ringler, P., Gu, L., Shen, Q.J., 2013. RNA-sequencing reveals previously unannotated protein- and microRNA-coding genes expressed in aleurone cells of rice seeds. Genomics.

Weng, J., Gu, S., Wan, X., Gao, H., Guo, T., Su, N., et al., 2008. Isolation and initial characterization of GW5, a major QTL associated with rice grain width and weight. Cell Res. 18 (12), 1199–1209.

Wu, B., Zhang, B., Dai, Y., Zhang, L., Shang-Guan, K., Peng, Y., et al., 2012. Brittle culm15 encodes a membrane-associated chitinase-like protein required for cellulose biosynthesis in rice. Plant Physiol. 159 (4), 1440–1452.

Xie, K., Yang, Y., 2013. RNA-guided genome editing in plants using a CRISPR-Cas system. Mol. Plant. 6 (6), 1975–1983.

Xing, Y., Zhang, Q., 2010. Genetic and molecular bases of rice yield. Annu. Rev. Plant Biol. 61, 421–442.

Xue, D.W., Qian, Q., 2007. Genetic basis and resources innovation of super rice breeding in China. J. Shenyang Agric. Univ. 38 (5), 667.

Yang, Y., Zhong, J., Ouyang, Y.D., Yao, J., 2013. The integrative expression and co-expression analysis of the AGO gene family in rice. Gene. 528 (2), 221–235.

Yoshida, H., Nagato, Y., 2011. Flower development in rice. J. Exp. Bot. 62 (14), 4719–4730.

Yu, J., Hu, S., Wang, J., Wong, G.K., Li, S., Liu, B., et al., 2002. A draft sequence of the rice genome (*Oryza sativa* L. ssp. indica). Science. 296 (5565), 79–92.

Zhang, Q., 2007. Strategies for developing Green Super Rice. Proc. Natl. Acad. Sci. USA. 104 (42), 16402–16409.

Zhang, D., Luo, X., Zhu, L., 2011. Cytological analysis and genetic control of rice anther development. J. Genet. Genomics. 38 (9), 379–390.

Zhao, Y., Zhou, D.X., 2012. Epigenomic modification and epigenetic regulation in rice. J. Genet. Genomics. 39 (7), 307–315.

Genome-wide DNA methylation in tomato

11

Rupert Fray[1] *and Silin Zhong*[2]
[1]Plant and Crop Sciences, University of Nottingham, UK; [2]State Key Laboratory of Agrobiotechnology, School of Life Sciences, The Chinese University of Hong Kong, Hong Kong

11.1 Introduction

DNA methylation is the biochemical process by which a methyl group is added to a DNA nucleotide base. Methylation of the fifth position of the DNA base cytosine, is a conserved epigenetic modification that is found in most plant, animal and fungal genomes (Feng et al., 2010; Law and Jacobsen, 2010). It is the best-characterized epigenetic mark and plays important roles in a range of biological processes, such as transposon silencing, genomic imprinting, X-chromosome inactivation, regulation of gene expression and maintenance of epigenetic memory. However, it should be noted that 5-methylcytosine is not the only form of DNA methylation, and many other nucleotides, or even different positions on the cytosine base could be targeted for methylation. For example, mammalian cells contain several oxidized forms of 5-methylcytosine such as 5-hydroxymethylcytosine, 5-formylcytosine and 5-carboxylcotisne generated by the enzyme Ten−eleven translocation (Tet) (Pastor et al., 2013). However, these methylated cytosines are unlikely to be present in plants, since a Tet homologue has not been found in any plant genome sequenced to date. In addition, bacterial cells also carry different methylated nucleotides such as N6-methyladenine and N4-methylcytosine, and their presence in plant and mammalian cells are still under debate (Ratel et al., 2006). In this chapter, we will focus on the 5-methylcytosine in the tomato nuclear genome. We will first review the current technologies for genome-wide DNA methylation analysis. Next, we will describe the unique methylation features of the tomato genome. At the end, we will discuss how DNA methylation variations regulate fruit development.

11.2 Methods to detect and quantify DNA methylation

11.2.1 Methylation-sensitive restriction enzymes

Bacteria have evolved a restriction-modification system to distinguish self and foreign DNA. They express DNA methyltransferases that methylate specific adenine

Applied Plant Genomics and Biotechnology.

or cytosine residues in their genome and protect the host from restriction enzyme cleavage. Methylation-sensitive restriction enzymes and their insensitive isoforms have long been utilized to detect DNA methylation in plant and animal genomes. For example, both HpaII and MspI recognize the same sequence CCGG. HpaII is unable to digest DNA when the second cytosine is methylated, while MspI is insensitive to the methylation status. By comparing their digestion pattern, the methylation level of the CG sites within the CCGG sequence context can be compared. Initially, this approach was restricted to the study of DNA methylation in given genomic regions, but in recent years, it has been coupled to microarray and next generation sequencing for genome-wide methylation analysis (Khulan et al., 2006). However, it is limited to methylcytosine sites within a given restriction enzyme-recognition sequence, and many methylation sites will be overlooked due to the lack of suitable restriction enzymes.

McrBC is a recently discovered methylation-dependent endonuclease. McrBC cuts DNA between two RmC (R = A or G) methylation sites in the form of $Pu^{m}C$ $(N_{40-3000})$ $Pu^{m}C$, while leaving the unmethylated DNA intact (Stewart and Raleigh, 1998; Sutherland et al., 1992). Hence, one could compare the methylation levels of a region of interest in different samples by using real-time PCR to quantify the unmethylated DNA that fails to be digested by McrBC. This method gained significant popularity due to its simple and inexpensive nature. McrBC digestion has also been successfully applied to genome-wide DNA methylation studies. After McrBC digestion, the unmethylated genomic DNA remains high-molecular weight, while the digested and hence methylated DNA is released as smaller fragments. These fragments can then be enriched and analysed using microarray or next generation sequencing approaches (Lippman et al., 2004). However, the major limitations for McrBC are the poor dynamic range and the lack of resolution, and it remains a semi-quantitative detection method.

11.2.2 Immuno-precipitation of methylated DNA

Methylated DNA regions could be enriched from a pool of fragmented genomic DNA by affinity purification using a monoclonal antibody that recognizes 5-methylcytosine. The enriched methylated DNA could be analysed by real-time PCR, microarray or deep sequencing (Keshet et al., 2006; Reynaud et al., 1992; Zilberman et al., 2007). An alternative approach is to affinity purify the methylated DNA using the methyl-binding domain of the naturally occurring CG methylation-binding proteins (Cross et al., 1994; Zhang et al., 2006). Since the antibody can recognize 5 mC in any sequence context, such approaches have overcome the limitation of restriction enzyme-based methods. It should be noticed that the antibody can only bind to exposed 5-mC in single-stranded DNA (Zhang et al., 2006; Zilberman et al., 2007). Hence, it is susceptible to the influence of the DNA secondary structure. In addition, all affinity-based methods rely on measuring the target 5 mC density. Therefore, a region with a few sparse 5-mC sites might not be differentiated from an unmethylated one.

11.2.3 Bisulphite sequencing

The current 'gold standard' for the analysis of DNA methylation is bisulphite sequencing. Conventional Sanger sequencing approaches could not detect DNA cytosine methylation, since methylation does not alter the basepairing capacity of cytosine. A ground breaking discovery was that sodium bisulphite could deaminate unmethylated cytosine residues to uracil in single-stranded DNA molecules, whilst methylated cytosines remain unchanged (Frommer et al., 1992). The bisulphite converted DNA can then be amplified with specific primers and sequenced by the Sanger method (Clark et al., 1994). Cytosine residues remaining in the resulting sequence represent previously methylated cytosines, while a cytosine to thymine substitution represents a bisulphite converted and hence unmethylated cytosine. Once multiple clones of the bisulphite converted DNA have been sequenced, the extent of methylation at each cytosine position can be calculated based on the number of mismatched bases (thymine) and matched bases (cytosine). Therefore, accurate data analysis relies on the premise that the DNA has been fully converted, with every unmethylated cytosine being deaminated to uracil. Since only cytosine residues on single-stranded DNA can be bisulphite converted, genomic regions less prone to denaturation are difficult to interpret as conversion may not be complete. If the conversion is incomplete, false positive results could arise, since unconverted unmethylated cytosines are interpreted as methylated ones. Besides the conversion rate, a key prerequisite for any bisulphite-based experiments is the knowledge of the reference DNA sequence, as methylation rate is determined by comparison of the sequencing result and the reference genome.

With the advance of next generation sequencing technologies, it is now possible to perform shotgun genome sequencing on bisulphite converted genomic DNA to measure the DNA methylation level at all cytosine positions in the genome. The first genome-wide DNA methylation map, which is often referred to as a 'methylome', has been generated by the bisulphite method coupled to Illumina shotgun genome sequencing (Cokus et al., 2008; Lister et al., 2008). For this method, genomic DNA has to be randomly fragmented by controlled sonication. The unmethylated C in these DNA fragment could then be bisulphite converted to U, and read by the next generation sequencing machine. The sequencing reads have to be aligned to both strands of the genome, since the Watson and Crick strands are no longer complimentary after bisulphite conversion. The methylation level of each cytosine positions could be calculated based on the number of converted and unconverted cytosines. The alignment and calculation are often performed using high performance computer servers and the software is freely available for academia research (Chen et al., 2010; Krueger and Andrews, 2011; Li and Li, 2009). Despite it is quantitative and having an unlimited dynamic range, the accuracy of the calculated methylation level depends on the sequencing depth of the cytosine position in question. Among the current genome-wide DNA methylation detection methodologies, whole genome bisulphite sequencing is the most comprehensive, providing single base resolution across the entire genome. It is also the most expensive, and the conventional wisdom is that a genome has to be sequenced 10–20 times per strand to achieve accurate results (Laird, 2010).

11.3 Distribution of 5-methylcytosine in tomato genome

11.3.1 Early studies on tomato genome DNA methylation

Studies of genome-wide DNA cytosine methylation in tomato date back over 20 years (Messeguer et al., 1991). Methylation-sensitive restriction enzyme restriction fragment length polymorphism (RFLP) analysis has shown that methylation pattern could be different among tomato cultivars, and the methylation polymorphisms were stably inherited and segregated in a Mendelian manner. High-performance liquid chromatography (HPLC) has also been used to estimate the global methylation level following the hydrolysis of genomic DNA. It was found that methylation levels varied in different tissues, and on average, 23% of the cytosines were methylated. Young tissues and protoplasts have less 5-methylcytosine (~20%) than the mature ones (~25%). Methylation level could also change in the same tissue at different developmental stages. For example, immature tomato fruits have a higher methylation level (30%) than the ripe fruits (~20%) (Teyssier et al., 2008).

11.3.1.1 Asymmetric CHH methylation in the PSY1 promoter

Since DNA replication could not copy epigenetic information such as the base modification, the newly synthesized DNA strand will initially lack cytosine methylation. If methylation exists on both strands of the DNA in the short palindromic sequence CG or CHG (where H = A, C or T), then after replication, the methylated strand acts to signal a maintenance methylase to initiate methylation at the homologous site on the newly synthesized strand. In this way, once a methylated site is established, it will be perpetuated through subsequent cell divisions – provided that the maintenance methylase is present. For this reason, methylation at sequences of the form CHH (where H represents any nucleotide other than G) was unexpected. However, early experiments indicated that methylation might exist at sites other than CG or CHG. In the case of the tomato phytoene synthase *PSY1* gene, analysis of numerous restriction digests of wild type and the insertion mutant, *yellowflesh*, by Southern analysis indicated that a single copy of the promoter region existed. However, when such analysis was carried out using the restriction enzyme *Hin*dIII, a different digestion pattern was observed that could only be interpreted as due to a subset of sites being resistant to cleavage (Fray, 1994). During ripening, increased digestion of this upstream *Hin*dIII site was observed. *Hin*dIII is sensitive to methylation within its target sequence (AAGCTT); it is now known that this *Hin*dIII-recognition site is indeed a target for CHH cytosine methylation and that the level of methylation at this site drops during fruit ripening.

11.3.1.2 Genome-wide DNA methylation in tomato fruit

The tomato fruit methylomes were generated by bisulphite genome sequencing (Zhong et al., 2013). A reference methylome was first built using pericarp tissue of the mature green tomato fruits at 39 days post anthesis (DPA). The mature green stage is a critical developmental time point when the fruit has attained its full size, embryo development is completed, the seeds have attained viability and the fruit is competent to produce and respond to the ripening hormone ethylene, yet it is several days prior to showing molecular or physiological signs of ripening initiation. The mature green fruit methylome had an average sequence depth of 23× per strand, and over 97% of the genome was covered. The average DNA methylation level determined by sequencing was 23.44%, which is close to the previously published HPLC results (Messeguer et al., 1991).

The DNA methylation sites could be classified into three sequence contexts: mCG, mCHG, mCHH (where H = A, C or T). In both plant and animal genomes, genebody methylation is primarily in the CG context, while all methylations could be found in transposons and repetitive regions. Unlike mammals for which cytosine methylation occurs mostly at CG sites and rarely at non-CG sites, CHH sites are much more common in plants. Among the 111,617,916 methylated cytosine sites found in the mature green tomato fruit genome, 72,806,375 (65%) of them are mCHH, while 19% and 16% are mCG and mCHG sites (Figure 11.1A). It should be noted each of the aforementioned 5-methylcytosine sites could be methylated to different levels. For example, the 5-methylcytosines in CG sequence context have the highest average methylation level (82.7%) compared to the mCHG (63.3%) and mCHH (25.3%) sites (Figure 11.1B). Within the CHG sequence context, the average methylation level for mCWG (W = A/T) is 81% genome-wide, while the CCG sites have a much lower methylation rate at 57%, suggesting that methylation in different sequence context could be established or maintained by different methyltransferases.

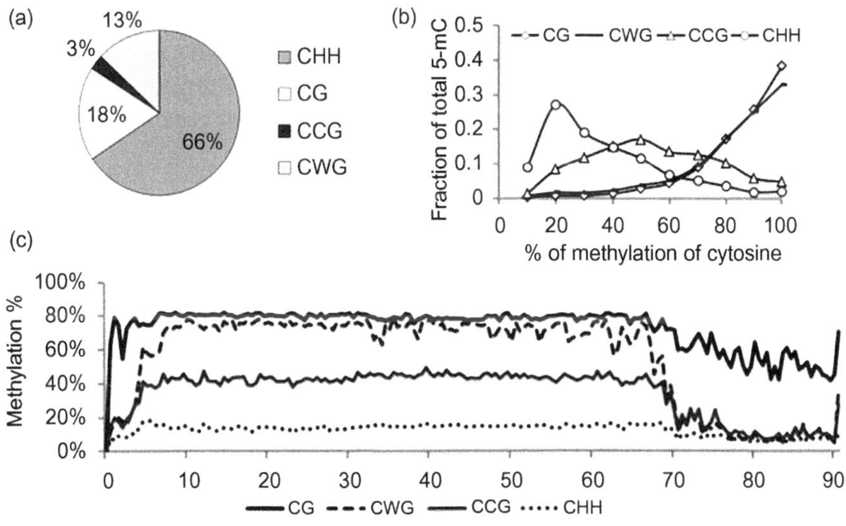

Figure 11.1 Genome-wide DNA methylation in tomato mature green fruit (a) fraction of 5-methylcytosines identified in each sequence context. (b) Distribution of the percentage methylation at each sequence context. The *y*-axis indicates the fraction of the total methylcytosines that are methylated to different levels (*x*-axis, 10 bins, from the lowest methylation at 0%–10% to the highest methylation level at 90%–100%). (c) Methylation rate of each sequence context in chromosome 1 (bin size = 500 kb).

11.3.2 DNA methylation and transposon

Almost all eukaryotic genomes contain transposons (TEs). Over half of the 900 Mb tomato genome is TEs, while it accounts for less than 15% of the small *Arabidopsis* genome (Sato et al., 2012). Since TEs are often constantly methylated, the genome-wide DNA methylation level and pattern are largely determined by TEs. Long terminal repeats (LTRs, $n = 691{,}933$) are the most abundant TE found in tomato and account for over 40% of the total genome where they tend to be concentrated in the heterochromatin. Short TEs such as the miniature inverted repeat transposable elements (MITEs) ($n = 200{,}281$), first described in maize and best characterized in the smaller rice genome, are more evenly distributed along the chromosomes and preferentially localize near the transcriptional start sites (TSSs) of genes (Jiang et al., 2003; Kuang et al., 2009; Sato et al., 2012). These promoter-located MITEs and their association with small RNA and gene expression will be discussed further in the following sections. All TEs including LTRs, MITEs and the less abundant short interspersed nuclear elements (SINEs, $n = 7{,}814$) and long interspersed nuclear elements (LINEs, $n = 1{,}409$) are heavily methylated in the tomato genome at all three sequence contexts (Figure 11.2), leading to a genome-wide methylation pattern closely matching the distribution of TEs (Figure 11.1C).

11.3.3 DNA methylation and small RNA

Plants have evolved a small RNA (sRNA)-mediated DNA methylation machinery to elicit locus-specific epigenetic silencing and it has been shown that methylation level correlates well with the sRNA abundance in *Arabidopsis* (Cokus et al., 2008; Lister et al., 2008). The tomato chromosome sRNA distribution pattern is opposite to that of the cytosine methylation and TEs, with gene-rich euchromatin having a higher sRNA density than heterochromatin (Zhong et al., 2013). For example, the average methylation rate in tomato chromosome 1 is 13.0% in euchromatin and

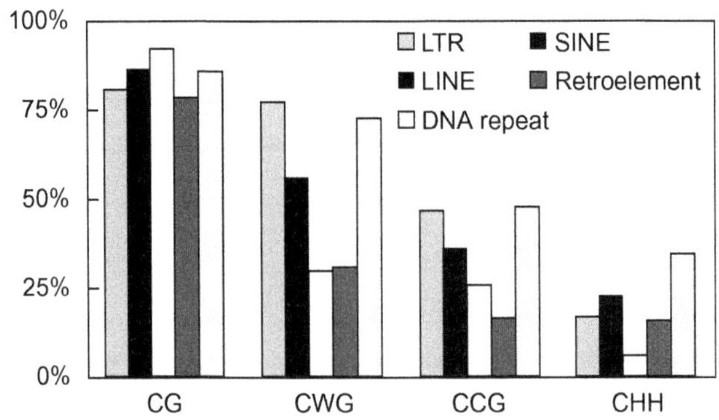

Figure 11.2 Methylation of repeats in the tomato genome. The average methylation rates in each sequence context of different types of repeats were shown. LINE, long interspersed nuclear element; LTR, long terminal repeat; SINE, short interspersed nuclear element.

25.6% in heterochromatin. In their sRNA loci, the average methylation levels are higher with 44.7% in euchromatin and 44.2% in heterochromatin. This supports the role of sRNA in RNA-direct DNA methylation in both euchromatin and heterochromatin. In addition, the higher DNA methylation level found in the sRNA-poor heterochromatin may well suggest that a sRNA-independent mechanism could be functional in the condensed heterochromatin regions with little or no transcription activity and lower sRNA density.

11.3.3.1 DNA methylation and gene expression

Epigenetic modifications such as DNA methylation are well known to regulate transcription to establish stable gene expression patterns during development (Feng et al., 2010; Law and Jacobsen, 2010; Smith and Meissner, 2013). It is often assumed that methylation represses gene expression and methylation level would negatively correlate with expression. However, recent genome-wide analyses have shown that average promoter methylation level does not always negatively associate with gene expression, instead, promoters of the intermediately expressed genes are most likely to be methylated (Zemach et al., 2010b). Similar pattern has been observed in tomato that the promoter and genebody CG sites were most likely to be methylated in the intermediately expressed genes (Figure 11.3). One distinction is

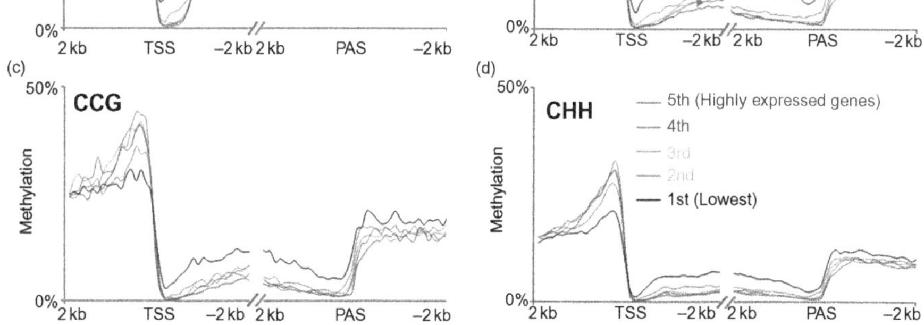

Figure 11.3 Association of DNA methylation with gene expression. Genes were classified into five groups based on their expression level in tomato fruit at the mature green stage (group 5: highest, group 1: lowest). Each group's average methylation levels in the region 2 kb upstream, genebody and the region 2 kb downstream is shown. (a) Distribution of CG methylation. (b) Distribution of CWG methylation. (c) Distribution of CCG methylation. (d) Distribution of CHH methylation. PAS, polyadenylation site; TSS, transcription start site. Bin size = 100 bp. (For interpretation of the references to colour in this figure, the reader is referred to the web version of this book.)

that near the TSSs, CG methylation level showed negative correlation with gene expression. On the other hand, both CHG and CHH methylation peaked at the 5′ end of genes and coincide with the presence of DNA repetitive region (MITEs) and 24 nt sRNA, while it has been reported that in the rice genome, only CHH methylation is associated with MITE and 24 nt sRNA (Zemach et al., 2010a).

DNA methylations in different sequence contexts or different regions of genes also have distinct association patterns with expression levels. Genes with the highest expression level have the lowest CHG and CHH methylation in the promoter (the region 2 kb upstream of the gene), but highest CHG and CHH methylation inside the genebody. This pattern of non-CG methylation corresponds well with MITE frequency and also closely resemble the 24 nt sRNA distribution pattern. It has been hypothesized that 24 nt sRNAs derived from MITEs could repress transcription in *Solaneceae* and rice (Jiang et al., 2003; Kuang et al., 2009), though a recent genome-wide survey of MITEs in rice suggests otherwise. Specifically, it has been reported that *mPINGs* were activated during plant stress and preferentially inserted into the 5′ upstream regions of genes with an apparent positive impact on downstream gene expression (Jiang et al., 2003; Kikuchi et al., 2003; Nakazaki et al., 2003). These could explain why the most actively transcribed genes in tomato have higher MITE frequency at their 5′ end.

11.3.3.2 DNA methylation and histone modification

It is well known that expressed and silenced genes are often associated with specific epigenetic marks (Cedar and Bergman, 2009). For example, promoters of active genes are often associated with H3K4me3, and H3K36me3 is a modification associated with the body of expressed genes, while Polycomb-silenced genes are associated with the repressive histone mark H3K27me3. Cluster analysis of tomato DNA methylation, sRNA and histone modifications within the 2 kb upstream and downstream regions of gene transcription start sites has shown that DNA methylation and sRNAs often co-localized, and they were depleted in the transcribed area of highly expressed genes. Inactive genes, particularity those associated with genebody H3K27me3, had higher levels of DNA methylation and sRNAs in both the 5′ and transcribed areas (Figure 11.4).

11.4 Genome-wide DNA methylation reprogramming during fruit ripening

11.4.1 Ethylene-dependent and -independent fruit ripening control

Fruits are seed dispersal vehicles for flowering plants usually derived from carpels or adjacent floral tissues. A fundamental question in biology is how flowering plants could coordinate the development of fruit and seed, and most importantly, prevent premature fruit ripening before seed development completes. The plant

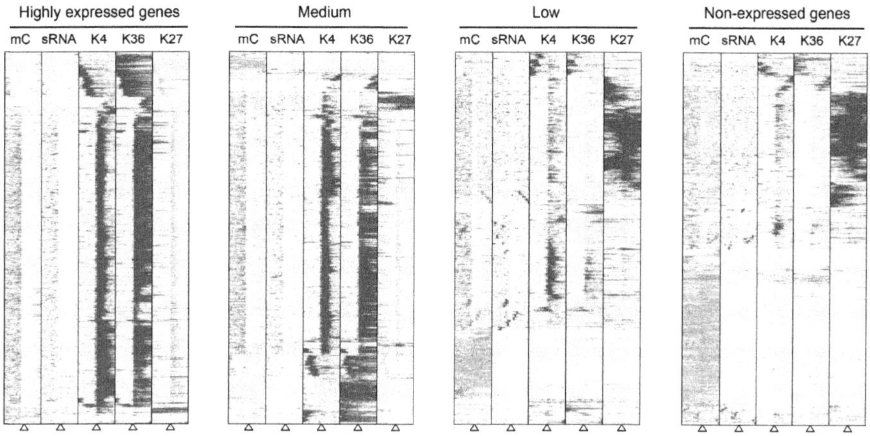

Figure 11.4 Cluster analysis of multiple epigenetics marks in gene region. Genes were first grouped by their expression level. Profiles of DNA methylation, sRNA density and histone marks in regions 2 kb upstream and downstream of each TSS were clustered. All density features were centred at the TSS indicated by the open triangle (red, high feature density; yellow, low; white, no data). K4, H3K4me3 density; K27, H3K27me3 density; K36, H3K36me3 density; mC, cytosine methylation rate; sRNA, sRNA density. (For interpretation of the references to colour in this figure legend, the reader is referred to the web version of this book.)

hormone ethylene has emerged as a master regulator of this process in climacteric fruits, because it can promote ripening in matured fruits, and inhibits ripening when its biosynthesis or signalling pathway is attenuated (Hamilton et al., 1990; Lin et al., 2009). However, the historic ripening model with ethylene as the master regulator has been challenged by the recent discovery of fruit-specific transcription factor mutants that block ripening in an ethylene-independent manner (Klee and Giovannoni, 2011). The term 'system 1 and 2 ethylene' was often used to explain the role of ethylene in both fruit ripening and other plant developmental processes (McMurchie et al., 1972). In this hypothesis, system 1 ethylene is produced by vegetative tissues and non-climacteric fruits at a basal level and is self-inhibitory, while the system 2 ethylene is produced by the climacteric fruits and is autocatalytic. Only after this transition is ethylene able to initiate ripening at the breaker stage, which is defined by dramatic changes in fruit colour, texture and aroma.

A rigid mechanism to repress system 2 ethylene and prevent premature seed release is a vital selection advantage to angiosperms. Being able to control this process is of great economical and agricultural importance. It has been shown that the recently discovered fruit transcription factors, such as RIN, NOR and CNR, are required for switching the fruit into a ripening-susceptible system 2 ethylene state precisely at the time when the seed become matured (Vrebalov et al., 2002; Manning et al., 2006). However, little is known about the exact molecular mechanism of this transition. It is tempting to hypothesize that the epigenome is involved in this stable regulatory machinery.

11.4.2 Genome-wide DNA methylation changes in tomato

DNA methylation is not static during plant development. Comparison of tomato epigenome profiles of four different fruit developmental stages revealed over 52,095 differentially methylated regions (DMRs). These DMRs are most enriched in the gene-dense euchromatin, and depleted in the transposon-dense heterochromatin (Zhong et al., 2013). This pattern is similar to the distribution of spontaneous transgenerational methylation variations found in *Arabidopsis* (Schmitz et al., 2011). However, the key difference is that the *Arabidopsis* transgenerational variations are located inside the genebody, while tomato DMR peaks at the region 5′ upstream, presumable in the promoter region.

On average, the tomato genome is losing methylation as the fruit matures (Zhong et al., 2013). It should be noted that global hypomethylation events have been observed in the other plants species. But it occurred mainly in TEs (Zemach et al., 2010a). In the tomato, loss-of-methylation tends to occur in the promoter region that is artificially defined as the region 2 kb upstream of gene (Figure 11.5). Both CG and CHG methylations decreased in this region during fruit ripening, with the ripened fruit at 52 DPA having the lowest methylation level. On the contrary, leaf tissue has the highest CG and CHG methylation, but the lowest CHH methylation. However, without the knowledge of the precise transcription factor-binding sites in these putative promoter regions, it is difficult to study the biological function of these methylation variations.

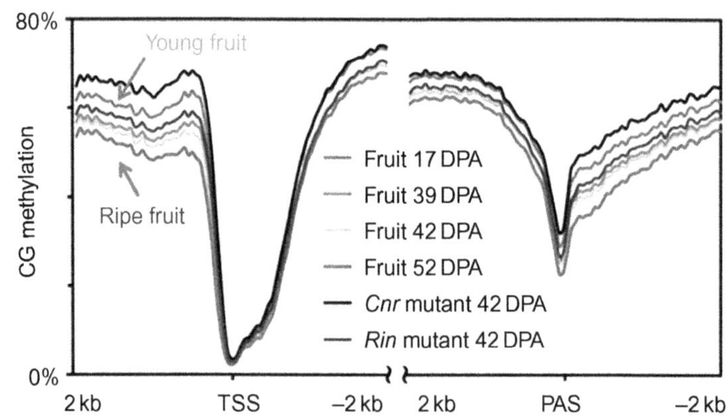

Figure 11.5 Decrease of CG methylation during fruit development ripening. The average methylation rate of tomato genes body and their 2 kb upstream and downstream regions are plotted (bin size = 200 bp). The wild-type tomato fruits from 17 to 52 DPA, as well as two transcription factor mutant (*rin* and *Cnr*) fruits at 42 DPA were shown. (For interpretation of the references to colour in this figure, the reader is referred to the web version of this book.)

11.4.3 Differential methylation associated with transcription factor-binding sites

Little is known about the role of epigenome in plant growth and development, but genome-wide DNA methylation changes have been extensively studied in mammalian models. Most recently, the human Encyclopaedia of DNA elements (ENCODE) project has systematically generated global DNA methylation profiles for a large collection of tissues and cells (Bernstein et al., 2012). They also use large-scale chromatin immuno-precipitation (ChIP-Seq) experiments to profile transcription factor binding in those samples, and found that DNA methylation level could be negatively correlated with the transcription factor binding.

Hence, for the DNA methylation variation sites found in tomato gene promoters, it is tempting to hypothesize that these changes might represent a large-scale reprogramming of transcriptional activity during fruit development, and the DMRs could harbour the transcription factor-binding sites that become accessible in ripening fruit. The MADS-box transcription factor RIPENING INHIBITOR (RIN) is one of the best-characterized regulators necessary for fruit ripening and its binding sites have been mapped by ChIP-Seq (Vrebalov et al., 2002; Zhong et al., 2013). It could directly bind to promoters of key ripening genes such as those associated with cell wall degradation, chlorophyll degradation, lycopene accumulation, ethylene and many fruit secondary metabolites. *RIN* loss-of-function fruits are unable to synthesize, nor respond to exogenous ethylene to initiate ripening, suggesting that *RIN* affects both ethylene biosynthesis and signalling. However, over-expression of RIN using 35S promoter could not force the immature fruit commit to early ripening, suggesting that additional factors are involved.

RIN could also target transcription factors that are involved in fruit ripening control, indicating that RIN is located in the centre of the transcription network in the ripening fruit. One of the most interesting finding is that RIN binds to its own promoter. This suggests that it could form a positive feedback loop that would enhance fruit ripening when initiated. This is of striking similarity to the hypothesis of system 2 ethylene that is self-autocatalytic. However, such a self-sustaining loop would represent a great danger for plants, as any leakiness of RIN expression could trigger ripening in immature fruit or non-fruit tissues, resulting in premature seed release.

One key observation from the tomato epigenome sequencing project is that the RIN-binding sites are frequently located near the DMRs in the tomato gene promoters. This suggests that DNA methylation could have been used to restrict RIN binding in immature fruits or non-fruit tissues in order to take advantage of the relatively stable epigenome as a final regulatory constraint over an irreversible developmental transition that could have severe negative consequences if deployed in the wrong place and at the wrong time.

A possible link between RIN binding and DNA methylation was suggested in a previous study that identified a natural epigenetic mutation in the tomato SQUAMOSA promoter binding protein-like (SPB)-box transcription factor gene colourless nonripening (*CNR*). *Cnr* is the only described epigenetic mutant with

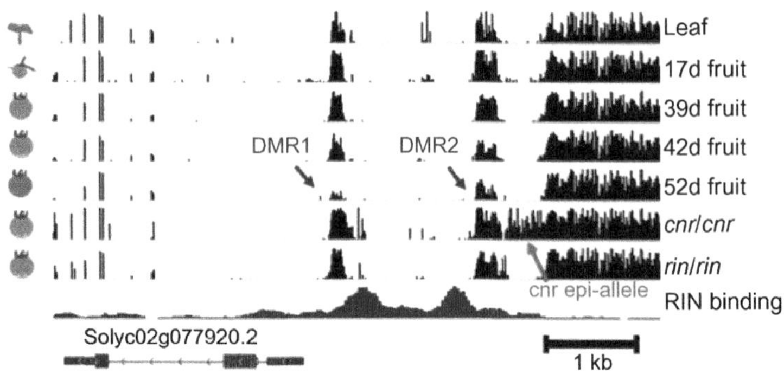

Figure 11.6 RIN-binding sites and DMRs in the *CNR* promoter. The *CNR* gene locus (Solyc02g077g20) is shown at the bottom (block: exon; line: intron; scale bar = 1 kb). Methylation level of wild-type tomato fruits at four developmental stages, as well as those of the *rin* and *Cnr* mutant fruit at 42 DPA were shown. The *Cnr* epi-allele are indicated by red arrow. The two DMRs gradually lose methylation during ripening in the *CNR* promoters, which are highlighted by blue arrows. The read density of RIN chromatin immunoprecipitation sequencing (ChIP-Seq) in the same region is shown at the bottom track. (For interpretation of the references to colour in this figure legend, the reader is referred to the web version of this book.)
Source: Figure adapted from Zhong et al., 2013.

pleiotropic defects in ripening (Manning et al., 2006). The mutant contains a hypermethylated region 2 kb upstream of the *CNR* gene that blocks its expression (Figure 11.6). RIN could bind to the *CNR* promoter in the wild-type fruit, and the binding sites are located next to the two DMRs in the *CNR* promoter. However, in the *Cnr* mutant, RIN failed to bind to the same hypermethylated sites (Martel et al., 2011; Zhong et al., 2013). Together with the observation that artificially demethylating the tomato genome could induce expression of RIN target genes and trigger premature ripening in small immature tomato fruits, it suggests that the methylation changes could be regulating whether transcription factors could interact with target gene promoter.

11.5 Conclusion

Epigenetic researches have been focused on model species such as *Arabidopsis* with relatively short life cycle and small genome, and it is often overlooked in crop species. There is a gradually growing interest in crop epigenome and using epigenetics to explain the enormous phenotypic diversity in crop species that could not be simply attributed to genetic variation. The developmental dynamics of DNA methylation during tomato fruit development clearly demonstrates that the epigenome could be a decisive factor in plant development. We suspect that such findings will soon be made in additional species besides tomato fruit ripening.

Acknowledgements

We gratefully acknowledge the State Key Laboratory of Agrobiotechnology in the Chinese University of Hong Kong for funding support. We thank Dr. Yun-Ru Chen for researching and assistance with the figures.

References

Bernstein, B.E., Birney, E., Dunham, I., Green, E.D., Gunter, C., Snyder, M., 2012. An integrated encyclopedia of DNA elements in the human genome. Nature. 489, 57–74.

Cedar, H., Bergman, Y., 2009. Linking DNA methylation and histone modification: patterns and paradigms. Nat. Rev. Genet. 10, 295–304.

Chen, P.Y., Cokus, S.J., Pellegrini, M., 2010. BS Seeker: precise mapping for bisulfite sequencing. BMC Bioinformatics. 11.

Clark, S.J., Harrison, J., Paul, C.L., Frommer, M., 1994. High-sensitivity mapping of methylated cytosines. Nucleic Acids Res. 22, 2990–2997.

Cokus, S.J., Feng, S., Zhang, X., Chen, Z., Merriman, B., Haudenschild, C.D., et al., 2008. Shotgun bisulphite sequencing of the *Arabidopsis* genome reveals DNA methylation patterning. Nature. 452, 215–219.

Cross, S.H., Charlton, J.A., Nan, X.S., Bird, A.P., 1994. Purification of Cpg Islands using a methylated DNA-binding column. Nat. Genet. 6, 236–244.

Feng, S., Jacobsen, S.E., Reik, W., 2010. Epigenetic reprogramming in plant and animal development. Science. 330, 622–627.

Fray R.G., 1994. Expression of foreign and endogenous genes in tomatoes. University of Nottingham (PhD thesis).

Frommer, M., Mcdonald, L.E., Millar, D.S., Collis, C.M., Watt, F., Grigg, G.W., et al., 1992. A genomic sequencing protocol that yields a positive display of 5-Methylcytosine residues in individual DNA strands. Proc. Natl. Acad. Sci. USA. 89, 1827–1831.

Hamilton, A.J., Lycett, G.W., Grierson, D., 1990. Antisense gene that inhibits synthesis of the hormone ethylene in transgenic plants. Nature. 346, 284–287.

Jiang, N., Bao, Z., Zhang, X., Hirochika, H., Eddy, S.R., McCouch, S.R., et al., 2003. An active DNA transposon family in rice. Nature. 421, 163–167.

Keshet, I., Schlesinger, Y., Farkash, S., Rand, E., Hecht, M., Segal, E., et al., 2006. Evidence for an instructive mechanism of de novo methylation in cancer cells. Nat. Genet. 38, 149–153.

Khulan, B., Thompson, R.F., Ye, K., Fazzari, M.J., Suzuki, M., Stasiek, E., et al., 2006. Comparative isoschizomer profiling of cytosine methylation: the HELP assay. Genome Res. 16, 1046–1055.

Kikuchi, K., Terauchi, K., Wada, M., Hirano, H.Y., 2003. The plant MITE mPing is mobilized in anther culture. Nature. 421, 167–170.

Klee, H.J., Giovannoni, J.J., 2011. Genetics and control of tomato fruit ripening and quality attributes. Ann. Rev. Genet. 45 (45), 41–59.

Krueger, F., Andrews, S.R., 2011. Bismark: a flexible aligner and methylation caller for Bisulfite-Seq applications. Bioinformatics. 27, 1571–1572.

Kuang, H., Padmanabhan, C., Li, F., Kamei, A., Bhaskar, P.B., Ouyang, S., et al., 2009. Identification of miniature inverted-repeat transposable elements (MITEs) and

biogenesis of their siRNAs in the Solanaceae: new functional implications for MITEs. Genome Res. 19, 42–56.
Laird, P.W., 2010. Principles and challenges of genome-wide DNA methylation analysis. Nat. Rev. Genet. 11, 191–203.
Law, J.A., Jacobsen, S.E., 2010. Establishing, maintaining and modifying DNA methylation patterns in plants and animals. Nat. Rev. Genet. 11, 204–220.
Li, Y.X., Li, W., 2009. BSMAP: whole genome bisulfite sequence MAPping program. BMC Bioinformatics. 10, 232.
Lin, Z.F., Zhong, S.L., Grierson, D., 2009. Recent advances in ethylene research. J. Exp. Bot. 60, 3311–3336.
Lippman, Z., Gendrel, A.V., Black, M., Vaughn, M.W., Dedhia, N., McCombie, W.R., et al., 2004. Role of transposable elements in heterochromatin and epigenetic control. Nature. 430, 471–476.
Lister, R., O'Malley, R.C., Tonti-Filippini, J., Gregory, B.D., Berry, C.C., Millar, A.H., et al., 2008. Highly integrated single-base resolution maps of the epigenome in *Arabidopsis*. Cell. 133, 523–536.
Manning, K., Tor, M., Poole, M., Hong, Y., Thompson, A.J., King, G.J., et al., 2006. A naturally occurring epigenetic mutation in a gene encoding an SBP-box transcription factor inhibits tomato fruit ripening. Nat. Genet. 38, 948–952.
Martel, C., Vrebalov, J., Tafelmeyer, P., Giovannoni, J.J., 2011. The tomato MADS-box transcription factor RIPENING INHIBITOR interacts with promoters involved in numerous ripening processes in a COLORLESS NONRIPENING-dependent manner. Plant Physiol. 157, 1568–1579.
McMurchie, E.J., McGlasson, W.B., Eaks, I.L., 1972. Treatment of fruit with propylene gives information about the biogenesis of ethylene. Nature. 237, 235–236.
Messeguer, R., Ganal, M.W., Steffens, J.C., Tanksley, S.D., 1991. Characterization of the level, target sites and inheritance of cytosine methylation in tomato nuclear DNA. Plant Mol. Biol. 16, 753–770.
Nakazaki, T., Okumoto, Y., Horibata, A., Yamahira, S., Teraishi, M., Nishida, H., et al., 2003. Mobilization of a transposon in the rice genome. Nature. 421, 170–172.
Pastor, W.A., Aravind, L., Rao, A., 2013. TETonic shift: biological roles of TET proteins in DNA demethylation and transcription. Nat. Rev. Mol. Cell Biol. 14, 341–356.
Ratel, D., Ravanat, J.L., Berger, F., Wion, D., 2006. N6-methyladenine: the other methylated base of DNA. Bioessays. 28, 309–315.
Reynaud, C., Bruno, C., Boullanger, P., Grange, J., Barbesti, S., Niveleau, A., 1992. Monitoring of urinary-excretion of modified nucleosides in cancer patients using a set of 6 monoclonal antibodies. Cancer Lett. 61, 255–262.
Sato, S., Tabata, S., Hirakawa, H., Asamizu, E., Shirasawa, K., Isobe, S., et al., 2012. The tomato genome sequence provides insights into fleshy fruit evolution. Nature. 485, 635–641.
Schmitz, R.J., Schultz, M.D., Lewsey, M.G., O'Malley, R.C., Urich, M.A., Libiger, O., et al., 2011. Transgenerational epigenetic instability is a source of novel methylation variants. Science. 334, 369–373.
Smith, Z.D., Meissner, A., 2013. DNA methylation: roles in mammalian development. Nat. Rev. Genet. 14, 204–220.
Stewart, F.J., Raleigh, E.A., 1998. Dependence of McrBC cleavage on distance between recognition elements. Biol. Chem. 379, 611–616.
Sutherland, E., Coe, L., Raleigh, E.A., 1992. Mcrbc – a multisubunit Gtp-dependent restriction endonuclease. J. Mol. Biol. 225, 327–348.

Teyssier, E., Bernacchia, G., Maury, S., Kit, A.H., Stammitti-Bert, L., Rolin, D., et al., 2008. Tissue dependent variations of DNA methylation and endoreduplication levels during tomato fruit development and ripening. Planta. 228, 391–399.

Vrebalov, J., Ruezinsky, D., Padmanabhan, V., White, R., Medrano, D., Drake, R., et al., 2002. A MADS-box gene necessary for fruit ripening at the tomato ripening-inhibitor (rin) locus. Science. 296, 343–346.

Zemach, A., Kim, M.Y., Silva, P., Rodrigues, J.A., Dotson, B., Brooks, M.D., et al., 2010a. Local DNA hypomethylation activates genes in rice endosperm. Proc. Natl. Acad. Sci. USA. 107, 18729–18734.

Zemach, A., McDaniel, I.E., Silva, P., Zilberman, D., 2010b. Genome-wide evolutionary analysis of eukaryotic DNA methylation. Science. 328, 916–919.

Zhang, X.Y., Yazaki, J., Sundaresan, A., Cokus, S., Chan, S.W.L., Chen, H.M., et al., 2006. Genome-wide high-resolution mapping and functional analysis of DNA methylation in *Arabidopsis*. Cell. 126, 1189–1201.

Zhong, S., Fei, Z., Chen, Y.R., Zheng, Y., Huang, M., Vrebalov, J., et al., 2013. Single-base resolution methylomes of tomato fruit development reveal epigenome modifications associated with ripening. Nat. Biotechnol. 31, 154–159.

Zilberman, D., Gehring, M., Tran, R.K., Ballinger, T., Henikoff, S., 2007. Genome-wide analysis of *Arabidopsis thaliana* DNA methylation uncovers an interdependence between methylation and transcription. Nature Genetics. 39, 61–69.

Recent application of biotechniques for the improvement of mango research

12

Mohammad Sorof Uddin[1] *and Qi Cheng*[2]
[1]Bangladesh Agricultural Research Institute, Gazipur, Bangladesh; [2]Biotechnology Research Institute of Chinese Academy of Agricultural Sciences, Beijing, People's Republic of China

12.1 Introduction

The mango (*Mangifera indica* L.; Anacardiaceae) is one of the most popular fruits of tropical and subtropical regions around the world, especially in Asia. Its popularity and importance can easily be realized by the fact that it is often referred to as the 'King of fruits' in the tropical world (Singh, 1996). There are many problems that affect mango production and reduce our target yield every year, including infections and insect–pest infestation (Uddin et al., 2012). Mango is a highly cross-pollinated plant and hand pollination is remarkably difficult. Most commercial cultivars have arisen from a selection of desirable types among naturally produced seedlings or as seedling selections from known mother trees (Mukherjee et al., 1968; Iyer and Degani, 1997; Karihaloo et al., 2003). Most of the superior clones are monoembryonic commercially cultivated cultivars including the 'Indian' and 'Floridians'. Propagation through sexual means does not ensure true to type plant reproduction. This leads to a long juvenile phase. Production problems are associated with both scion and rootstock. Scion cultivar problems include a biennial bearing habit, large tree size, susceptibility to major diseases and pests, short shelf life and some physiological disorders such as vegetative malformation, floral malformations and development of spongy tissue (Iyer and Degani, 1997). Mangoes are one of the most cultivated tropical fruits in the world, both for fresh and industrial consumption. At present the demand for mangoes in developed countries is rising. However, the vast majorities of mangoes are sold in local markets and consumed domestically (Santoso, 2000). Mango is renowned for combating nutritional disorders (Griesbach, 2003). Each part of the tree has a number of functions. A bark infusion can be a remedy for mouth infections in children (Bally, 2006). Furthermore, there are other items obtained from the trees themselves, e.g. gum, tannin and yellow dye (Narasimha-Char et al., 1979). The fruit has a high nutritional value and represents an important source for vitamins A and C.

Applied Plant Genomics and Biotechnology.

It is also a good source of potassium and β-carotene. Many mango hybrids resulted from hybridization are gaining ground owing to their novel characteristic features. However, there is an acute shortage of this material. Most of the important varieties are not amenable to hi-tech cultivation practices and do not meet the requirements of modern horticultural production systems. They do not have precocity in bearing, ability for dwarfing, regularity in bearing with high yield, resistance to major diseases or insects–pests and physiological disorders. Also they lack the ability to have good keeping quality. In addition, the world mango trade is narrowed to a great extent owing to the rapid perishable nature of fruits (Lizada, 1993). Mangoes are climacteric fruits and long distance transportation is sometimes a problem. Anthracnose caused by *Colletotrichum* spp. is regarded as one of the single most significant production and post-harvest problems (Dodd et al., 1998). There is a long-felt need to develop a variety combining most of the desirable horticultural traits. According to Singh (1996), an ideal mango variety should be a dwarf, regular bearer with medium size fruit (250–300 g). Additionally, it should be highly tolerant of various fungal and bacterial diseases. The fruits should have stable pleasant flavour combined with good keeping quality. The need for such an ideal phenotype cannot be met by conventional breeding. Conventional breeding of woody perennials like mango is difficult owing to their long juvenile phase, self incompatibility, low fruit set, heavy fruit drop, single seed per fruit, high degree of cross-pollination, polyembryony, polyploidy and heterozygous nature. As well as this there is only meagre information on inheritance of important qualitative traits. The application of molecular techniques in the breeding programme would expedite the development of desired cultivars. Employing biotechnology to correct genetic flaws of existing varieties could be of importance.

12.2 Origin and distribution

According to history the emperor Akbar, who reigned in northern India from 1556–1605, planted an orchard of a hundred thousand mango trees (Mukherjee, 1997). Because of the phytogeographical distribution of related species, the fossil records and the presence of plenty of wild and cultivated varieties in India, it was stated that the region of mango origin was most likely Indo-Burma. They then gradually spread to the tropical and subtropical regions of the world and most likely exported to other countries and continents (Singh, 1968; Kostermans and Bompard, 1993; Yonemori et al., 2002). Mukherjee (1949) reported that the genus *Mangifera* originated in Burma, Siam, Indo-China and the Malayan peninsula but the mango itself originated in the Assam–Burma region which includes the area now called Bangladesh. Jagarlamudi et al. (2011) reported that *M. indica* L. originated in a region that included the north-eastern part of India (Assam), the western part of Myanmar and Bangladesh (Chittagong Hill Tracts). India is thought to be the primary centre of diversity along with its status as the centre of mango origin. Presently, India harbours more than 1,000 mango cultivars and represents the biggest mango germ pool in the world. Australia, Bangladesh, People's Republic of China, the United States, Israel and Thailand are the other regions that maintain such healthy mango germ pools (Tomar et al., 2011).

12.3 Economic importance

Mango plays a major role in the global trade as it constitutes approximately 50% of all tropical fruits produced worldwide, equivalent to 5.5% of all fruits produced globally (Jedele et al., 2003). The world's total mango production has increased over the years, from about 24.4 megatonnes (MT) in 1999 (FAOSTAT, 2000) to 33.8 MT in 2008 (FAO, 2009). The major producers are Asia with about 74%, followed by Latin America and the Caribbean with 16%, Africa with 10% and Europe and Oceania with less than 1% (Galán Sauco, 2004; FAO, 2009). According to FAOSTAT (2010), India, China, Thailand, Mexico, Pakistan, Brazil, the Philippines, Indonesia, Nigeria and Vietnam are the leading mango producing countries of the world. These 10 countries cover 85% of the entire world production (Gunjate, 2009). Importation of processed mango such as canned mangoes, mango flavoured beverages and processed mango pulp has also increased in previous years (de Almeida et al., 2000). The major importer countries are USA, China, Netherlands, UAE, France, Malaysia, UK, Saudi Arabia, Germany, Singapore and major exporter countries are Mexico, the Philippines, Pakistan, Brazil, India, Netherlands, Peru, Guatemala, France, Haiti (CIA World Factbook, 2011). In 1998, the total value of mango exportation was about US$375.5 million and the total exported volume was 510000 tons, compared with a production of 23–28 MT. This implies that only a small quantity of production was exported and consequently there is a possibility to increase the export market. The main characteristics of international markets are that the price is established at the import market. Consumer profit is also an important variable that determines mango demand and it is important that consumers are given information about alternative forms of consumption (de Almeida et al., 2000).

12.4 Cytology

Present information on the cytology of the mango is quite limited. Only *Mangifera* species, i.e. *M. indica*, *M. caesia*, *M. sylvatica*, *M. odorata* *M. zeylanica* and *M. caloneura*, have been studied. It was found that the number of chromosomes was $2n = 40$ and $n = 20$ (Mukherjee, 1950; Roy and Visweswariya, 1951). The somatic chromosomes have lengths from 0.4–2.0 μ. There are eleven chromosome types that vary for at least one chromosome between different cultivars. There are a large number of somatic chromosomes with high numbers of nucleolar chromosomes resulting in a regular pairing and disjunctions of chromosomes during meiosis, as well as an absence of any multivalent formation and good fertility that can be linked to the polyploidy nature of mango (Mukherjee, 1963). The appearance of different degrees of variation indicates mangoes are polyploidy and have a hybrid nature. In production of new varieties it is important to consider natural hybridization because of high compatibility between varieties, resulting from large morphological similarity of chromosomes (Mukherjee, 1963; Bompard, 1993; Iyer and Degani, 1997).

12.5 Molecular biotechniques applied on mango

Molecular biotechnology has been applied to the mango plant for various aspects in different mango growing regions. Out of more than 1,000 known varieties of mangoes, only 350 are of commercial importance. The original wild mangoes had small fruits with little fibrous flesh and it is believed that natural hybridization occurred between *M. indica* and *M. sylvatica* in South Asia. Selection for better quality has been carried out for 4,000–6,000 years and vegetative propagation for 400 years (Morton, 1987).

12.5.1 Development of DNA extraction method

Recent development of an easy and efficient protocol for genomic DNA extraction makes obtaining DNA possible all year round. High quality genomic DNA is the first step in the development of DNA-based markers for fingerprinting and genetic diversity of crops, including mango, a woody perennial. Poor-quality genomic DNA hinders the successful application of analytical DNA-based tools. The standard protocols for DNA extraction are not suitable for mature mango leaves since the extracted genomic DNA often contains secondary metabolites that interfere with analytical applications. Recently, a simple, easy and efficient method was developed (Uddin et al., 2014). In this method 0.4 M glucose was used during extraction with an additional step to remove polyphenols, polysaccharides and secondary metabolites from genomic DNA extracted from young or mature leaf tissue, then a modified traditional cetyl trimethyl ammonium bromide (CTAB) method was applied. Glucose avoids contamination and browning by polyphenolics that improved DNA quality, compare to the traditional CTAB method. There are a few protocols for mango genomic DNA extraction mostly used for young and fresh leaves (Dellaporta et al., 1983; Doyle and Doyle, 1990; Davis et al., 1995). However, these leaf types are not always available on a mango tree. Moreover, vegetative growth occurs only 3–4 times a year, on an individual stem, depending upon the cultivar and growth conditions (Davenport and Nunez-Elisea, 1997). To conclude, by this modified protocol availability of genomic DNA is ensured year round. This new protocol was also able to extract high quality genomic DNA from other woody perennials such as walnut, guava, lychee, pear, grape and sugarcane.

12.5.2 Molecular markers for genetic diversity analysis

The assessment of genetic diversity among cultivars is potentially an important tool for a mango breeding purpose, since it can provide breeders with the means for analysing variation available in germplasm collections. Initially, traditional methods, which combine agronomic and morphological characteristics, were used to characterize the mango germplasm. However, the classification can be compounded by the influence of the environment on these vegetative characteristics which can lead to continuous variation and a high degree of plasticity. This often does not reflect their true genetic diversity. From the last two decades both genetic and DNA markers have been developed that are environmentally insensitive. Generally, these markers were used to identify dissimilarities between individuals in a population of the same species or between different species at genomic level. Each molecular marker has the advantage of being abundant, phenotypically neutral, showing absence of epistasis and not being influenced by the developmental stage or tissue of the plant or environmental conditions (Mohapatra, 2007). Many molecular markers are nowadays utilized for numerous purposes, e.g. characterization of germplasm, varietal identification and clonal fidelity testing, assessment of genetic diversity, validation of genetic relationship and marker-assisted selection (Hoogendijk and Williams, 2001). Differentiation of cultivars through morphological features is inefficient and inaccurate. This problem is further compounded by the perennial nature of the crop plant (Rahman et al., 2007). Molecular approaches offer an efficient alternative tool to conventional breeding (Krishna and Singh, 2007). The use of molecular markers, which comprise isozyme and DNA markers, can be used for cultivar identification. Another promising application could be marker-aided selection (MAS) to expedite the breeding programme. There are two types of molecular markers – isozyme markers and DNA markers – applied on mango. Each has its own advantages and disadvantages.

12.5.2.1 Isozyme marker

At first isozyme markers were used by Gan et al. (1981) to note the genetic variation in mango cultivars. Later Degani et al. (1990) worked on different enzymes, namely GPI (Glucose-6-phosphate isomerase-EC 5.3.1.9), TPI (triosephophate isomerases-EC 5.3.1.1) and LAP (leucine aminopeptidase-EC 3.4.11.1), to identify 6 loci with 17 allelomorphs in 41 mango cultivars. Jitanawongse and Changtragoon (2000) also used several enzyme systems to identify mango hybrids and true hybrids resulting from hybridization using eleven isozyme systems.

12.5.2.2 Restriction fragment length polymorphism

Restriction fragment length polymorphisms (RFLPs) were developed by Botstein et al. (1980), which uses restriction enzymes that cut the DNA molecule at specific sites, called restriction sites, resulting in different fragments of variable lengths. After separation by electrophoresis, fragments are transferred to nitrocellulose or nylon filters through southern blotting, followed by hybridization with radioactively labelled DNA probes and visualization using photographic film (Varshney et al., 2004).

A study by Eiadthong et al. (1999), on 13 *Mangifera* species, classified the species into two groups based on eight informative mutation sites detected by four endonuclease enzymes. The monomorphic group of 11 *Mangifera* species formed a cluster with *A. occidentale*. This study used a combination of two types of molecular markers, RFLP and amplified fragment length polymorphism (AFLP).

12.5.2.3 Random amplified polymorphic DNA

In random amplified polymorphic DNA (RAPD) analysis the sequence of the fragment to be amplified is unknown. Primers are drawn with random sequences of about 10 bp and the technique is used in organisms where the DNA sequence is unknown (Williams et al., 1990). RAPD analysis has the advantage of being neutrally selective, as it does not use radioisotopes, is able to use DNA of low quality and primers are more accessible than when using the RFLP technique. However, disadvantages include a limited detection of polymorphisms, a low resolution profile, which may result in low bands, and detection of only the dominating allelomorphs. It was found that RAPD, due to low annealing temperatures, are less reproducible than other techniques (Williams et al., 1990; Kapteyn and Simon, 2002).

Schnell and Knoght (1993) used nine *Mangifera* species to determine genetic relationships using RAPDs. The classification of species based on RAPD data was different compared to classification based on phenotyphic characteristics. This was the first study involving molecular markers in *Mangifera*.

Twenty-five mango accessions were analysed using RAPDs for the identification of cultivars and validation of genetic relationships in *M. indica* (Schnell et al., 1995). Eighty random decamer primers were used and 28 of these gave polymorphisms. This study included a maternal half-sib (MSH) family. RAPD data was used to create simple matching coefficients that were analysed phonetically and by means of principle co-ordinate analysis (PCA). The randomly selected accessions were scattered with no apparent pattern while the MSH clustered together in both the phenetic dendogram and the PCA.

The RAPD technique was used to investigate two species of bush mango (*Irvinga gabonensis* [Aubry-Lecomte ex O'Rorke] Baill. and *I. wombolu* Vermoesen) from central/west Africa. Significant genetic integrity was detected and no confirmation of hybridization was seen. Results of the study indicated two different species, despite morphological similarity and previous misidentification as being the same species. This study also confirmed that the RAPD technique can be applied to species with diminutive genetic diversity information available (Lowe et al., 2000). Forty genotypes from the Brazilian Research Institute (EMBRAPA) were analysed using 13 primers that produced 176 reproducible RAPD markers. Of the 176 markers, 116 were polymorphic, detecting 65.9% polymorphism. The author concluded that RAPD analysis showed efficient differences to determine genotype polymorphism in mango germplasm (de Sousa and Costa Lima, 2004).

Furthermore, RAPD analysis was used for the identification of molecular markers linked to differential flowering behaviour of mangoes in the Andaman and Nicobar Islands (Damodaran et al., 2007). The study reported that specific bands in the range 200–300 bp, amplified by the primers OPX9, OPX10, OPF4 and OPC2, were found only in multiple-flowering open pollinated clones of Neelam and Banganapali, while the same was absent in single flowering clones and varieties. RAPD markers were used by Marcela et al. (2009) for the analysis of diversity among six population of Colombian mango.

12.5.2.4 Amplified fragment length polymorphism

AFLP is a polymerase chain reaction (PCR)-based method, which is similar to RAPD analysis and can be performed on genomes of any crop and complexity. It is a universal and multi locus marker and applies PCR amplification of restriction fragments from total double-digested genomic DNA, under highly stringent conditions. AFLP analysis utilizes six- (EcoRI, PstI, HindIII) and four-base (Msel or TaqI) cutters for template preparation. Following digestion, adapters are added to the restricted DNA to create primer annealing sites. The initial PCR step uses primers with a single selective nucleotide and reduces the whole complexity of the combination up to 16-fold, allowing the target sequence to become the predominant species. Products from the first PCR are used as templates for a second amplification that uses three selective nucleotides on the 3′-end of each primer. Other enzyme systems substitute six-base for eight-base cutting enzymes, such as Sse83871 or its isochizomer SdaI or SbfI (Mohle and Schwarz, 2004). The AFLP technique results in predominant amplification of those restriction fragments that have a rare cutter sequence on the one end and a fragment cutter sequence on the other end. The basis for using two restriction enzymes is as follows:

- The frequent will produce small DNA fragments that will amplify well and are in the best size range for separation on denaturing gels (sequence gels).
- The number of fragments to be amplified is decreased by using the rare cutter and this limits the number of selective nucleotides desired for selective amplification.
- The use of two restriction enzymes makes it possible to label one strand of the double stranded PCR products that prevents the incidence of doublets on the gels due to different mobility of the two strands of the amplified fragment.
- Using two different restriction enzymes gives the most agility in tuning the number of fragments to be amplified.
- A large number of different fingerprints can be created by the diverse combinations of a small number of primers (Vos et al., 1995).

A study using 31F_1 progenies from crosses between ‘Alphonso’ and ‘Palmer’ led to the construction of maps for each cultivar that was useful for analysing correlations of traits like fruit size, shape and colour (Phumichai et al., 2000). AFLP analysis was demonstrated to be useful for identification of mango cultivars and rootstocks (Kashkush et al., 2001). The authors reported genetic relationships and diversity within *Mangifera* species with no differences between morphological and molecular data in this study. Hence AFLP analysis can be considered an applicable and effective tool in taxonomic analysis (Phumichai et al., 2000). A study was done to clarify the effectiveness of AFLP markers for the identification of accessions in four *Mangifera* species that are important in Malaysia and to explore the genetic relationship and diversity among these *Mangifera* species for the basic knowledge of *Mangifera* breeding. They concluded that AFLP is robust, useful and an appropriate tool for identifying *Mangifera* species and for detecting genetic relationships between the four species tested (Yamanaka et al., 2006).

12.5.2.5 Simple sequence repeat

Simple sequence repeats (SSRs) are widely used as a versatile tool in plant breeding programs, as well as in evolutionary studies, because of their high ability for showing diversity among cultivars (Adato et al., 1995; Mhameed et al., 1996; Levi and Rowland, 1997). SSRs are also known as micro-satellites. According to Holton (2001) SSRs are about 1–6 nucleotides and 1–4 nucleotides DNA sequences long (Chaters et al., 1996). The advantages are that they are dispersed and plentiful in all genomes with elevated levels of polymorphism compared to other molecular markers. As a disadvantage, SSR analysis is an expensive and time-consuming process especially when the creation of a library is needed. For many crops, to construct a high resolution linkage map, using only SSR markers is expensive, but it is usually more reasonable to combine SSR and AFLP analysis. Other advantages of SSR include co-dominant inheritance, analytical simplicity and its transferability (Weber, 1990; He et al., 2003).

In Thailand a study was done to identify mango cultivars and evaluate their genetic variation using SSR anchored primers. Results indicated that two Thai mango cultivars were found to be far distant in their genetic relationship from the other cultivars. Seven cultivars were in the same group as two Florida cultivars, one Philippine cultivar and one Indonesian cultivar. Four non-Thai cultivars were divided into two groups, each group enclosed Indian cultivars. The analysis did not present evident distinction between the polymorphic and monoembryonic seed races (Eiadthong et al., 1999).

Viruel et al. (2005) reported on the development of a set of 16 SSRs for mango using two genomic libraries enriched with CT repeats from DNA extracted from 'Tommy Atkins'. The analysis of 28 mango genotypes using these 16 SSRs showed three main groups using both cluster analysis and PCA indicating similar distribution of the genotypes. Cultivars were grouped according to their geographical origin and pedigree history. There are two main types of mangoes (monoembryonic and polyembryonic) which were clearly differentiated.

Honsho et al. (2004) isolated and characterized new SSRs in mango to identify 36 cultivars from different places, namely Thailand, Australia, the United States and Taiwan. An AC genomic library was created using the mango variety 'Irwin'. SSR alleles indicated high frequencies and tended to be shared by Thailand cultivars, whereas rare alleles were found in cultivars from other regions. This could have been due to the similar genetic background in 29 of the 36 cultivars from

Thailand. SSR analysis shows great potential for mango improvement and can be performed for variety identification, validation of parentages and estimation of genetic variation in existing populations and characterization of rootstocks (Brettell et al., 2002).

12.5.2.6 Inter-simple sequence repeats

Amplification of inter-simple sequence repeats (ISSRs) is a relatively novel technique and has proven to be a powerful, rapid, simple, reproducible and inexpensive way to assess genetic diversity or to identify closely related cultivars in many species including fruit trees. The ISSR marker system detects polymorphisms in inter-microsatellite DNA regions without any prior sequence knowledge (Zietkiewicz et al., 1994). Primers are based on a repeat sequence, often with a degenerate 3′ anchor, and amplify the sequence between two micro-satellites resulting in a large number of amplification products per primer being produced, providing high reproducibility at a low cost. ISSR markers have been widely used for cultivar identification in many species including banana, sorghum, *Arabidopsis*, sunflower and potato (Bornet and Branchard, 2001; Godwin et al., 1997). ISSR markers have also been used for cultivar identification, genetic diversity analysis and validation of mango genotypes. ISSRs have been applied in identification of mango varieties in different mango growing regions including Australia (Gonzalez et al., 2002), China (He et al., 2005, 2007; Luo et al., 2011) and India (Pandit et al., 2007; Gajeria et al., 2011; Tomar et al., 2011; Samal et al., 2012; Srivastava et al., 2012; Damodaran et al., 2012; Uddin et al., 2014). Amplification pattern showing ISSR-PCR were applied to different mango cultivars (Figure 12.1).

12.5.2.7 Start codon targeted primers

Start codon targeted (SCoT) primers were used for genetic diversity analysis, cultivars identification and to understand relationships among the cultivars. The SCoT primers were designed to amplify from the short conserved region surrounding the ATG translation start codon (Joshi et al., 1997; Sawant et al., 1999; Collard and Mackill, 2009; Xiong et al., 2009). In the last few years, a few researchers started using SCoT primers on mango and this proved to be a noble marker system which was able to preferentially detect polymorphisms in coding sequences (Luo et al., 2010, 2011, 2012).

12.5.2.8 Developments of CAPS markers

Cleaved amplified polymorphic sequence (CAPS) markers have been developed recently for the identification of true hybrids in F1 progeny. It could be that a shorter time will be required for confirmation of F1 hybrids. Shudo et al. (2013) applied these markers in mango for the confirmation of F1 progeny.

12.5.2.9 18S rRNA gene sequence

18S ribosomal RNA is a part of ribosomal RNA and the structural RNA for the small component of eukaryotic cytoplasomic ribosomes. It is one of the basic components of

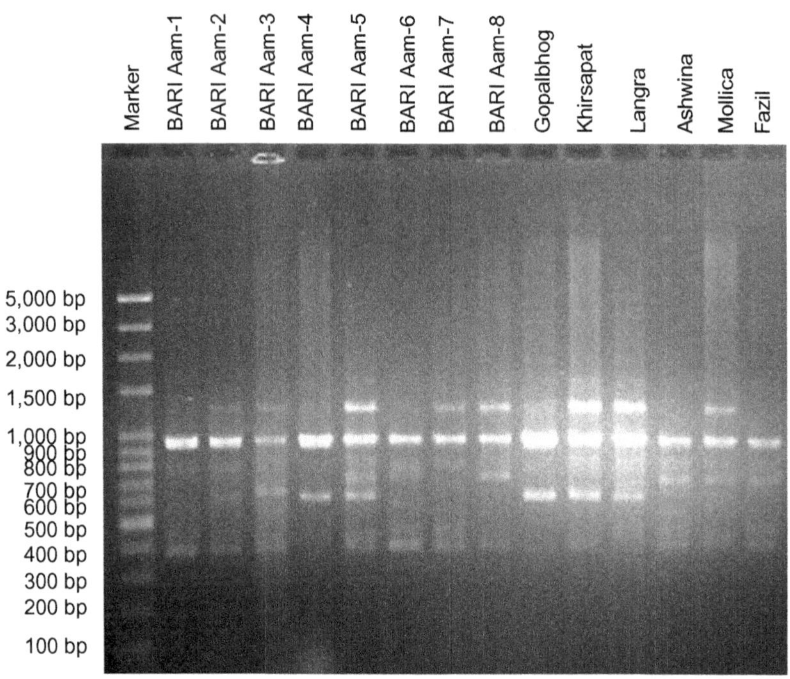

Figure 12.1 ISSR-PCR amplification in different mango varieties.

all eukaryotic cells. The genes coding for 18S rRNA are referred to as 18SrDNA. The small subunit 18SrRNA gene is one of the most frequently used genes in phylogenetic studies and is an important marker for the random target PCR in environmental biodiversity screening. Recently a phylogenetic tree was constructed using mangoes' 18SrDNA gene sequences that has been successfully analysed for mango genetic diversity and relationships. Previously phylogenetic relationships of *Mangifera* species was carried out by using Internal Transcribed Spacer (ITS) region of rDNA sequences. This showed the hybrid nature of the origin of *M. indica* (Yonemori et al., 2002; Roy and Chattopadhyay, 2011). The method characterized the Guti cultivar based on sequence information of the 18S rRNA gene and was able to construct a phylogenetic relationship with other species. Recent amplification of 18SrRNA gene sequence from different mango cultivars is shown in Figure 12.2.

12.6 Application of RAA method on mango

Nucleic acid amplification is essential to most nucleic acid testing strategies. The current techniques require sophisticated equipment or complex experimental procedures. Due to this, their uptake has been limited to the outside laboratories. A continuous electricity supply is expensive and comparatively difficult to maintain in

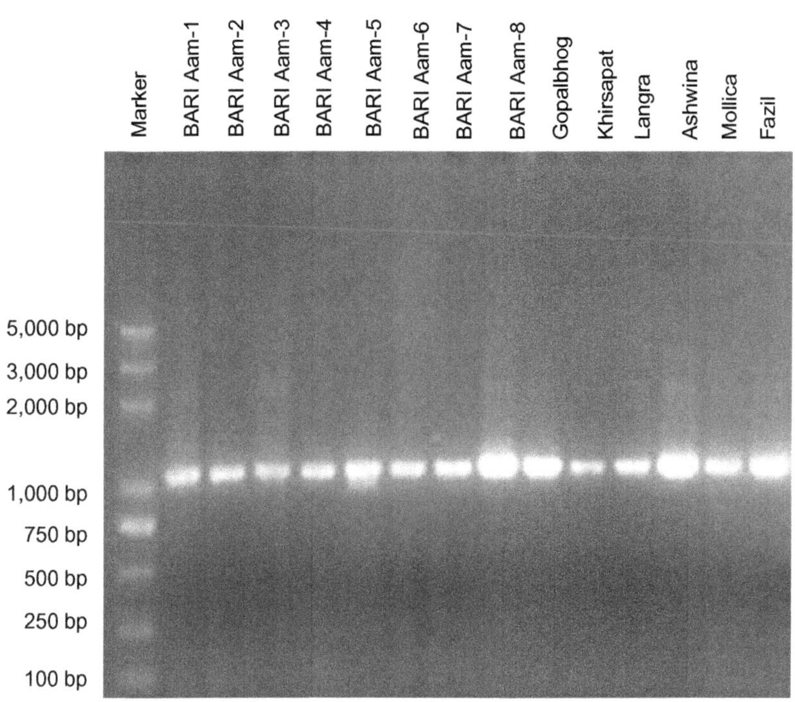

Figure 12.2 18SrRNA-PCR amplification in different mango varieties.

some regions. However, it is the pre-requisite for the operation of sophisticated equipment. Recombinase-aid amplification method (RAA) is the most recent isothermal amplification technology in which the classical thermal stable enzyme has been replaced by recombinase, a DNA-binding protein and DNA polymerase. The RAA reaction can be carried out at optimized 37°C or under room temperature without any heating assistance (Lu et al., 2010). This rapid RAA method has great application potential in tropical and subtropical mango growing regions, for field use or point of care diagnosis. It was proved that this method successfully amplified mango genomic DNA with a 20–28 bp primer length (Figure 12.3).

12.7 Problems in mango improvement using biotechnology

Micro-propagation of mango has not met with the commercial success as obtained in other fruit crops like pineapple, banana and strawberry. This is due to many problems associated with it, including latent microbial infection, excessive polyphenol exudation and early explant necrosis. Biotechnology could resolve some of the most serious problems of the mango industry. Moreover, molecular methods are

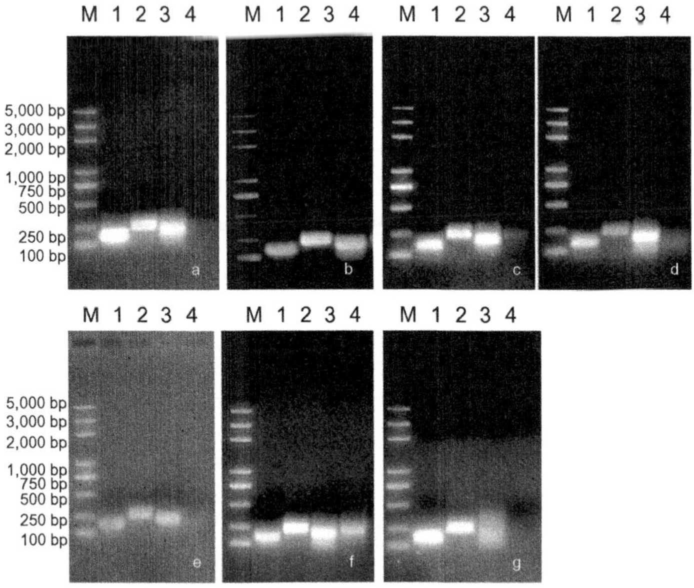

Figure 12.3 18S rRNA-RAA (Recombinase-Aid Amplification) assays applied on different fruit crops amplified with different base pairs primers. DNA marker is Trans 2K plus (100, 250, 500, 750, 1,000, 2,000, 3,000 and 5,000 bp) and a, b, c, d, e, f and g represent mango, walnut, pear, lychee, guava, grape and sugarcane, respectively, and lane 1, 2, 3 and 4 represent 28, 24, 20 and 18 base pairs primer length amplification.

useful for taxonomical characterization, as well as to understand the regulation and expression of important traits/genes, along with understanding genetic diversity analysis and relationships.

12.8 Conclusion and direction of future research

In the last few decades, biotechnological research of mango is in progress in different mango growing regions around the world. There have been a few studies carried out on micro-propagation, cloning of useful genes and genetic transformation. A few researchers are trying to develop a mango genomic resource to aid in the discovery of genes controlling consumer and grower traits, including fruit quality and tree architectures to increase the future efficiency in mango breeding. Presently, most of research is based on molecular marker-assisted breeding. Biotechnology holds several promises for mango improvement. Tissue culture techniques like anther and ovary culture can be exploited for raising homozygous lines. Likewise, genetic transformation to raise stable transformants for different characters is gradually been explored. Genetic markers are of special significance as they can aid in conventional breeding approaches in *Mangifera* spp.

Considerable success has been achieved in the development of regeneration protocols in several mango cultivars. Transformation of mango through repetitive somatic embryogenesis has also successfully been accomplished. Despite the successful regeneration of different genotypes, the conversion rate of somatic embryos into normal plantlets remains low. Future research must be focused on enhancing the conversion frequency of somatic embryos into normal plantlets, and regeneration of plantlets from shoot/nodal segments. Most of the important mango varieties which dominate the world mango trade, such as Haden, Kent, Sensation, Alphonso and BARI Aam-7, have large canopies that disqualify them from being included in the concept of high density planting. Introduction of dwarfing gene(s) from Indian cultivars such as Amrapali, Kerala and Manjeera, could be a solution to induce dwarfing in otherwise vigorous cultivars.

Another major problem confounding cultivation of many mango varieties is the occurrence of alternate bearing or irregular bearing. This problem could be alleviated by the introduction of flower-meristem-activity AGAMOUS-LIKE 20 (AGL 20), APETALA1 (AP1) and LEAFY (LFY) (Blazquez and Weigel, 2000) genes from *Arabidopsis*. AGL20 plays a pivotal role in floral evocation by integrating signals from several different pathways involving both environmental and internal cues (Borner et al., 2000). Once activated, AGL20 triggers the expression of LFY, and LFY which turns on the expression of AP1 (Simon et al., 1996). Likewise, the advances made in the field of biotechnology can also meet the challenges of abiotic and biotic stresses. Rapid strides in molecular biology and in other aspects of biotechnology have opened up new approaches in plant breeding. Biotechnology is an important and invaluable asset to the breeders which hold the greater promise to revolutionize the mango industry by development of altered varieties intended to serve the specific purpose through precise genetic manipulation, which was so far unachievable through conventional breeding. Without resorting to these new technologies mango breeding will continue to be a slow process.

References

Adato, A., Sharon, D., Lavigne, U., Hillel, I., Gazit, S., 1995. Application of DNA fingerprints for identification and genetic analysis of mango *Mangifera indica* genotypes. J. Am. Soc. Hort. Sci. 120, 259–264.

Bally, I.S.E., 2006. Species profiles for Pacific Island agroforestry. Permanent Agriculture Resources (PAR). <http://www.traditionaltree.org> (accessed 30.04.08.).

Blazquez, M.A., Weigel, D., 2000. Integration of floral inductive signals in *Arabidopsis*. Nature. 2000, 889–892.

Bompard, J.M., 1993. The genus *Mangifera* rediscovered: the potential contribution of wild species to mango cultivation. Acta Hort. 341, 69–77.

Borner, R., Kampmann, G., Chandler, J., Gleissener, R., Wisman, E., et al., 2000. A MADS domain gene involved in the transition to flowering in *Arabidopsis*. Plant J. 24, 591–599.

Bornet, B., Branchard, M., 2001. Non-anchored inter simple sequence repeat (ISSR) markers: reproducible and specific tools for genome fingerprinting. Plant Mol. Biol. Rep. 19, 209–215.

Botstein, D., White, R.L., Skolnick, M., Davis, R.W., 1980. Construction of a genetic linkage map in man using restriction fragment length polymorphism. Am. J. Human Genet. 32, 314–331.

Brettell, R., Coulson, M., Gonzalez, A., 2002. Development of DNA markers (ISSRs) in mango. Acta Hort. 575, 139–143.

Chaters, Y.M., Robertson, A., Wilkinson, M.J., Ramsay, G., 1996. PCR analysis of oilseed rape cultivars (*Brassica napus* L. sp. *oleifera*) using 5′-anchored simple sequence repeat (SSR) primers. Theor. Appl. Genet. 92, 442–447.

CIA World Factbook, 2011. http://www.mapsofworld.com/world-top-ten/mango-exporting-countries.html.

Collard, B.C.Y., Mackill, D.J., 2009. Start codon targeted (SCoT) polymorphism: a simple, novel DNA marker technique for generating gene-targeted markers in plants. Plant Mol. Biol. Rep. 27, 86–93.

Damodaran, T., Medhi, R.P., Kapil Dev, G., Damodaran, V., Rai, R.B., et al., 2007. Identification of molecular markers linked with differential flowering behavior of mangoes in Andaman and Nicobar Islands. Curr. Sci. 92, 1054–1056.

Damodaran, T., Kannan, R., Ahmed, I., Srivastava, R.C., Rai, R.B., Umamaheshwari, S., 2012. Assessing genetic relationships among mango (*Mangifera indica* L.) accessions of Andaman Islands using inter simple sequence repeat markers. N. Z. J. Crop Hort. Sci. 40, 229–240.

Davenport, T.L., Nunez-Elisea, R., 1997. Reproductive physiology. In: Litz, R.E. (Ed.), The Mango: Botany, Production and Uses. CAB International, Wallingford, UK, pp. 69–146.

Davis, T.M., Yu, H., Haigis, K.M., Mcgowan, P.J., 1995. Template mixing: a method of enhancing detection and interpretation of co dominant RAPD markers. Theor. Appl. Genet. 91, 582–588.

de Almeida, C.O., Souza, J.S., Mendes, L.N., Pereira, R.J., 2000. Aspects socioeconomics. In: Silva, J.M. (Ed.), Manga Producao. EMBRAPA, Brasilia, p. 11.

Degani, C., El-Batsri, R., Gazit, S., 1990. Enzyme polymorphism in mango. J. Am. Soc. Hort. Sci. 115, 844–847.

De Sousa, V.A., Lima, P.C., 2004. Genetic variability in mango genotypes detected by RAPD markers. Acta Hort. 645, 303–310.

Dellaporta, S.L., Wood, J., Hicks, J.B., 1983. A plant mini-preparation; version 2. Plant Mol. Biol. Rep. 1, 19–21.

Dodd, J.C., Prusky, D., Jeffris, P., 1998. Fruits diseases. In: Litz, R.E. (Ed.), The Mango: Botany Production and Uses, second ed. CAB International, Wallingford, UK, pp. 257–280.

Doyle, J.J., Doyle, J.L., 1990. Isolation of plant DNA from fresh tissue. Focus. 12, 13–15.

Eiadthong, W., Yonemori, K., Sugiora, A., Utsunomiya, N., Subhadrabandhu, S., 1999. Identification of mango cultivars of Thailand and evaluation of their genetic variation using the amplified fragments by simple sequence repeat (SSR) anchored primers. Sci. Hort. 82, 57–66.

FAO, 2009. Food outlook global market analysis. Tropical fruit. <http://www.fao.org/docrep/012/ak341e14.htm> (accessed 27.05.10.).

FAOSTAT, 2000. FAO Statistics. Food and Agriculture Organization of the United Nations, Rome, Italy.

FAOSTAT, 2010. FAO Statistics. Food and Agriculture Organization of the United Nations, Rome, Italy.

Gajeria, H.P., Tomar, R.S., Patel, S.V., Viradia, R.R., Golakiya, B.A., 2011. Comparison of RAPD and ISSR markers for genetic diversity analysis among different endangered *Mangifera indica* genotypes of Indian gir forest region. J. Plant Biochem. Biotechnol. 20, 217–223.

Galán Sauco, V., 2004. Mango production and world market: current situation and future prospects. Acta Hort. 645, 107–116.

Gan, Y.Y., Zaini, S., Idris, A., 1981. Genetic variation in the vegetative propagated mango *Mangifera indica*. Pertanika. 4, 53–63.

Godwin, I.D., Aitken, E.A.B., Smith, L.W., 1997. Application of inter simple sequence repeat (ISSR) markers to plant genetics. Electrophoresis. 18, 1524–1528.

Gonzalez, A., Coulson, M., Brettel, R., 2002. Development of DNA markers (ISSRs) in mango. Acta Hort. (ISHS). 575, 139–143.

Griesbach, J., 2003. Mango Book Kenya. World Agroforestry Centre, Nairobi, Kenya, pp. 4–6.

Gunjate, R.T., 2009. Advances in mango culture in India. Acta Hort. 820, 69–78.

He, C., Poysa, V., Yu, K., 2003. Development and characterization of simple sequence repeats (SSR) markers and their use in determining relationships among *Lycopersicon esculentum* cultivars. Theor. Appl. Genet. 106, 363–373.

He, X.H., Li, Y.R., Guo, Y.Z., Tang, Z.P., Li, R.B., 2005. Genetic analysis of 23 mango cultivars collection in Guangxi province revealed by ISSR. Mol. Plant Breed. 3, 829–834.

He, X.H., Li, Y.R., Guo, Y.Z., Ou, S.J., Li, R.B., 2007. Identification of closely related mango cultivars by ISSR. Guihaia. 27, 44–47.

Holton, T.A., 2001. Plant genotyping by analysis of microsatellites. In: Henry, R. (Ed.), Plant Genotyping: The DNA Fingerprinting Plants. Centre for Plant Conservation Genetics, Southern Cross University. CABI International, Lismore, Australia, pp. 15–28.

Honsho, C., Nishiyama, K., Eiadthong, S., Yonemori, K., 2004. Isolation and characterization of new microsatellites markers in mango (*Mangifera indica*). Mol. Ecol. Notes. 5, 152–154.

Hoogendijk, M., Williams, D., 2001. Characterizing the genetic diversity of home garden crops: some examples from Americas. In: Second International Home Gardens Workshop, 17–19 July 2001, Witzenhausen, Germany, pp. 34–40.

Iyer, C.P.A., Degani, C., 1997. Classical breeding and genetics. In: Litz, R.E. (Ed.), The Mango: Botany Production and Uses. CAB International, Wallingford, UK, pp. 49–68.

Jagarlamudi, S., Rosaiah, G., Kurapati, R.K., Pinnamaneni, R., 2011. Molecular identification of mango (*Mangifera indica* L.) var. totupura. Bioinformation. 5, 405–409.

Jedele, S., Hau, A.M., Oppen, M.V., 2003. An analysis of the world markers for mangoes and its importance for developing countries. Paper was presented in the Conference on International Agricultural Research for Development on 8–10 October in Gottingen, Germany.

Jintanawongse, S., Changtragoon, S., 2000. Identification of cultivars and certification of hybrids in mango (*Mangifera indica* L.) by isozyme gene marker. Acta Hort. 509, 177–184.

Joshi, C.P., Zhou, H., Huang, X., Chiang, V.L., 1997. Context sequences of translation initiation codon in plants. Plant Mol. Biol. 35, 993–1001.

Kapteyn, J., Simon, J.E., 2002. The use of RAPDs for assessment of identity, diversity and quality of Echinacea. In: Whipkey, J.J. (Ed.), Trends in New Crops and New Uses. ASHS Press, Alexandria, Egypt, pp. 509–513.

Karihaloo, J.L., Dwivedi, Y.K., Archak, S., 2003. Analysis of genetic diversity of Indian mango cultivars using RAPD markers. J. Hort. Sci. Biotech. 78, 198–289.

Kashkush, K., Fang, J., Tomer, J., Hillel, J., Lavi, U., 2001. Cultivar identification and genetic map of mango *Mangifera indica*. Euphytica. 122, 129–136.

Kostermans, A.J., Bompard, J.M., 1993. The Mangoes their Botany, Nomenculture, Horticulture and Utilization. Academic Press, London, p. 233.

Krishna, H., Singh, S.K., 2007. Biotechnological advances in mango (*Mangifera indica* L.) and their future implication in crop improvement – a review. Biotecnol. Adv. 25, 223–243.

Levi, A., Rowland, L.J., 1997. Identifying blueberry cultivars and evaluating their genetic relationships using randomly amplified polymorphic DNA (RAPD) markers. HortScience. 32, 1105–1108.

Lizada, C., 1993. Mango. In: Seymour, G., Taylor, J., Tucker, G. (Eds.), Biochemistry of Fruit Ripening. Chapman & Hall Inc., New York, NY, pp. 255–266.

Lowe, A.C., Gillies, M., Wilson, J., Dawson, I.K., 2000. Conservation genetics of bush mango from central/west Africa: implications from random amplified polymorphic DNA analysis. Mol. Ecol. 9, 831–841.

Lu, B., Cheng, H.R., Yan, Q.F., Huang, Z.J., Shen, G.F., et al., 2010. Recombinase-aid amplification: a novel technology of *in vitro* rapid nucleic acid amplification. Sci. Sinica Vitae. 40, 983–988.

Luo, C., He, X.H., Chen, H., Ou, S.J., Gao, M.P., 2010. Analysis of diversity and relationships among mango cultivars using Start Codon Targeted (SCoT) markers. Biochem. Syst. Ecol. 38, 1176–1184.

Luo, C., He, X.H., Chen, H., Ou, S.J., Gao, M.P., Brown, J.S., et al., 2011. Genetic diversity of mango cultivars estimated using SCot and ISSR markers. Biochem. Syst. Ecol. 39, 676–684.

Luo, C., He, X.H., Chen, H., Ou, S.J., Gao, M.P., 2012. Genetic relationship and diversity of *Mangifera indica* L.: revealed through SCoT analysis. Genet. Res. Crop Evol. 59, 1505–1515.

Marcela, D.M., Ingrid, S.G., Manuel, R.G., Elizabeth, H.J., 2009. Analysis of diversity among six populations of Colombian mango (*Mangifera indica* L. cvar. Hilacha) using RAPDs markers. Electron. J. Biotechnol. 12, 1–8.

Mhameed, S., Sharon, D., Hillel, J., Lahav, E., Kaufman, D., et al., 1996. Level of heterozygisity and mode of inheritance of variable number of tandem repeat loci in avocado. J. Am. Soc. Hort. Sci. 121, 778–782.

Mohapatra, T., 2007. Molecular markers in crop improvement: current applications and future prospects. In: Keshavachandran, R., Nazeem, P., Girija, D., John, P., Peter, K. (Eds.), Recent Trends in Horticultural Biotechnology. New India Publishing, New Delhi, India, pp. 1–10.

Mohle, V., Schwarz, G., 2004. Genotyping tools in plant breeding from restriction fragment length polymorphism to single nucleotide polymorphisms. In: Lorz, H., Wei, G. (Eds.), Molecular Marker Systems in Plant Breeding and Crop Improvement. Biotechnology in Agriculture and Forestry, Springer-Verlag, Berlin, Heidelberg, Germany, pp. 23–38.

Morton, J.F., 1987. Fruits of Warm Climates. Florida Flair Books, Miami, FL, pp. 221–239.

Mukherjee, S.K., 1949. The mango and its wild relatives. Sci. Culture. 15, 5–9.

Mukherjee, S.K., 1950. Mango: its allopolyploid nature. Nature. 166, 196–197.

Mukherjee, S.K., 1963. Cytology and breeding of mango. Punjab Hort. J. 3, 107–115.

Mukherjee, S.K., 1997. Introduction: Botany and importance. In: Litz, R.E. (Ed.), The Mango: Botany, Production and Uses. CAB International, Wallingford, UK, p. 8.

Mukherjee, S.K., Singh, R.N., Majumder, P.K., Sharma, D.K., 1968. Present position regarding breeding of mango in India. Euphytica. 17, 462–467.

Narasimha-Char, B.L., Atchyuta-Ramayya, D., Azeemoddin, G., Thirumala-Rao, S.D., 1979. J. Oil Technol. Assoc. India. 11, 94.
Pandit, S.S., Mitra, S., Giri, A.P., Pujari, K.H., Patil, B.P., Narayan, D.J., et al., 2007. Genetic diversity analysis of mango cultivars using inter simple sequence repeat markers. Curr. Sci. 93, 1135–1141.
Phumichai, J.C., Bapprasert, C., Chunwongse, C., Sukonsawan, S., 2000. Molecular mapping of mango cultivars "Alphonso" and "Palmer". Acta Hort. 509, 193–207.
Rahman, M.L., Rabbani, M.G., Siddique, M.N.A., Rahman, M.A., Garvey, E.J., Rahaman, E. H.M.S., 2007. Molecular characterization of 28 mango germplasm using RAPD. Plant Tissue Cult. Biotech. 17, 71–77.
Roy, B., Visweswariya, S.S., 1951. Cytogenetics of Mango and Banana. Report of Maharashtra Association for Cultivation of Science, Pune, India.
Roy, S.C., Chattopadhyay, A., 2011. Evaluation of genetic diversity in mango germplasm resources using RAPD markers and characterization of cultivar Guti based on 18S rRNA gene sequence. Indian J. Genet. Plant Breed. 71, 254–261.
Samal, K.C., Jena, R.C., Swain, S.S., Das, B.K., Chand, P.K., 2012. Evaluation of genetic diversity among commercial cultivars, hybrids and local mango (*Mangifera indica* L.) genotypes of India using cumulative RAPD and ISSR markers. Euphytica. 185, 195–213.
Santoso, P., 2000. Mango production and industry in Indonesia. Acta Hort. 509, 51–58.
Sawant, S.V., Singh, P.K., Gupta, S.K., Madnala, R., Tuli, R., 1999. Conserved nucleotide sequences in highly expressed genes in plants. J. Genet. 78, 123–131.
Schnell, R.J., Knight Jr., R.J., 1993. Genetic relationships among *Mangifera* spp. Based on RAPD markers. Acta Hort. 341, 86–92.
Schnell, R.J., Ronning, C.M., Knoght, R.J.J., 1995. Identification of cultivars and validation of genetic relationship in *Mangifera indica* L. using RAPD markers. Theor. Appl. Genet. 90, 269–274.
Shudo, A., Tarora, K., Makishi, Y., Ichi, R., Takahashi, K., et al., 2013. Development of CAPS markers and their application in breeding for mango (*Mangifera indica* L.). Euphytica. 190, 345–355.
Simon, R., Igeno, M.I., Coupland, G., 1996. Activation of floral meristem identity genes in Arabidopsis. Nature. 384, 303–310.
Singh, H., 1996. Mango. ICAR, New Delhi, India.
Singh, L.B., 1968. The Mango Botany, Cultivation and Uses. Leonard Hill Books, London, p. 438.
Srivastava, N., Bajpai, A., Chandra, R., Muthukumar, M., Srivastava, M.K., 2012. Comparison of PCR based marker systems for genetic analysis in different cultivars of mango. J. Environ. Biol. 33, 159–166.
Tomar, R.S., Gajeria, H.P., Viradiya, R.R., Pater, S.V., Golakiya, B.A., 2011. Phylogenetic relationship among mango (*Mangifera indica* L.) landraces of Saurashtra based on DNA fingerprinting. J. Hort. Forest. 3, 379–385.
Uddin M.S., 2013. Application of molecular techniques for the improvement of mango production. (Doctoral thesis). Chinese Academy of Agricultural Sciences, Beijing, People's Republic of China.
Uddin, M.S., Feng, J.X., He, X.H., Li, S.W., Cheng, Q., 2012. Gummosis disease of mango and its control. Chinese Forest. Sci. Tech. 11, 25–32.
Uddin, M.S., Xin-Hua, He., SunWen, L.I., Jaime, A., Da Silva, T., Cheng, Qi., 2014. An Improved method to extract DNA from Mango. Biologia. 69, 133–138.

Varshney, A., Mohapatra, T., Sharma, R.P., 2004. Molecular mapping and marker assisted selection of traits for crop improvement. In: Srivastava, A., Narula, S. (Eds.), Plant Biotechnology and Molecular Markers. Anamaya, New Delhi, India, p. 289.

Viruel, M.A., Escribano, P., Barbieri, M., Ferri, M., Hormaza, J.I., 2005. Figerprinting, embryo type and geographic differentiation in mango (*Mangifera indica* L.) with microsatellites. Mol. Breed. 15, 383–393.

Vos, P., Hogers, R., Bleeker, M., Reijans, M., Van de Lee, T., et al., 1995. AFLP: a new techniques for DNA fingerprinting. Nucleic Acids Res. 23, 4407–4414.

Weber, J., 1990. Informativeness of human $(dC\text{-}dA)_n$ $(dG\text{-}dT)_n$ polymorphisms. Genomics. 7, 524–530.

Williams, J.K., Kubelic, R.A., Livak, K., Rafalski, J.A., Tingey, S.V., 1990. DNA polymorphisms amplified by arbitrary primers are useful as genetic markers. Nucleic Acids Res. 18, 6531–6535.

Xiong, F.Q., Tang, R.H., Chen, Z.L., Pan, L.H., Zhuang, W.J., 2009. SCoT. A novel gene targeted marker technique based on the translation start codon. Mol. Plant Breed. 7, 635–638.

Yamanaka, N., Hasran, M., Xu, D.H., Tsunematsu, H., Idris, S., et al., 2006. Genetic relationship and diversity of four *Mangifera* species revealed through AFLP analysis. Genet. Res. Crop Evol. 53, 949–954.

Yonemori, K., Honsho, C., Kanzaki, S., Eiadthong, W., Sugiura, A., 2002. Phylogenetic relationships of *Mangifera* species revealed by ITS sequences of nuclear ribosomal DNA and a possibility of their hybrid region. Plant Syst. Evol. 231, 59–75.

Zietkiewicz, E., Rafalski, A., Labuda, D., 1994. Genome fingerprinting by simple sequence repeat (SSR) anchored polymerase chain reaction amplification. Genomics. 20, 176–183.

Cotton genomics and biotechnology

13

Hao Juan and Sun Yuqiang
College of Life and Environmental Science, Hangzhou Normal University, Hangzhou, Zhejiang, People's Republic of China

13.1 Introduction

Cotton (*Gossypium*) is one of the most important textile crops economically. It includes approximately 50 species distributed in a variety of areas worldwide, mainly under subtropical and tropical environmental conditions. There are four species that have independently been domesticated for their fibres, two each in Africa–Asia and the Americas. A parallel level of cytogenetic and genomic diversity has arisen during the global radiation of the genus, leading to the evolution of eight groups of diploid ($n = 13$) species (genome groups A-G and K) (Wendel and Cronn, 2003). In addition to textile manufacturing, cotton and cotton by-products provide raw materials used to produce a wealth of consumers-based products, foodstuffs, livestock feed, fertilizer and paper. The production, marketing, consumption and trade of cotton-based products further stimulate the economy, and cotton is the number one value-added crop producing an annual global textile mills market value of US$630.6 billion in 2011. Approximately 90% of cotton's value resides in the fibre (lint), yet yield and fibre quality have declined, especially over the last decade. This downward trend has been attributed to general erosion in genetic diversity of cotton varieties, and an increased vulnerability of the crop to environmental stress (Bowman et al., 1996).

For the critical need to increase diversity in the gene pool, cotton improvement programs are increasingly turning to the application of many approaches to breeding and germplasm utilization. There has long been interest in the genetic improvement of this valuable crop species. Conventional breeding and, more recently, biotechnological approaches, including tissue culture, genetic transformation and somatic hybridization, are presently being applied to cotton breeding. The combination of biotechnology and conventional breeding will become the main way of cotton breeding. Traditional plant breeding techniques together with classic biotechnological approaches have been widely used to increase crop yields by selecting improved varieties that are more productive and resistant to diseases and pathogens. Unfortunately, some important traits, such as resistance to insect pests and to some herbicides, appear to be absent from the genetic pools of cultivars. It is necessary to create new germplasms or utilize wild species and kindred species. Cotton biotechnology, embodied as cotton cell engineering and genetic engineering, is also a very important approach to assist cotton breeding.

Applied Plant Genomics and Biotechnology.

13.2 Cotton genomics

The cotton species (*Gossypium*) are divided into 5 tetraploid ($A_1A_1D_1D_1$ to $A_5A_5D_5D_5$, $2n=4x=52$) and nearly 45 diploid ($2n=2x=26$) species and is believed to have originated from the same ancestor about 5–10 million years ago. The tetraploid cotton species are thought to have formed by an allopolyploidization event with an A-genome species as the maternal parent and a D-genome species as the pollen-providing parent approximately 1–2 million years ago. Cotton species have huge and complex genomes. The haploid genome sizes of cotton were approximately 885 Mb for *Gossypium raimondii*, 1.7 Gb for *Gossypium arboreum* and 2.4 Gb for *Gossypium hirsutum*, respectively (Figure 13.1) (Paterson, 2009). Surely, decoding cotton genomes will provide a major source of candidate genes important for the genetic improvement of cotton quality and productivity, and also improve understanding of the functional and agronomic significance of polyploidy and genome size variation within the *Gossypium* genus (Chen et al., 2007).

Wang et al. produced *G. raimondii* genome using a next-generation Illumina paired-end sequencing strategy. A total of 78.7 Gb of next-generation Illumina paired-end reads was generated by sequencing genome shotgun libraries that

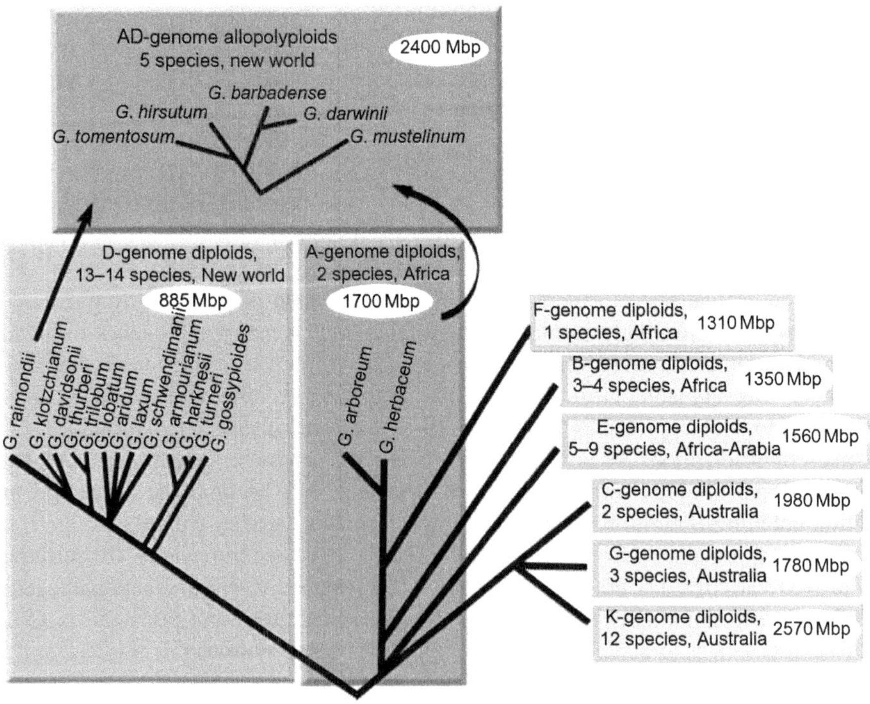

Figure 13.1 Evolutionary history of *Gossypium*.
Source: From Paterson (2009).

covered 103.6-fold of the 775.2 Mb assembled *G. raimondii* genome. The assembly consisted of 4,715 scaffolds and 41,307 contigs, which accounted for about 88.1% of the estimated *G. raimondii* genome. Of the total genome content, introns, exons, DNA transposable elements, long terminal repeats and other repeat sequences accounted for 6.9%, 6.4%, 4.4%, 42.6% and 13.0%, respectively. As previously reported for *Zea mays* (*Z. mays*) most *G. raimondii* chromosomes and transposable elements were distributed mainly in gene-poor regions. A total of 40,976 protein-coding genes were identified in the *G. raimondii* genome, with 92.2% (37,780 of 40,976) of these further confirmed by transcriptome sequencing data. The *G. raimondii* genome had a similar exon number per gene, a higher gene number and a lower mean gene density compared with the *Arabidopsis thaliana* genome. Yet it possessed a core set of 9,525 gene families in common similar to *A. thaliana, T. cacao* and *Z. mays*. *G. raimondii* and *T. cacao* were found to belong to a common subclade and probably diverged from a common ancestor approximately 33.7 million years ago, according to the examination of 745 single-copy gene families from 9 sequenced plant genomes. It is also indicated that the paleohexaploidization event occurred in a common progenitor between 115.4 and 146.1 million years ago before their speciation into the two present-day species 33.7 million years ago. Evidence of the hexaploidization event was observed. Wang et al. (2012) identified 2,355 syntenic blocks in the *G. raimondii* genome and found that approximately 40% of the paralogous genes were present in more than one block, which suggests that cotton genome has undergone substantial chromosome rearrangement during its evolution. Many genes involved in cotton fibre initiation and elongation were exhumed by qualitative analysis of transcriptional differences.

Almost simultaneously, *G. raimondii* genome sequencing and analysis had also been published that mainly focused on the analysis of repeated polyploidization of genomes and the evolution of spinnable cotton fibres (Paterson *et al.*, 2012). They speculated that the cotton lineage experienced an abrupt 5-6-fold ploidy increase shortly after its divergence from the common ancestor shared with *T. cacao* approximately 60 million years ago. Tracing to this 5-6-fold ploidy increase, paralogous genes show a single peak of synonymous nucleotide substitution values. The pairwise cytological similarity among A-genome chromosomes suggests the most recent event was a duplication. Allopolyploidy conferred about 30-36-fold duplication of ancestral angiosperm genes in cottons such as *G. hirsutum* and *G. barbadense* and reunited divergent *Gossypium* genomes at least 1–2 million years ago. It was considered that paleopolyploidy could accelerate cotton mutation rates. The complexity of a Malvaceae-specific lades of MYB transcription family was increased by paleopolyploidy, which may have led to the differentiation of epidermal cells into fibres rather than mucilages of other Malvaceae. Among 8 R1R2R3, 204 R2R3 and 194 heterogeneous MYB transcription factors in *G. raimondii*, there are six members in subgroup 9 known only in Malvaceae, including a possible ‘fibre clade’ different from the *Arabidopsis* GL1-like subgroup 15 related to trichome and root hair development. These genes were predominantly expressed in the early stages of fibre development, and 50% of subgroup 9 genes have higher expression in elite cultivated tetraploid cottons than the wild tetraploids. During the domestication process, changes in gene expression can improve adaptability to against environmental stress. There are a total of 300 genes encoding nucleotide-binding site (NBS) proteins involved in pest and disease resistance in *G. raimondii*, and at least 15 cellulose synthase (CESA) and 35 cellulose-synthase-like (CSL) genes required for cellulose synthesis and synthesis of cell wall matrix polysaccharides that surround cellulose microfibrils. Spinnable *G. herbaceum* A and non-spinnable *G. longicalyx* F genomes were sequenced and compared with the outgroup *G. raimondii* D genome in order to clarify the evolution of spinnable nascent fibres. The sequence of *G. hirsutum* cultivar revealed many non-reciprocal DNA exchanges between subgenomes, which might have brought about phenotypic innovation and emergent properties. By comparison to diploid progenitors and outgroups, more opportunities were obtained for dissecting phenotypic innovation and emergent properties of other polyploids, particularly angiosperms.

As cotton research moving to a 'post-genomic' era, the many diverse internet resources become available to the international research community for cotton genomic research. The Cotton Genome Project (CGP), initiated and performed by Institute of Cotton Research of CAAS and accompanied by BGI. The CGP is mainly focused on cotton sequencing and functional analysis (http://cgp.genomics.org.cn/page/species/index.jsp). It provides BLAST and MAPVIEW of genome, CDS and protein in *G. raimondii.* CottonGen (*http://www.cottongen.org*) is a new cotton community genomics, genetics and breeding database being developed to enable basic, translational and applied research in cotton. It is being built using the open-source Tripal database infrastructure. CottonGen consolidates and expands the data from CottonDB and the Cotton Marker Database (CMD), providing enhanced tools for easy querying, visualizing and downloading research data including gene, germplasm, marker, QTL and related publications.

13.3 Cotton fibre function genomics

Cotton is one of the most important economic crops in the world. Its main product, cotton fibres, are single-celled trichomes from individual epidermal cells on the outer integument of the ovules. As the most important natural raw materials used in the textile industry, cotton fibres serve as an excellent model for studying single-cell development. So, the main objective of the current cotton research is to clarify the molecular mechanisms of cotton fibre development. Cotton fibre development consists of four overlapping stages: initiation (from 3 days before anthesis to 3 days post anthesis [DPA]), elongation/primary cell wall (PCW) synthesis (2–20 DPA), secondary cell wall (SCW) biosynthesis (15–45 DPA) and dehydration and maturation (45–50 DPA) (Figure 13.2) (Haigler et al., 2012). Fibre initiation is a key factor for yield formation; elongation determines both yield and quality; and SCW biosynthesis mainly impact fibre quality. Biotechnology is the only way to improve the efficiency of cotton production and cotton fibre yield and quality.

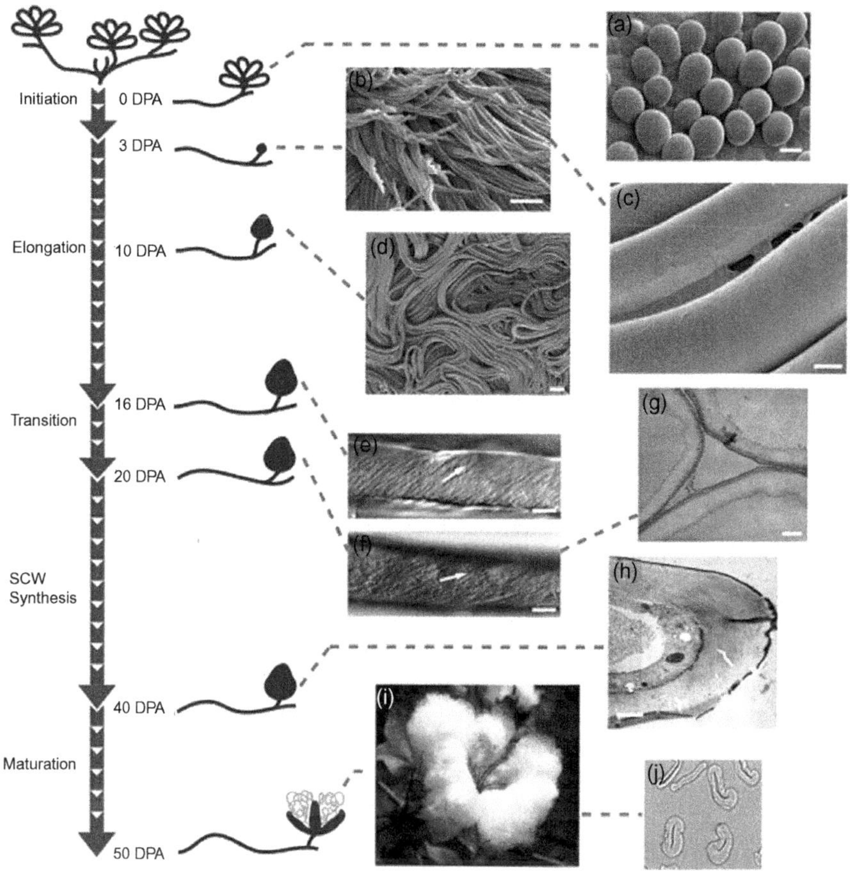

Figure 13.2 A representation of the progression of cotton fibre development.
Source: From Haigler et al. (2012).

With the development of functional genomics and next-generation sequencing technology, it becomes easier to locate the gene in the cotton Chr14 (Han et al., 2006). Two fibre elongation-related genes *GhNLP* and *GhSAMS* were mapped on Chr19 and Chr14, respectively (Zhang et al., 2008). A high-density linkage map is a prerequisite for cotton genome research and genetic breeding. Currently, four higher density maps have been released (Guo et al., 2008; John et al., 2012; Rong et al., 2004; Yu et al., 2011), and SSR markers play an irreplaceable role in the construction of these maps. With the reduced cost of sequencing, SNP and IDP mark has been developing rapidly. The rich EST sequences in cotton public databases provide a valuable resource for the development of EST-SNP markers. The CMD (http://www.cottonmarker.org/) provides centralized access to all 17,448 publicly available SSRs and 312 mapped cotton RFLP sequences containing SSRs.

Currently, single nucleotide polymorphisms computationally mined in cotton ESTs (eSNPs) from the NCBI dbEST database have been included. The standardized panel screened data is available for many of the microsatellites which consists of 12 diverse genotypes selected from cultivated and exotic cottons.

Mature cotton fibre cells are extremely long, up to 3–5 cm, and are occupied in major part by a secondary wall that consists mainly of cellulose (>90%) and some minor noncellulosic carbohydrates, such as xyloglucans, pectic polysaccharides, xylans, glucomannans and glucans (Meinert and Delmer, 1977; Huwyler et al., 1979). Owing to its exceptional cell length and simple SCW composition, the cotton fibre provides an excellent model for studies of plant cell elongation and cell wall biogenesis (Kim and Triplett, 2001). During recent years, significant progress has been made in large-scale identification of genes and proteins involved in fibre development, particularly those related to fibre elongation (Arpat et al., 2004; Shi et al., 2006; Gou et al., 2007), and a few fibre elongation-related genes have been structurally or functionally characterized (Ruan et al., 2003; Li et al., 2005; Zhang et al., 2011).

A R2R3 MYB transcription factor GhMYB109 was isolated from cotton (*G. hirsutum* L.) fibres, which is highly homologous with AtGL1 and AtWER. This gene was specifically expressed in initial and elongating fibre cells. In the antisense transgenic cotton plants, fibre initiation was delayed, fibre elongation was blocked and the expression of genes related to ethylene synthesis was inhibited (Pu et al., 2008). Another R2R3 MYB transcription factor, GaMYB2, regulates fibre cell differentiation and initiation, which is predominantly expressed in early fibre development. Heterologous expression of *GL1::GaMYB2* in *gl1* plants can rescue the mutant phenotype in >90% of the transformants. Constitutive overexpression of *GaMYB2* can also induce the production of seed trichomes (Wang et al., 2004). The promoter of *GaMYB2* was fused to the GUS reporter gene to analyse the expression patterns in cotton, *Arabidopsis* and tobacco, respectively. It showed that the *GaMYB2* promoter exhibited activities not only in developing fibre cells and trichomes of other aerial organs in cotton, but also in *Arabidopsis* trichomes and tobacco glandular secreting trichomes (Shangguan et al., 2008). Four putative homologues of *Arabidopsis* WD-repeat gene *TTG1* were isolated from *G. raimondii* by a combination of PCR-based and library screening techniques. Two of them, *GhTTG1* and *GhTTG3*, were able to restore trichome formation in the *Arabidopsis ttg1* mutant and complement the anthocyanin defect in the *Matthiola incana ttg1* mutant (Humphries et al., 2005). GaHOX1, a class-IV HD-ZIP gene family member, with 66% amino acid sequence identities to *Arabidopsis* GL2, was predominantly expressed in cotton fibre cells during initiation and early elongation stages. In the induction of the promoter PGL2, GaHOX1 could restore the trichome development to the wild-type plant level in the *gl2-2* mutant plants. While the wild-type plants transformed with 35S::GL2 emerged a *gl2*-mutant-like phenotype with reduced viability (Guan et al., 2008). Another putative class-IV HD-ZIP transcription factor, GhHD1, was identified. The expression level of *GhHD1* in the ovule epidermis of fibreless mutants was lower than in wild-type ovules. RNAi transgenic lines were almost completely hairless and showed a delay in fibre initiation, but had other normal phenotypes such as plant height, morphology, flowering and seed setting. Vectors containing a constitutive S7 and a seedcoat-specific FBP7 promoter were transformed into cotton, respectively. Both of the overexpression lines exhibited significant increases of the number of fibre initials relative to their wild-types (Walford et al., 2012). Wu et al. (2006) also isolated three important transcription factors (GhMYB25, GhMYB25-like and GhHD1) through microarray analysis. Overexpression of *GhMYB25* in cotton increased cotton fibre and leaf trichome initiation. Suppression of *GhMYB25* resulted in delayed fibre initiation, shorter fibres and dramatically reduced trichome numbers on petioles, petals and leaves (Machado et al., 2009). Silencing of *GhMYB25-like* abolished fibre development similar to the fibreless mutant, without affecting the development of trichomes on petioles, petals and leaves (Walford et al., 2011). Overexpression of *GhHD1* increased fibre initiation, but did not affect the leaf trichomes, while reducing the transcripts of *GhHD1* delayed the timing of fibre initiation and decreased trichome formation (Walford et al., 2012).

Following the initiation, the cotton fibres enter into the rapid elongation phase immediately. Developing cotton fibre processes two important stages: cell elongation and secondary wall formation. By about 2 weeks, cotton fibre cells can stretch from 10–15 μm to 2.5–3.0 cm. Thus more studies get some preliminary results and showed kinetics supporting the fibre cell polarity elongation. Rapid fibre elongation is driven by high turgor with a highly extendable PCW. The key to generate turgor is some accumulation of ions and salts and the formation of large central vacuole. Ruan et al. (2001) investigated the gating of fibre plasmodesmata by confocal imaging of the membrane-impermeant fluorescent solute carboxyfluorescein and analysed the expression of the plasma membrane sucrose, K^+ transporters and cell wall expansion in elongating cotton fibres at selected stages of development. The dynamic mode controlling rapid fibre elongation was proposed as plasmodesmata is initially open in the early stage of fibre elongation (0–9 DPA), and at this stage the fibre cell elongation mainly depends on the maximal expression of *GhEXP1*. But plasmodesmata is closed at 10 DPA and re-opened at 16 DPA. At 10–15 DPA, sucrose and K^+ are transported into fibre cells to decrease intracellular osmolality and increase turgor in order for cell elongation. The estimated turgor is decreased rapidly in fibres at 16 DPA. Fibre cell elongation is terminated by the combination of loss of higher turgor and increased cell wall rigidity. The transcript of *GhEXP1* is dramatically reduced to undetectable levels at 20 DPA. For the role of sucrose in the initiation and elongation of the single-celled fibres, Ruan et al. (2003) transformed a sucrose synthase gene (*Sus*) suppression construct into cotton and analysed the early fibre development on 0–3 DPA. Suppressing the expression of *Sus* by more than 70% in the ovule epidermis resulted in a fibreless phenotype. In those ovules, the fibre initials were fewer and collapsed or shrunken. The level of *Sus* inhibition correlated strongly with the degree of suppression of fibre initiation and elongation, which demonstrated that *Sus* played a rate-limiting role in the initiation and elongation of the cotton fibres. A novel sucrose synthase gene (*SusA1*) from *G. hirsutum* line 7235 was described. Suppression of *GhSusA1* reduced fibre quality and decreased the boll size and seed weight, but overexpression of *GhSusA1* increased fibre length and strength and led to elevated seedling biomass (Jiang et al., 2012). GhMADS11 is a novel MADS-box protein, which is preferentially expressed in elongating fibres. Overexpression of this gene in fission yeast could promote cell elongation significantly (Li et al., 2011). A class I TCP transcription factor (GbTCP) from the normalized cDNA library of sea-island cotton fibre (−2 to 25 DPA) (*G. barbadense*) was preferentially expressed in the elongating cotton fibre. Suppression of *GbTCP* expression resulted in shorter fibres and lower fibre quality possibly by affecting fibre elongation at the rapid elongation stage (10–15 DPA), but did not show changes on the plant architecture and fibre initiation. Ectopic overexpression of *GbTCP* in *Arabidopsis* resulted in more and longer root hairs. In *G. hirsutum*, WLIM1a localises in the cytosol and nucleus and moves into the nucleus in response to hydrogen peroxide and preferentially expressed during the elongation and secondary wall synthesis stages in developing fibres. Overexpression of WLIM1a in cotton led to significant changes in fibre length and secondary wall structure. Compared with the wild type, fibres of WLIM1a-overexpressing plants grew longer and formed a thinner and more compact SCW, which contributed to improved fibre strength and fineness. WLIM1a gene in fibre development in *G. hirsutum* has dual roles; it acts as an actin bundler to facilitate elongation of fibre cells and also functions as a transcription factor to activate expression of Phe ammonia lyase–box genes involved in phenylpropanoid biosynthesis to build up the SCW (Han et al., 2013). Currently there are mainly two hypotheses about cotton fibre elongation mode. Rapidly elongating fibre cells seem to expand via a diffuse-growth mode. For example, organelle zonation did not occur in the apical part of fast-growing tips of elongating cotton fibre cells by scanning electron microscopy and transmission electron microscopy (Seagull, 1990). In addition, the newly deposited cellulose microfibrils and cortical microtubules were transversely oriented with respect to the growth axis in cotton fibre cells (Tiwari and Wilkins, 1995). However, another hypothesis was proposed on the basis of the currently available results, cotton fibre cells elongate via linear cell-growth mode by the combination of both tip-growth and diffuse-growth modes (Qin and Zhu, 2011).

Plant hormones play an extremely important role in cotton fibre development. *In vitro* ovule culture is an effective tool to study hormonal regulation on fibre development and was established by Beasley and Ting (1973) as early as in the 1970s. They found that IAA and GA could promote fibre development, but kinetin (KT) and abscisic acid (ABA)-inhibited fibre development. Overexpression of the IAA biosynthetic gene *iaaM* in cotton fibres, driven by FBP7 promoter, increased IAA levels in the epidermis of cotton ovules at the fibre initiation stage. The number of lint fibres was increased substantially in a 4-year field trial. The lint percentage of the transgenic cotton was higher in transgenic plants than in wild-type plants, resulting in an increase of >15% in lint yield. But no significant change in fibre length was observed (Zhang et al., 2011). Xiao et al. (2010) cloned a GA synthetic pathway key enzyme gene *GhGA20ox1-3* from cotton fibres. qRT-PCR analysis revealed that *GhGA20ox1* was preferentially expressed in elongating fibres. The endogenous GA content was consistent with the expression level of this gene. Overexpression of *GhGA20ox1* in cotton increased endogenous GA levels in fibres and ovules with GA over-production phenotypes. The number of fibre initials per ovule and fibre lengths was increased compared with the control plants. Upregulation of the *GhGA20ox1* promoted fibre initiation and elongation by regulating GA levels (Xiao et al., 2010).

Ethylene biosynthesis was one of the most significant upregulated biochemical pathways during fibre development through sequencing by sequencing 36,000 cDNAs from 5 to 10 DPA cotton ovules and subsequent microarray analysis (Shi et al., 2006). The expression of three 1-Aminocyclopropane-1-Carboxylic Acid Oxidase ACO1-3 was increased significantly during the fibre elongation phase and reduced in the *fl* mutant. Amounts of gaseous ethylene release were closely related to the expression of *ACO*. Treatment with increasing concentrations of ethylene *in vitro* ovule culture increased the fibre length. Accordingly, treatment with the ethylene biosynthetic inhibitor AVG reduced the fibre length dramatically. A steroid 5a-reductase GhDET2 was cloned from developing fibre cells and predominantly expressed during the fibre initiation and rapid elongation stage. Suppression of *GhDET2* inhibited both fibre initiation and elongation. Treating cultured ovules with steroid 5a-reductase inhibitor reduced fibre elongation. Furthermore, seed coat-specific expression of *GhDET2* increased fibre cell number and length significantly (Luo et al., 2007). JA was an inhibitor of ovule and fibre development *in vitro*. Continuous exogenous JA treatment inhibited fibre cell elongation and had a dose-dependent effect (Tan et al., 2011). GbTCP analysis also indirectly revealed that jasmonic acid played an important role in cotton fibre initiation and elongation (Hao et al., 2012).

In fibre development, there are many unfavourable environmental conditions including biotic and abiotic stresses which restrict the yield of cotton. Several endogenous phytohormones, such as ethylene (ET), salicylic acid (SA), JA, ABA and brassinosteroids (BR) activated different defence pathways related to a complex defence signaling network. Crosstalk among the plant hormones provides plants with a powerful ability to finely regulate the immune response (Bari and Jones, 2009). Among these stresses, *Verticillium* wilt is a devastating disease worldwide, reducing the quality and yield of the fibres by up to 30%. An important sesquiterpene phytoalexin gossypol, which is specific to cotton, plays a crucial role in the defence against the invasion of insects and pathogens (Luo et al., 2001). Through sequencing, a number of genes involved in disease resistance, such as pathogen-related proteins, aerobic metabolism enzymes, ethylene biosynthesis and response genes have been identified from the resistant cotton cultivar *G. barbadense* L. variety 7124 by suppression subtractive hybridization (Xu et al., 2011a). Using the RNA-Seq method and histochemistry, a number of genes involved in lignin metabolism were identified and revealed that lignin metabolism had a crucial role in the resistance to the wilt fungus *Verticillium dahliae* (Xu et al., 2011b). A comparative proteomic analysis was performed on Mock and *V. dahliae* inoculated roots of *G. barbadense* L. variety 7124, which identified a total of 188 differentially expressed proteins by mass spectrometry analysis. These proteins could be divided into 17 biological processes through gene ontology annotation, most of which were implicated in secondary metabolism, reactive oxygen burst and phytohormone signalling pathways. Furthermore, the roles of three classes of genes, involved in BR signalling, JA signalling and gossypol metabolism, were characterized by employing virus-induced gene silencing (VIGS). BRs, JA and gossypol, act as important factors in the cotton disease resistance to *V. dahliae*, which provides new insights into the molecular basis of cotton defence against *V. dahliae* (Gao et al., 2013). The productivity of cotton is also adversely affected by drought stress. Genome-wide transcriptome analysis was used to identify differentially expressed genes at various stages of fibre development under drought stress. A large number of genes, including transcription factors (NAC, WRKY, C2H2 and AP2-EREBP), ion transporters, osmoprotectants, heat shock proteins and pathways participated in hormone biosynthesis and signalling were up-regulated during fibre elongation. This study showed that drought had relatively less impact on fibre initiation but had profound effect on fibre elongation (Padmalatha et al., 2012).

13.4 Cotton biotechnology

'Genetically modified cotton' refers to the use of cotton varieties with transgenic or genetically modified (GM) traits, which have made a significant contribution in the dramatic reduction in insecticides applied to cotton crops. Other benefits of GM cotton are increased populations of beneficial insects and wildlife in cotton fields, reduced pesticide run off, improved farm worker and neighbour safety, a decrease in labour and fuel usage, improved soil quality, reduced production costs, increased yield, reduced risks and further opportunities to grow cotton in areas of high pest infestation.

Cotton incorporating transgenic traits has a sound track record of safe and successful use in China, America and Australia, with no adverse incident recorded in over 15 years of growing these varieties. Insect-resistant cotton contains one or two genes from the naturally occurring soil bacterium *Bacillus thuringiensis* (Bt) and gives the plant an in-built tolerance to the *Helicoverpa* caterpillar. When the caterpillar ingests a small part of the cotton plant, the Bt protein disrupts the caterpillar's digestive system and it dies. Bt is a naturally occurring soil organism that produces insecticidal proteins and Bt sprays have been safely used for over 50 years in agriculture. Three types of transgenic cotton were produced by two different genes from the naturally occurring soil bacterium *Bacillus thuringiensis* (Bt), or from the soil bacterium called *Agrobacterium tumefaciens*, or from the soil microorganism *Streptomyces hygroscopicus*. The use of transgenic cotton is often used in a combination of natural controls and pest-specific chemistry to further reduce pesticide use.

Herbicide tolerant cotton can reduce the amount of soil cultivation and herbicide required on cotton crops to control weeds and facilitates healthier soils through less soil disruption and reductions in residual herbicides. Over the last decade new cotton varieties released have contained new features such as improved fibre quality, disease resistance, maturity and regional adaptability – research is being undertaken to develop varieties that require less water and/or are drought tolerant.

Cottonseed oil doesn't require GM food labelling. This is because when cottonseed is crushed to make oil, the oil is separated from the Bt and other transgenic proteins. Cottonseed oil from a transgenic cotton plant variety contains no GM material.

Recent advances in plant tissue culture technology offer novel and valuable means for cultivar improvement. Embryo rescue allows the development of many interspecific and intergeneric cotton hybrids between wild diploid cotton species and diploid or tetraploid cultivars to create interspecific and intergeneric crosses that would normally produce seeds that are aborted. Regeneration through somatic embryogenesis is preferred over organogenesis for probable single-cell origin of the somatic embryo (Merkle et al., 1995) and efficient *in vitro* techniques for the regeneration of plantlets from cotton are continually being developed (Firoozabady and DeBoer, 1993; Kumria et al., 2003; Price and Smith, 1979; Sun et al., 2006). Methods also have been developed for plant regeneration from cotton protoplasts (Finer and Smith, 1982; Peeters et al., 1994; Sun et al., 2006).

In cotton biotechnology, a breakthrough was made in production of somatic hybrids by protoplast fusion in cotton between *G. hirsutum* and wild species *G. klotzschianum* (Sun et al., 2004), then many somatic hybrids between cultivars and wild species were reported (Sun et al., 2005, 2011; Yu et al., 2012). The wild species included *G. klotzschianum*, *G. davidsonii*, *G. stocksii*, *G. bickii* and *G. trilobum*. The somatic hybrids provide many intermediate materials to create many introgression lines and transfer the specific gene/genes from wild species to cultivars.

Biotechnology is the present and the future of cotton. It is also very essential to identify genes, isolate important agronomic traits and create novel germplasms for breeding. The technology is already providing increased yields on less land with less water, soil and other resources. And scientific advances in biotechnology promoting the cotton traditional breeding process are coming at an ever-faster rate, portending a future of full cotton sustainability.

List of abbreviations

ABA Abscisic acid
AVG Aminoethoxyvinylglycine
BGI Beijing Genomics institution
BR brassinosteroids
CAAS the Chinese Academy of Agricultural Sciences
CDS Coding sequence
CESA Cellulose synthase
CGP Cotton Genome Project
CMD Cotton Marker Database
CSL Cellulose-synthase-like
DPA days post anthesis
EST Expressed sequence tag

ET ethylene
GA Gibberellins
GM genetically modified
GUS β-glucuronidase
IAA Indole-3-acetic acid
IDP Insertion-deletion length polymorphism
KT kinetin
PCW primary cell wall
QTL Quantitative trait loci
RFLP Restriction Fragment Length Polymorphism
SA salicylic acid
SCW secondary cell wall
SNP Single Nucleotide Polymorphisms
SSR Simple Sequence Repeat
VIGS virus-induced gene silencing

References

Arpat, A.B., Waugh, M., Sullivan, J.P., Gonzales, M., Frisch, D., Main, D., et al., 2004. Functional genomics of cell elongation in developing cotton fibers. Plant Mol. Biol. 54, 911–929.

Bari, R., Jones, J.D., 2009. Role of plant hormones in plant defence responses. Plant Mol. Biol. 69, 473–488.

Beasley, C., Ting, I.P., 1973. The effects of plant growth substances on *in vitro* fiber development from fertilized cotton ovules. Am. J. Bot. 60, 130–139.

Bowman, D.T., May, O.L., Calhoun, D.S., 1996. Genetic base of upland cotton cultivars released between 1970 and 1990. Crop Sci. 36, 577–581.

Chen, Z.J., Scheffler, B.E., Dennis, E., Triplett, B.A., Zhang, T., Guo, W., et al., 2007. Toward sequencing cotton (*Gossypium*) genomes. Plant Physiol. 145, 1303–1310.

Finer, J.J., Smith, R.H., 1982. Isolation and culture of protoplasts from cotton (*Gossypium Klotzschianum* Anderss) callus cultures. Plant Sci. Lett. 26, 147–151.

Firoozabady, E., DeBoer, D.L., 1993. Plant regeneration via somatic embryogenesis in many cultivars of cotton (*Gossypium hirsutum* L.). In Vitro Cell Dev. Biol. Plant. 29, 166–173.

Gao, W., Long, L., Zhu, L., Xu, L., Gao, W., Sun, L., et al., 2013. Proteomic and virus-induced gene silencing (VIGS) analyses reveal that Gossypol, Brassinosteroids and Jasmonic acid contribute to the resistance of cotton to *Verticillium dahliae*. Mol. Cell. Proteomics. 12, 3690–3703.

Guan, X., Li, Q., Shan, C., Wang, S., Mao, Y., Wang, L., et al., 2008. The HD-Zip IV gene GaHOX1 from cotton is a functional homologue of the *Arabidopsis* GLABRA2. Physiol. Plant. 134, 174–182.

Guo, W., Cai, C., Wang, C., Zhao, L., Wang, L., Zhang, T., 2008. A preliminary analysis of genome structure and composition in *Gossypium hirsutum*. BMC Genomics. 9, 314.

Haigler, C.H., Betancur, L., Stiff, M.R., Tuttle, J.R., 2012. Cotton fiber: a powerful single-cell model for cell wall and cellulose research. Front. Plant Sci. 3, 104.

Han, L., Li, Y., Wang, H., Wu, X., Li, C., Luo, M., et al., 2013. The dual functions of WLIM1a in cell elongation and secondary wall formation in developing cotton fibers. Plant Cell. 25 (11), 4421–4438.

Han, Z., Chu, Y., Guo, W., Zhang, T., 2006. Divergence and mapping of SNPs of the *FbL2A* gene for cotton fiber development. J. Agric. Biotechnol. 14, 360–364.

Hao, J., Tu, L., Hu, H., Tan, J., Deng, F., Tang, W., et al., 2012. GbTCP, a cotton TCP transcription factor, confers fibre elongation and root hair development by a complex regulating system. J. Exp. Bot. 63, 6267–6281.

Humphries, J.A., Walker, A.R., Timmis, J.N., Orford, S.J., 2005. Two WD-repeat genes from cotton are functional homologues of the *Arabidopsis thaliana* TRANSPARENT TESTA GLABRA1 (TTG1) gene. Plant Mol. Biol. 57, 67–81.

Huwyler, H.R., Franz, G., Meier, H., 1979. Changes in the composition of cotton fibre cell walls during development. Planta. 146, 635–642.

Jiang, Y., Guo, W., Zhu, H., Ruan, Y.L., Zhang, T., 2012. Overexpression of *GhSusA1* increases plant biomass and improves cotton fiber yield and quality. Plant Biotechnol. J. 10, 301–312.

John, Z.Y., Kohel, R.J., Fang, D.D., Cho, J., Van Deynze, A., Ulloa, M., et al., 2012. A high-density simple sequence repeat and single nucleotide polymorphism genetic map of the tetraploid cotton genome. G3: Genes|Genomes|Genetics. 2, 43–58.

Kim, H.J., Triplett, B.A., 2001. Cotton fiber growth in planta and in vitro. Models for plant cell elongation and cell wall biogenesis. Plant Physiol. 127, 1361–1366.

Kumria, R., Sunnichan, V.G., Das, D.K., Gupta, S.K., Reddy, V.S., Bhatnagar, R.K., et al., 2003. High-frequency somatic embryo normal plants in cotton (*Gossypium hirsutum*) through metabolic stress. Plant Cell Rep. 21 (7), 635–639.

Li, X.B., Fan, X.P., Wang, X.L., Cai, L., Yang, W.C., 2005. The cotton ACTIN1 gene is functionally expressed in fibers and participates in fiber elongation. Plant Cell. 17, 859–875.

Li, Y., Ning, H., Zhang, Z., Wu, Y., Jiang, J., Su, S., et al., 2011. A cotton gene encoding novel MADS-box protein is preferentially expressed in fibers and functions in cell elongation. Acta Biochim. Biophys. Sin. 43 (8), 607–617.

Luo, M., Xiao, Y., Li, X., Lu, X., Deng, W., Li, D., et al., 2007. GhDET2, a steroid 5alpha-reductase, plays an important role in cotton fiber cell initiation and elongation. Plant J. 51, 419–430.

Luo, P., Wang, Y.H., Wang, G.D., Essenberg, M., Chen, X.Y., 2001. Molecular cloning and functional identification of (+)-delta-cadinene-8-hydroxylase, a cytochrome P450 mono-oxygenase (CYP706B1) of cotton sesquiterpene biosynthesis. Plant J. 28, 95–104.

Machado, A., Wu, Y., Yang, Y., Llewellyn, D.J., Dennis, E.S., 2009. The MYB transcription factor GhMYB25 regulates early fibre and trichome development. Plant J. 59, 52–62.

Meinert, M.C., Delmer, D.P., 1977. Changes in biochemical composition of the cell wall of the cotton fiber during development. Plant Physiol. 59, 1088–1097.

Merkle, S.A., Parrot, W.A., Flinn, B.S., 1995. Morphogenic aspects of somatic embryogenesis. In Vitro Embryogenesis Plants. 20, 155–203.

Padmalatha, K., Dhandapani, G., Kanakachari, M., Kumar, S., Dass, A., Patil, D., et al., 2012. Genome-wide transcriptomic analysis of cotton under drought stress reveal significant down-regulation of genes and pathways involved in fibre elongation and up-regulation of defense responsive genes. Plant Mol. Biol. 78, 223–246.

Paterson, A.H., 2009. Genetics and genomics of cotton. Plant Genetics/Genomics. Springer Science + Business Media, New York, NY.

Paterson, A.H., Wendel, J.F., Gundlach, H., Guo, H., Jenkins, J., Jin, D., et al., 2012. Repeated polyploidization of *Gossypium* genomes and the evolution of spinnable cotton fibres. Nature. 492, 423–427.

Peeters, M.C., Willems, K., Swennen, R., 1994. Protoplast-to-plant regeneration in cotton (*Gossypium hirsutum* L. cv. Coker312) using feeder layers. Plant Cell Rep. 13 (3–4), 208–211.

Price, H.J., Smith, R.H., 1979. Somatic embryogenesis in suspension cultures of *Gossypium klotzschianum* Anders. Planta. 145, 305–307.

Pu, L., Li, Q., Fan, X., Yang, W., Xue, Y., 2008. The R2R3 MYB transcription factor GhMYB109 is required for cotton fiber development. Genetics. 180, 811–820.

Qin, Y.-M., Zhu, Y.-X., 2011. How cotton fibers elongate: a tale of linear cell-growth mode. Curr. Opin. Plant Biol. 14, 106–111.

Rong, J., Abbey, C., Bowers, J.E., Brubaker, C.L., Chang, C., Chee, P.W., et al., 2004. A 3347-locus genetic recombination map of sequence-tagged sites reveals features of genome organization, transmission and evolution of cotton (*Gossypium*). Genetics. 166, 389–417.

Ruan, Y.L., Llewellyn, D.J., Furbank, R.T., 2001. The control of single-celled cotton fiber elongation by developmentally reversible gating of plasmodesmata and coordinated expression of sucrose and K+ transporters and expansin. Plant Cell. 13, 47–60.

Ruan, Y.L., Llewellyn, D.J., Furbank, R.T., 2003. Suppression of sucrose synthase gene expression represses cotton fiber cell initiation, elongation, and seed development. Plant Cell. 15, 952–964.

Seagull, R., 1990. The effects of microtubule and microfilament disrupting agents on cytoskeletal arrays and wall deposition in developing cotton fibers. Protoplasma. 159, 44–59.

Shangguan, X.X., Xu, B., Yu, Z.X., Wang, L.J., Chen, X.Y., 2008. Promoter of a cotton fibre MYB gene functional in trichomes of *Arabidopsis* and glandular trichomes of tobacco. J. Exp. Bot. 59 (13), 3533–3542.

Shi, Y.H., Zhu, S.W., Mao, X.Z., Feng, J.X., Qin, Y.M., Zhang, L., et al., 2006. Transcriptome profiling, molecular biological, and physiological studies reveal a major role for ethylene in cotton fiber cell elongation. Plant Cell. 18, 651–664.

Sun, Y.Q., Zhang, X.L., 2006. Somatic hybrids between *Gossypium hirsutum* L. (4x) and *G. davidsonii* Kellog (2x) produced by protoplast fusion. Euphytica. 151 (3), 393–400.

Sun, Y.Q., Zhang, X.L., Nie, Y.C., Guo, X.P., Jin, S.X., Liang, S.G., 2004. Production and characterization of somatic hybrids between upland cotton (*Gossypium hirsutum*) and wild cotton (*G. klotzschianum* Anderss) via electrofusion. Theor. Appl. Genet. 109, 472–479.

Sun, Y.Q., Zhang, X.L., Nie, Y.C., Guo, X.P., 2005. Production of fertile somatic hybrids of *Gossypium hirsutum* + *G. bickii* and *G. hirsutum* + *G. stockii* via protoplast fusion. Plant Cell Tiss. Organ. Cult. 83, 303–310.

Sun, Y.Q., Liu, S.M., Wang, Y., Jones, B.J., Wang, H.Z., Zhu, S.J., 2011. An interspecific somatic hybrid between upland cotton (*G. hirsutum* L. cv. ZDM-3) and wild diploid cotton (*G. klotzschianum* A.). Plant Cell Tiss. Organ. Cult. 106, 425–433.

Tan, J., Tu, L., Deng, F., Wu, R., Zhang, X., 2011. Exogenous jasmonic acid inhibits cotton fiber elongation. J. Plant Growth Regul. 31 (4), 599–605.

Tiwari, S.C., Wilkins, T.A., 1995. Cotton (*Gossypium hirsutum*) seed trichomes expand via diffuse growing mechanism. Can. J. Bot. 73, 746–757.

Walford, S.A., Wu, Y., Llewellyn, D.J., Dennis, E.S., 2011. GhMYB25-like: a key factor in early cotton fibre development. Plant J. 65, 785–797.

Walford, S.A., Wu, Y., Llewellyn, D.J., Dennis, E.S., 2012. Epidermal cell differentiation in cotton mediated by the homeodomain-leucine zipper gene, GhHD-1. Plant J. 71 (3), 464–478.

Wang, K., Wang, Z., Li, F., Ye, W., Wang, J., Song, G., et al., 2012. The draft genome of a diploid cotton *Gossypium raimondii*. Nat. Genet. 44, 1098–1103.

Wang, S., Wang, J.W., Yu, N., Li, C.H., Luo, B., Gou, J.Y., et al., 2004. Control of plant trichome development by a cotton fiber MYB gene. Plant Cell. 16, 2323–2334.

Wendel, J.F., Cronn, R.C., 2003. Polyploidy and the evolutionary history of cotton. Adv. Agron. 78, 139–186.

Wu, Y., Machado, A.C., White, R.G., Llewellyn, D.J., Dennis, E.S., 2006. Expression profiling identifies genes expressed early during lint fibre initiation in cotton. Plant Cell Physiol. 47, 107–127.

Xiao, Y.H., Li, D.M., Yin, M.H., Li, X.B., Zhang, M., Wang, Y.J., et al., 2010. Gibberellin 20-oxidase promotes initiation and elongation of cotton fibers by regulating gibberellin synthesis. J. Plant Physiol. 167, 829–837.

Xu, L., Zhu, L., Tu, L., Guo, X., Long, L., Sun, L., et al., 2011a. Differential gene expression in cotton defence response to *Verticillium dahliae* by SSH. J. Phytopathol. 159, 606–615.

Xu, L., Zhu, L., Tu, L., Liu, L., Yuan, D., Jin, L., et al., 2011b. Lignin metabolism has a central role in the resistance of cotton to the wilt fungus *Verticillium dahliae* as revealed by RNA-Seq-dependent transcriptional analysis and histochemistry. J. Exp. Bot. 62 (15), 5607–5621.

Yu, X.S., Chu, B.J., Liu, R.E., Sun, J., Brian, J., Wang, H.Z., et al., 2012. Characteristics of fertile somatic hybrids of *G. hirsutum* L. and *G. trilobum* generated via protoplast fusion. Theor. Appl. Genet. 125, 1503–1516.

Yu, Y., Yuan, D., Liang, S., Li, X., Wang, X., Lin, Z., et al., 2011. Genome structure of cotton revealed by a genome-wide SSR genetic map constructed from a BC1 population between *Gossypium hirsutum* and *G. barbadense*. BMC Genomics. 12, 15.

Zhang, M., Zheng, X., Song, S., Zeng, Q., Hou, L., Li, D., et al., 2011. Spatiotemporal manipulation of auxin biosynthesis in cotton ovule epidermal cells enhances fiber yield and quality. Nat. Biotech. 29, 453–458.

Zhang, Y., Zhu, Y., Guo, W., 2008. Cloning, characterization and mapping of two genes *ghsams* and *ghnlp* related with fiber development in *gossypium hirsutum* l. Sci. Agric. Sin. 41 (9), 2581–2588.

Virus technology for functional genomics in plants

14

Cheng Qin[1], Qi Zhang[1], Meiling He[1], Junhua Kong[1], Bin Li[1], Atef Mohamed[2,3], Weiwei Chen[1], Pengcheng Zhang[1], Xian Zhang[1], Zhiming Yu[1], Tongfei Lai[1], Nongnong Shi[1], Toba Osman[2] and Yiguo Hong[1,3]
[1]Research Centre for Plant RNA Signalling, College of Life and Environmental Sciences, Hangzhou Normal University, Hangzhou, People's Republic of China; [2]Department of Botany, Faculty of Agriculture, Fayoum University, Fayoum, Egypt; [3]School of Life Sciences, University of Warwick, Coventry, UK

Acronyms

ACC 1-aminocyclopropane-1-carboxylate
ACO1 ACC oxidase 1
CaLCV cabbage leaf-curl geminivirus
FT FLOWERING LOCUS T
LeHB-1 tomato homeobox protein 1
Mr VIGS microRNA-mediated virus-induced gene silencing
PVT plant virus technology
PVX potato virus X
RISC RNA-induced silencing complex
RNAi RNA interference
Sir VIGS siRNA-mediated virus-induced gene silencing
VbMS virus-based microRNA silencing
VIGC virus-induced gene complementation
VIGE virus-induced genome editing
ViTGS virus-induced transcriptional gene silencing
VRMA virus-based RNA mobility assay
TRV tobacco rattle virus

14.1 Introduction

Whole genome sequences and their annotations for the model dicot *Arabidopsis* and the moncot rice have predicted 30,000–40,000 protein-coding genes in the respective genomes. There are also numerous no-protein-coding genes coding for microRNAs, non-coding long and small RNAs. In plants, the high throughput and vast capacity of next-generation sequencing have speeded up whole genome

Applied Plant Genomics and Biotechnology.

sequences of an increasing number of higher plant and crop species including *Arabidopsis*, rice, maize, potato and tomato. These data together with bioinformatic predictions, as well as transcriptome and proteome analyses of the expression patterns of thousands of unknown genes, will certainly allow identification of a large number of putative protein- and no-protein-coding genes without designated biological functions. Traditionally, to define gene function often involves forward and reverse genetics analyses (Robinson and Parkin, 2009). Both approaches involve mutant analysis that provides a genetic link between genotype and phenotype in order to assign gene function. In plants, mutants can either be identified in natural populations or induced by chemical and physical mutagenesis, T-DNA insertion, transposon integration, RNAi-mediated knockdown or genome editing (Baltes et al., 2014). However, these genetics tools are largely restricted to model plants such as *Arabidopsis*, tobacco or rice. Definite proof of gene function often involves complementation in a mutant background, which is usually achieved through stable transformation. Moreover, transformation is time consuming and technically challenging in many crops of economic importance. Thus to elucidate the biological function of genes predicted from whole genome sequences in order to bridge the knowledge gap from gene-to-function is still a very challenging task.

Plant virus technology (PVT) such as small interfering RNA (siRNA)-mediated virus-induced transcriptional (ViTGS) and posttranscriptional gene silencing (Sir VIGS), microRNA-mediated VIGS (Mr VIGS), virus-based RNA mobility assays (VRMAs) and virus-based ectopic gene expression, is an important forward and reverse genetic approach to study gene function in plants. PVT does not require transformation and it has attracted attention because it is an effective tool to define gene function in plants. In particular, virus-induced gene silencing (VIGS) has become a rapid reverse- and forward-genetics tool in functional genomics since the early development and refinement of VIGS vectors based on plant viruses including *Tobacco mosaic virus*, *Potato virus X* (PVX) and *Tobacco rattle virus* (TRV) (Lindbo et al., 1993; Kumagai et al., 1995; Ruiz et al., 1998; Liu et al., 2002). VIGS has now been applied to more than 30 plant species across the angiosperm phylogeny (Becker and Lange, 2009; Senthil-Kumar and Mysore, 2011). It should be noted that viruses are pathogens and virus infection often causes disease symptoms. This could lead to a complex interpretation of phenotypic alternations in plants.

However, PVT possesses still several important advantages for application in plant functional genomics. First, PVT is an easy and quick technique to investigate gene function in plants. It can result in 'gain-of-function' or 'loss-of-function' phenotypes, thus providing a biological link between phenotype and genotype. Second, PVT possesses a relatively high efficiency for phenotypic complementation or creating mutant phenotypes. Third, PVT does not require time-consuming and labour-intensive stable transformation. Therefore, it can often lead to exciting findings within a short period of time. Fourth, the feasibility of PVT is applicable to most of plant and crop species if not all. This is particularly useful in plants and crops that are difficult to transform.

14.1.1 Virus-based transient gene expression system (Lin et al., 2008; Li et al., 2009)

Virus-based transient gene expression system has long been used for production of proteins such as recombinant subunit vaccines and pharmaceutical proteins in model plants and crops. Indeed many plant DNA and RNA viruses have been engineered to express heterologous genes in systemically infected plant tissues with high levels of foreign production (Porta and Lomonossoff, 2002). These transient systems are also used to express pathogenic genes and examine their associated phenotypes, thus revealing gene functions in plant–pathogen interactions (Van Wezel et al., 2001; Hong et al., 2003). It becomes obvious that such virus-based transient gene expression system can be applied to over-express endogenous genes in order to investigate their functions in plant development and growth.

For instance, the *Arabidopsis FLOWERING LOCUS T* (*FT*) gene encodes the mobile florigen that is required to induce flowering and viral transient expression of this gene promotes floral induction (Li et al., 2009). This was achieved by cloning the *Arabidopsis FT*-coding sequence into the PVX-based gene expression vector. PVX/FT has the capacity to produce a functional FT protein once a plant is treated with this recombinant virus. Indeed, after treated with PVX/FT, short-day *Nicotiana tabacum* cv. Hick's Maryland Mammoth tobaccos were induced to flowering under non-inducing long-day conditions while control plants infected with PVX/mFT carrying a mutated non-translatable FT mRNA remained vegetative. Viral delivery of wild-type and mutated *FT* RNA was readily detected and free FT protein expressed only from PVX/FT but not from PVX/mFT was also detectable in systemically infected leaves. The fact that PVX/FT could induce flowering but PVX/mFT could not, shows that the *Arabidopsis FT* RNA alone is not sufficient to initiate flowering but that its protein product expressed from PVX/FT is absolutely essential for floral induction. Similar experiments using different virus-based vector systems demonstrate that viral transient expression of homologous or indigenous *FT* genes can induce early flowering in vegetable crops and fruit trees, such as cucurbit and apple (Lin et al., 2007; Yamagishi et al., 2014).

Another example is to use virus-based transient gene expression system to study how ethylene biosynthesis gene expression is regulated in order to modulate tomato development and fruit ripening (Lin et al., 2008). Ethylene is required for climacteric fruit ripening. Inhibition of ethylene biosynthesis genes, ACC (1-aminocyclopropane-1-carboxylate) synthase and ACC oxidase, prevents or delays ripening, but it is not known how these genes are controlled during normal tomato development and fruit ripening. LeHB-1, a previously uncharacterized tomato homeobox protein, was shown by gel retardation assay to interact with the promoter of *LeACO1*, an ACC oxidase gene expressed during ripening. Inhibition of *LeHB-1* mRNA accumulation via VIGS greatly reduced *LeACO1* mRNA levels and inhibited fruit ripening. However, ectopic overexpression of *LeHB-1* from a virus vector to developing flowers triggered altered floral organ morphology, including production of multiple flowers within one sepal whorl, fusion of sepals and petals and conversion of sepals into carpel-like structures that grew into fruits and ripened. These unexpected findings suggest that *LeHB-1* is not only involved in the control of fruit ripening via regulation of *LeACO1* gene expression but also plays a critical role in floral organogenesis and tomato development (Lin et al., 2008).

14.1.2 VIGC – virus-induced gene complementation (Zhou et al., 2012; Kong et al., 2013)

A viral transient expression system is useful to express endogenous genes for rapid induction of specific phenotypes in wild-type background plant. However, the simultaneous expression of an endogenous gene from a plant genome can complicate the interpretation of the biological relevance of the transient expression of the same gene from a virus vector. Thus, it would be of advantage to use virus vectors to express plant genes in a mutant background to lead phenotypic complementation, thus as a 'gain-of-function' tool to define gene functions and bridge knowledge gaps between genotypes and phenotypes in plants.

To exploit whether virus technology could be indeed used in 'gain-of-function' assay, we have developed a virus-induced gene complementation (VIGC) system from a modified PVX vector. Viral ectopic overexpression of the MADS-box transcription factor (TF) gene *LeMADS-RIN*, involving no stable tomato transformation, was able to complement non-ripening phenotype in the tomato *Ripening inhibitor* (*rin*) mutant (Zhou et al., 2012). Here green fruits were needle-injected with either recombinant PVX RNA transcripts of PVX/LeMADS-RIN (wild-type), PVX/mLeMADS-RIN (non-sense-mutated), or the empty PVX vector through the carpopodium (pedicel), immature (approximately 3–4 weeks after anthesis) or mature stages. All mock-treated fruits and fruits injected with PVX or PVX/mLeMADS-RIN remained non-ripe, showing characteristics of the *rin* mutant phenotype. However, fruits injected with PVX/LeMADS-RIN appeared to ripen at 2–3 weeks after virus-injection, as evidenced by developing red sectors on the fruit. Thus, expression of LeMADS-RIN from PVX/LeMADS-RIN can functionally complement the *rin* mutant.

To further test the usefulness of the VIGC technique, we expressed the SBP-box LeSPL-CNR TF in the tomato *Colourless non-ripening* (*Cnr*) mutant fruit and found that viral delivery of a functional *LeSPL-CNR* gene caused complementation of the *Cnr* mutant (Kong et al., 2013; Manning et al., 2006). Moreover, comparative gene expression analysis of ripening and non-ripening tissues at the precisely equivalent growing stage collected from same individual fruits, revealed that LeMADS-RIN is a master TF in a transcriptional network in modulating tomato fruit ripening. This model is supported by the fact that LeMADS-RIN directly interacts with promoters of several key ripening-associated TFs (Fujisawa et al., 2012; Martel et al., 2011; Zhou et al., 2012).

14.1.3 Sir VIGS and Mr VIGS – virus-induced gene silencing (Tang et al., 2010; Chen et al., 2014)

Traditional VIGS is a powerful virus-based short interfering RNA-mediated RNA silencing technique dubbed Sir VIGS for plant functional genomics (Becker and

Lange, 2009; Senthil-Kumar and Mysore, 2011). Sir VIGS has been widely used in many plant species, owing to its capability of effective and rapid gene 'knock-down'. However, a drawback for Sir VIGS is the off-target silencing and it can cause nonspecific mRNA degradation or translation repression. This disadvantage can be overcome to a certain extent by microRNA (miRNA)-based VIGS, known as Mr VIGS, in which only one known mature miRNA can be expressed from a specifically designed miRNA precursor by a virus vector, leading to highly specific silencing of target genes (Tang et al., 2010; Chen et al., 2014). Mr VIGS is developed based on the fact that miRNAs have been shown to regulate gene expression by RNA silencing in various organisms.

A Cabbage leaf-curl geminivirus (CaLCV) vector is modified to express artificial and endogenous miRNAs in plants. Indeed this Mr VIGS system was effective to silence the expression of a wide range of endogenous genes in *N. benthamiana*. For instance, silencing of *SGT1* led to the loss of *N*-mediated resistance to *Tobacco mosaic virus*. Ectopic expression of endogenous *miR156* and *miR165* but not their mutant forms via Mr VIGS resulted in earlier abnormal developmental phenotypes. Thus, the CaLCV-based miRNA expression system can be utilized not only to specifically silence genes involved in general metabolism and defence but also to investigate the function of endogenous miRNAs in plants (Tang et al., 2010). More recently a second Mr VIGS system was developed, which uses a TRV-based vector (Chen et al., 2014).

14.1.4 VbMS – virus-based microRNA silencing (Sha et al., 2014)

MiRNAs are genome-encoded 20–24-nucleotide small RNAs that act as post-transcriptional regulators in eukaryotes. In plants, mature miRNAs are excised from the primary miRNA transcript by RNase III-like endoribonucleases 1 and are selectively loaded into RNA-induced silencing complex (RISC). The RISC can then cause either target mRNA degradation or translational repression directed by the miRNA with complete or partial complementarity to the target transcript. MiRNAs play pivotal roles in various biological processes across kingdoms. Many plant miRNAs have been experimentally identified or predicted by bioinformatics mining of small RNA databases. However, their functions remain largely unidentified due to the lack of effective genetic tools.

Plant viral vectors have been widely used for transient gene expression and for gene silencing in plants (Senthil-Kumar and Mysore, 2011). These viral vectors have the potential to become an attractive and quick approach to uncover miRNA function in plants, especially in those unsuited for genetic transformation. Indeed, we have recently reported a virus-based miRNA silencing (VbMS) system that can be used for functional analysis of plant miRNAs. VbMS is performed through TRV-based expression of miRNA target mimics to silence *endogenous* miRNAs. For example, VbMS of either miR172 or miR165/166 caused developmental defects in *N. benthamiana* while VbMS of miR319 reduced the complexity of tomato compound leaves. Therefore, TRV-based VbMS is a powerful tool to silence endogenous miRNAs and to dissect their functions in different plant species (Sha et al., 2014).

14.2 VRMA – virus-based RNA mobility assay (Li et al., 2009, 2011)

RNA trafficking plays an important role in systemic signalling that controls plant development and defence against pathogen infection. Hundreds of RNA transcripts have been recently identified in phloem, suggesting phloem-mobile RNAs may act as long-distance signalling molecules in plants. Indeed, systemic movement of a homeobox fusion transcript and gibberellic acid-insensitive RNA regulates leaf architecture (Haywood et al., 2005; Kim et al., 2001), a non-cell-autonomous mobile RNA represents a long-distance signal that modulates potato tuber formation (Banerjee et al., 2006), and siRNAs are components of intercellular and systemic mobile signals for innate RNA silencing defence (Dunoyer et al., 2005). RNA trafficking is also critical for plant viruses and viroids to establish systemic infection. It has been demonstrated that an RNA motif directs long-distance trafficking of a small naked RNA viroid (Zhong et al., 2007). Moreover, a short RNA sequence is found to be involved in cell-to-cell movement of a plant viral RNA (Lough et al., 2006), and replication-independent viral RNA can move over long distances in plants (Gopinath and Kao, 2007).

In floral induction, the mobile florigen is encoded by the *Arabidopsis FT* gene. *FT* transcribes mRNA in the leaf, but its encoded FT protein functions in the shoot apices where flowers develop. The *Arabidopsis* FT protein and its orthologues have been shown to be involved in long-distance signalling in floral induction (Jackson and Hong, 2012). However, whether *FT* mRNA is also capable of systemic spread remains to be demonstrated. This question is now addressed by a VRMA.

The novel RNA mobility assay vectors are based on two distinct movement-defective RNA viruses, PVX and *Turnip crinkle virus* (Li et al., 2009, 2011). The mobility of the *FT* RNA is reflected by its ability to recover cell-to-cell and long-distance spread of both movement-defective VRMA vectors throughout plants. In VRMAs, not only does *FT* RNA move over long distances but, remarkably, it also facilitates the systemic spread of heterologous green fluorescent protein (GFP) mRNA and different viral RNAs in plants. The *FT* RNA movement does not rely on the expression of the FT protein. Moreover, VRMAs also reveal that the *FT* RNA mobility is determined by a *cis*-acting element localized within nucleotides 1–102 of the *FT* mRNA coding sequence. Viral ectopic expression of non-translatable *FT* RNA induces early flowering in the short-day *N. tabacum* Maryland Mammoth tobacco (Li et al., 2011). Thus VRMAs has proved to be a powerful tool to demonstrate that a plant self-mobile RNA molecule can mediate long-distance trafficking of heterologous RNAs and raise the possibility that *FT* RNA, along with the FT protein, may be involved in the spread of the floral stimulus throughout the plant (Jackson and Hong, 2012).

14.3 Conclusion

In this article, we summarized some latest development of plant virus-based techniques including VIGC, Mr VIGS, VbMS and VRMA and their great applications to study floral induction, fruit ripening, miRNA and mRNA signalling in plant development. Plant viruses can be developed as user-friendly and unique tools for investigation into roles that plant genes play in various developmental processes. Further elucidation of molecular mechanisms about how viruses hijack plant cell replication and gene expression machinery and how they maintain a delicate balance between causing diseases and host cellular antiviral defence may bring novel ideas for developing new virus-based techniques to address fundamental questions in plant sciences.

References

Baltes, N.J., Gil-Humanes, J., Cermak, T., Atkins, P.A., Voytas, D.F., 2014. DNA replicons for plant genome engineering. Plant Cell. 26, 151–163.

Banerjee, A.K., Chatterjee, M., Yu, Y., Suh, S.-G., Miller, W.A., Hannapel, D.J., 2006. Dynamics of a mobile RNA of potato involved in a long-distance signaling pathway. Plant Cell. 18, 3443–3457.

Becker, A., Lange, M., 2009. VIGS – genomics goes functional. Trends Plant Sci. 15, 1–4.

Chen., W., Zhang, Q., Kong, J., Hu, F., Li, B., Wu, C., et al., 2014. Chapter 11 MR VIGS: microRNA-based virus induced gene silencing in plants. In: Mysore, K.S., Muthapa, S.-K. (Eds.), Methods in Molecular Biology, Plant Gene Silencing: Methods and Protocols. Humana Press, Springer.

Dunoyer, P., Himber, C., Voinnet, O., 2005. DICER-LIKE 4 is required for RNA interference and produces the 21-nucleotide small interfering RNA component of the plant cell-to-cell silencing signal. Nat. Genet. 37, 1356–1360.

Fujisawa, M., Shima, Y., Higuchi, N., Nakano, T., Koyama, Y., Kasumi, T., et al., 2012. Direct targets of the tomato-ripening regulator RIN identified by transcriptome and chromatin immunoprecipitation analyses. Planta. 235, 1107–1122.

Gopinath, K., Kao, C.C., 2007. Replication-independent long-distance trafficking by viral RNAs in *Nicotiana benthamiana*. Plant Cell. 19, 1179–1191.

Haywood, V., Yu, T.-S., Huang, N.-C., Lucas, W.J., 2005. Phloem long-distance trafficking of GIBBERELLIC ACID-INSENSITIVE RNA regulates leaf development. Plant J. 42, 49–68.

Hong, Y., Stanley, J., Van Wezel, R., 2003. Novel system for the simultaneous analysis of geminivirus DNA replication and plant interactions in *Nicotiana benthamiana*. J. Virol. 77, 13315–13322.

Jackson, S.J., Hong, Y., 2012. Systemic movement of FT mRNA and a possible role in floral induction. Front. Plant Sci. 3, 127.

Kim, M., Canio, W., Kessler, S., Sinha, N., 2001. Developmental changes due to long-distance movement of a homeobox fusion transcript in tomato. Science. 293, 287–289.

Kong, J., Chen, W., Shen, J., Qin, C., Lai, T., Zhang, P., et al., 2013. Virus-induced gene complementation in tomato. Plant Signal Behav. 4, 8.

Kumagai, M.H., Donson, J., Della-Cioppa, G., Harvey, D., Hanley, K., Grill, L.K., 1995. Cytoplasmic inhibition of carotenoid biosynthesis with virus-derived RNA. Proc. Natl. Acad. Sci. USA. 92, 1679–1683.

Li, C., Zhang, K., Zeng, X., Jackson, S., Zhou, Y., Hong, Y., 2009. A cis element within flowering locus T mRNA determines its mobility and facilitates trafficking of heterologous viral RNA. J. Virol. 83, 3540–3548.

Li, C., Gu, M., Shi, N., Zhang, H., Yang, X., Osman, T., et al., 2011. Mobile FT mRNA contributes to the systemic florigen signalling in floral induction. Sci. Rep. 1, 73.
Lin, M.K., Belanger, H., Lee, Y.J., Varkonyl-Gasic, E., Kaoka, K., Miura, E., et al., 2007. FLOWERING LOCUS T protein may act as the long-distance florigenic signal in the cucurbits. Plant Cell. 19, 1488–1506.
Lin, Z.F., Hong, Y.G., Yin, M.G., Li, C.Y., Zhang, K., Grierson, D., 2008. A tomato HD-Zip homeobox protein, LeHB-1, plays an important role in floral organogenesis and ripening. Plant J. 55, 301–310.
Lindbo, J.A., Silva-Rosales, L., Proebsting, W.M., Dougherty, W.G., 1993. Induction of a highly specific antiviral state in transgenic plants: implications for regulation of gene expression and virus resistance. Plant Cell. 5, 1749–1759.
Liu, Y., Schiff, M., Dinesh-Kumar, S.P., 2002. Virus-induced gene silencing in tomato. Plant J. 31, 777–786.
Lough, T.J., Lee, R.H., Emerson, S.J., Forster, R.L., Lucas, W.J., 2006. Functional analysis of the 5′ untranslated region of potexvirus RNA reveals a role in viral replication and cell-to-cell movement. Virology. 351, 455–465.
Manning, K., Tor, M., Poole, M., Hong, Y., Thompson, A.J., King, G.J., et al., 2006. A naturally occurring epigenetic mutation in a gene encoding an SBP-box transcription factor inhibits tomato fruit ripening. Nat. Genet. 38, 948–952.
Martel, C., Vrebalov, J., Tafelmeyer, P., Giovannoni, J.J., 2011. The Tomato MADS-Box transcription factor RIPENING INHIBITOR interacts with promoters involved in numerous ripening processes in a COLORLESS NONRIPENING-dependent manner. Plant Physiol. 157, 1568–1579.
Porta, C., Lomonossoff, G.P., 2002. Viruses as vectors for the expression of foreign sequences in plants. Biotechnol. Genet. Eng. Rev. 19, 245–291.
Robinson, S.J., Parkin, I.A.P., 2009. Bridging the gene-to-function knowledge gap through functional genomics. In: Somers D.J., Langridge, P., Gustafson, J.P. (Eds.), Plant Genomics Methods and Protocols. Methods in Mol. Biol. 513, 153–173.
Ruiz, M.T., Voinnet, O., Baulcombe, D.C., 1998. Initiation and maintenance of virus-induced gene silencing. Plant Cell. 10, 937–946.
Senthil-Kumar, M., Mysore, K.S., 2011. New dimensions for VIGS in plant functional genomics. Trends Plant Sci. 16, 656–665.
Sha, A., Zhao, J., Yin, K., Tang, Y., Wang, Y., Wei, X., et al., 2014. Virus-based microRNA silencing in plants. Plant Physiol. 164, 36–47.
Tang, Y., Wang, F., Zhao, J., Xie, K., Hong, Y., Liu, Y., 2010. Virus-based microRNA expression for gene functional analysis in plants. Plant Physiol. 153, 632–641.
Van Wezel, R., Liu, H., Tien, P., Stanley, J., Hong, Y., 2001. Gene C2 of the monopartite geminivirus tomato yellow leaf curl virus from China virus encodes a pathogenicity determinant that is localised in the nucleus. Mol. Plant-Microbe Interact. 14, 1125–1128.
Yamagishi, N., Kishigami, R., Yoshikawa, N., 2014. Reduced generation time of apple seedlings to within a year by means of a plant virus vector: a new plant-breeding technique with no transmission of genetic modification to the next generation. Plant Biotechnol. J. 12, 60–68.
Zhong, X., Tao, X., Stombaugh, J., Leontis, N., Ding, B., 2007. Tertiary structure and function of an RNA motif required for plant vascular entry to initiate systemic trafficking. EMBO J. 26, 3836–3846.
Zhou, T., Zhang, H., Lai, T., Qin, C., Shi, N., Wang, H., et al., 2012. Virus-induced gene complementation reveals a transcription factor network in modulation of tomato fruit ripening. Sci. Rep. 2, 836.

PARP proteins, NAD, epigenetics, antioxidative response to abiotic stress

15

Palmiro Poltronieri[1] *and Masanao Miwa*[2]
[1]CNR-ISPA, National Research Council of Italy, Lecce, Italy; [2]Nagahama Institute of Bioscience and Technology, Nagahama, Japan

15.1 Introduction

In most organisms, from higher eukaryotes to plants, chemical and oxidative damage activates poly-ADP-ribose polymerases (PARPs), also known as ADP-Ribose Transfer proteins (ADPRTs) that are involved in cell protection in response to the stress signal.

PARP genes are found exclusively in eukaryotes. In mammals, 18 PARPs has been described, linked with processes such as DNA damage repair, transcriptional regulation, chromatin modifications and regulation of nucleoli and centrosome functions (Kanai et al., 2003). PARP proteins are characterized by the PARP-domain motive. PARPs modify their target proteins post-translationally by adding ADP-ribose polymers (pAR-ylation) to glutamic acid and aspartic acid residues. The ADP-ribose moiety originates from nicotinamide adenine dinucleotide (NAD+), cleaved into the ADP-ribose and nicotinamide. This degradation of NAD+ relates PARP activity and energy homoeostasis that in certain case leads to cell death.

Plants have up to three PARP proteins. The red and green algae do not encode members of this family or encode only one or two representatives. However, there are several types of PARP-domain proteins, grouped for the presence of a NAD-binding domain.

In this review several cases are presented regarding the involvement of these enzymes in covalent modification of proteins, in signalling pathways, as recruiters of protein complexes and in activation of transcription factor (TF) complexes involved in biotic and abiotic stress response, as well as in NAD depletion, culminating with localized cell death (hypersensitive response, HR).

15.1.1 PARPs in DNA repair

Mammalian PARP1 and PARP2 are localized in the nucleus. They bind to DNA, sensing DNA single- and double-strand breaks through their zinc fingers in the N-terminal domain. Double-stranded DNA and DNA break is recognized by Zn1,

Applied Plant Genomics and Biotechnology.

while Zn2 binds to single-stranded DNA (Dantzer and Santoro, 2013). Zinc finger 3 (Zn3) has the ability to bind to single-stranded RNA, and may have an indirect role in RNA-dependent activation. PARP activation leads to recruitment of DNA repair proteins, release of histones and chromatin opening. In the repair of DNA single-strand breaks, such as those formed by exposure to ionizing radiation and alkylating agents, XRCC1 protein interacts with DNA ligase III and PARP to participate in the base excision repair pathway. The interaction between PARP and XRRCC1 occurs through the BRCA1 carboxy-terminal (BRCT) domain found in several DNA damage repair and cell cycle checkpoint proteins (Masson et al., 1998), bringing together DNA repairing complexes and DNA polymerase β (Menegazzi et al., 1991). In PARP1, a homodimerization domain resides within its Leucine zipper region. PARP1 proteins catalyse the covalent attachment of ADP-ribose units from NAD+ to a limited number of proteins, among them are histones H1 and H2B, and several DNA-binding proteins regulating DNA replication, repair and transcription (Miwa et al., 1995; Zhang et al., 2013). The C-terminal region contain the PARP domain responsible for binding to NAD+ and competitive inhibitors such as nicotinamide, 3-amino benzamide and 3-methoxy benzamide, exploited in the adjuvant treatment of certain types of cancers (Rouleau et al., 2010). Poly-ADP-ribose (pADPr) is a branched polymer containing from many to hundreds of ADP-ribose units. The automodification domain in the central part of the enzyme is subject to addition of ADPr units to several different glutamic acid residues. In turn, the displacement and dissociation of PARP from DNA breaks occurs due to repulsion of negative charges. A rapid turnover of poly-ADP-ribose polymers is required and is obtained by the concerted activity of PARP and poly-ADP-ribose glycohydrolase (PARG) (Ohashi et al., 2003). The synthesis of pADPr (covalent modification of proteins) and the reverse modification, the release of pADPr from proteins, is a very fast reaction, thanks to the colocalization of PARP and PARG. Poly-ADP-ribose (pADPr) when produced in high quantity for a prolonged time leads to the consumption of intracellular NAD+ and ATP used to restore the NAD pool. Recently a new role has been proposed for unbound pADPr polymer in the activation and recruitment of pADPR-binding proteins (Gagné et al., 2008). This study on pADPr interacting proteins has unravelled the presence of many proteins and signalling pathway involved in pADPr signalling. The study revealed the presence of a new structure (pADPr-binding zinc finger, PBZ) found in DNA repair and checkpoint proteins, which mediates specific pADPr binding (Ahel et al., 2008). Proteins found possessing the pADPr-binding motif include DNA excision repair proteins (ERCC-2, ERCC-6); DNA polymerase subunit α; DNA primase small subunit; DNA replication licensing factors (MCM3, MCM5, MCM7); DNA topoisomerase 2-β; mitochondrial DNA topoisomerase I; SWI/SNF-related matrix-associated actin-dependent regulator of chromatin subfamily A member 5 (SMCA5); centromere protein T (CENP-T); the DNA damage checkpoint response protein HUS1; several histone metabolism-associated proteins, such as the histone acetyltransferases MYST3, MYST4, p300 and PCAF (26) or the pADPr-binding protein DEK (Gagné et al., 2008).

15.1.2 Functions of PARP proteins beyond poly(ADP-ribosyl)ation and DNA repair

HsPARP1 has been shown to function in gene expression non-enzymatically, both as a TF/coregulator and at the chromatin level (Schreiber et al., 2006). For instance, HsPARP1 functions as a co-activator of NF-KB but its enzymatic activity is not required (Hassa and Hottiger, 2008). Even for the PARP enzymes for which poly (ADP-ribosyl)ation activity has been demonstrated, functions has been determined that do not depend on such activity. HsPARP1 can bind directly to regulatory sequences, impacting transcriptional activity, as has been shown for the *CXCL1* promoter (Nirodi et al., 2001) or by cooperation with other proteins that mediate the DNA binding, as shown in the *COX-2* promoter region (Lin et al., 2011). In addition, it can bind to nucleosomes and promote compaction of chromatin (Kim et al., 2004; Wacker et al., 2007).

Human PARP9 (HsPARP9) does not have enzymatic activity, is inducible by interferon and is able to increase the expression of interferon-stimulated genes (Juszczynski et al., 2006), suggesting a role in host defence against viruses. Another enzymatically inactive PARP, HsPARP13, interacts with viral RNA and recruits factors to degrade that RNA (Chen et al., 2008; Gao et al., 2002; Zhu and Gao, 2008). HsPARP13 is also able to induce type I interferon genes by associating with the RIG-I viral RNA receptor in a ligand-dependent manner, promoting oligomerization of this protein. This stimulates ATPase activity of RIG-I and enhancement of NF-KB signalling (Hayakawa et al., 2011). PARP10 has been shown to be a mono ADP-ribosyl transferase (ADPRT). Vyas and colleagues performed a systems-level analysis in human somatic cells and HeLa cells of each PARP protein and the PAR polymer, examining localization and expression throughout the cell cycle. Researchers examined the knock-down phenotype of each PARP and performed follow up analyses to help elucidate function (Vyas et al., 2013). They identified new physiological functions for the PARP family, including regulation of cell viability, of cellular membrane structures and of actin cytoskeleton. Finally, they analysed the function of PARP14, a member of MacroPARPs (with macro domain), finding that PARP14 is a component of focal adhesion complexes regulating the strength and stability of cellular attachment to substrate. The protein module 'macro domain' was found to recognize monomeric, polymeric or both forms of ADP-ribose (Dani et al., 2009). Following the determination of the crystal structure of the ADP-ribose—macro-domain complex, several scientists have exploited selected macro domains for interaction with free ADP-ribose-like molecules and mono-ADP-ribosylated proteins. These studies has led to the identification of glyceraldehyde 3-phosphate dehydrogenase (GAPDH), protein disulphide isomerase (PDI) GRP78/BiP, vimentin, glutamate dehydrogesase (GDH), the elongation factor 1—2 and HNRPH1.

15.1.3 PARP role and activity in the cell reprogramming

Cellular identity is an almost immutable programme: when cells became differentiated, they cannot revert by themselves to a stem-cell state. In differentiated cells, genes typically expressed in other cell types are repressed. Stem-cell pluripotency genes are inactive in differentiated cells. However, cell transformation by introduction of few TFs (Oct4, Sox2 and Klf4) can reverse such lineage restrictions, inducing many differentiated cell types to become iPSCs.

Gene expression is controlled by two classes of protein: TFs and epigenetic regulators. TFs act by binding directly to DNA, whereas epigenetic regulators can influence gene expression in various ways, e.g. by altering histones (and local packaged DNA complexity). Researchers recently focused on two epigenetic regulators, PARP1 and ten-eleven translocation-2 (Tet2), that stimulate the expression of dormant pluripotency genes in fibroblasts, thereby initiating the reprogramming of these cells into iPSCs (Doege et al., 2012). The study showed that, during cell reprogramming, PARP1 and Tet2 induce the removal of a repressing mark (H3K27me3) from histones, as well as the addition of an activating mark (H3K4me2) with gene activation effects, in the pluripotency genes *Nanog* and *Esrrb*, days before the genes became active. Oct4 binds to *Parp1* in stem cells and exogenously added Oct4 induced the expression of PARP protein (Doege et al., 2012). Some fundamental epigenetic machinery, such as the Trithorax (TrxG) H3K4 methyltransferase protein complex, might be broadly required to activate the expression of alternative-lineage genes for any reprogramming event.

15.1.4 NAD pool in cell physiology and plant stress response

The metabolic processes that occur in living cells involve oxidation–reduction (redox) chemistry sustained by redox compounds such as glutathione, ascorbate and pyridine nucleotides. Among these redox carriers, NAD + is a cofactor and energy-bearing compound in cellular oxidations along catabolism and is therefore essential for plant growth and development. In addition to its redox role, there is now understanding that NAD is a signal molecule controlling crucial functions, such as primary and secondary carbon metabolism. Recent studies using integrative *-omics* approaches combined with molecular pathology have shown that manipulating NAD biosynthesis and recycling lead to an alteration of metabolites pools and developmental processes, and changes in the resistance to various pathogens.

NAD^+ is essential to ensure metabolic energy production in mitochondria. Indeed, the tricarboxylic acid (TCA) pathway generates reducing power as NADH, which is subsequently used by oxidative phosphorylation to yield ATP. Mitochondria have an adenine nucleotide translocator/mitochondrial substrate carrier (NDT2) allowing influx of NAD + and acting as an antiport coupled to ADP. Plastids also have a NAD carrier, NDT1. *NDT1* and *NDT2* genes are highly expressed in *Arabidopsis* tissues under high metabolic activities. Recently, a peroxisomal NAD^+ carrier (PXN) has been shown required for optimal fatty acid degradation and for providing the peroxisomes with coenzymes A and NAD^+. Plants exposed to harsh environmental conditions (abiotic and biotic stress) usually face secondary oxidative damage and NAD/NADP homoeostasis is crucial for cell survival. When mitochondrial NADPH pools are depleted, there is an increase in reactive oxygen species (ROS) content, DNA damage and lipid peroxidation. Plant cells can repair oxidative damage with numerous NADPH-dependent antioxidant systems such as thioredoxin and glutathione reductases.

ADPr degradation is also involved in the restoration of NAD pool. Reactions dependent on NAD-binding enzymes are affected by compounds with competitive binding, such as nicotinamide, and trigonelline, an alkaloid formed by the methylation of the nitrogen atom of nicotinamide, a product of niacin metabolism occurring in many plants.

15.1.5 NUDIX hydrolases, pADPr and NAD recycling

In plants, ADPr homoeostasis involves NUDIX (nucleoside diphosphates linked to some moiety X) hydrolases. NUDIX hydrolases (NUDX or NUDT, in *Arabidopsis*) cleave pADPr and NAD to produce nicotinamide. In *Aspergillus nidulans*, NdxA increases acetylation levels of histone H4. The fungal sirtuin SirA uses NAD(+) as a cosubstrate to deacetylate the lysine 16 residue of histone H4 and represses target gene expression. The impaired acetylation of histone and secondary metabolite synthesis in the NdxA-deficient strain were restored by inhibiting SirA, showing that SirA mediates NdxA-dependent regulation. Thus, NdxA controls total levels of NAD(+)/NADH and negatively regulates sirtuin function and chromatin structure (Shimizu et al., 2012).

In *Arabidopsis*, several NUDX genes affect plant response to stress. Depending on the NUDX considered, the effective enzymatic activity can either be NAD-hydrolysing or ADPr-hydrolysing and might influence NADH/NAD+ ratio or ADPr contents. AtNUDX2 protein has pyrophosphatase activities towards both ADP-ribose and NADH. AtNUDX7 is involved in the response to oxidative stress (Ishikawa et al., 2010) while AtNUDX6 positively regulates salicylic acid signalling and NPR1 TF activity (Ishikawa et al., 2009). Further work is needed on the NUDX family to better understand the substrate specificity of these enzymes and the particular context in which each of them is involved.

15.1.6 Plant PARPs and PARP-domain containing proteins

So far nine proteins with a PARP signature were identified in *Arabidopsis*, of which two have high similarity to PARP1. Of the three *Arabidopsis* PARPs (PARP1-3), PARP1 and PARP2 are assigned to tolerance of abiotic and biotic stress (Adams-Phillips et al., 2010), but they have been implicated also in developmental processes (Lamb et al., 2012). PARP1 and PARP2 are associated with DNA repair and transcriptional regulation. PARP3, a more recently identified PARP, is induced by several abiotic stresses and developmental processes, e.g. during seed development. PARPs also affect the level of abscisic acid (ABA) and ABA-related signalling. Down-regulation of PARP activity increases resistance against abiotic stresses such as temperature, excessive light and drought and this resistance is correlated with a reduced poly(ADP-ribose) level (Briggs and Bent, 2011).

How reduced PARP activity leads to the described effects is under investigation. One possibility is through altered NAD+ metabolism (Pétriacq et al., 2013): lower

NAD+ consumption may lead to a lower need for the highly energy-dependent recycling pathway. This allows the NAD+ and ATP pools to be maintained and avoids the link to cell death pathways where NAD+ and ATP depletion are major inductive signals. Another possibility is that PARP regulates key stress signalling pathways at the transcriptional level.

15.1.7 PARP-domain proteins: radical induced cell death-1 and SRO1 in stress-induced morphogenetic response

PARP-domain proteins contain a NAD-binding domain similarly to PARP proteins. Six PARP-domain containing proteins, such as Radical Induced Cell Death 1 (RCD1), have been studied for the effects found in mutants accumulating oxidative stress.

Although first identified in *Arabidopsis thaliana*, these proteins are found throughout land plants and consist of two subgroups (Citarelli et al., 2010; Jaspers et al., 2010). *Arabidopsis thaliana* RCD1, the first member of the SRO family identified, may be active without pADPr synthesis (Jaspers et al., 2010), while wheat RCD1 produce poly-ADP-ribose (Liu et al., 2014). This enzyme promotes the accumulation of ROS, enhancing the activity of NADPH oxidase and the expression of NAD(P)H dehydrogenase, in conjunction with the suppression of alternative oxidase expression. Moreover, it promotes the activity of ascorbate–GSH cycle enzymes and GSH–peroxidase cycle enzymes, which regulate ROS content and cellular redox homoeostasis.

RCD1 and SRO1 proteins confer enhanced resistance and tolerance to hydrogen peroxide and superoxide anions, assessed through phenotypic effects of mutant plants. The SRO (Similar to Radical cell death One) family is characterized by PARP catalytic domains. However, the catalytic domains within this group show variability and this observation may not be applicable to all SRO family members (Citarelli et al., 2010). SRO family members inhibit accumulation of ROS, which contributes to stress-induced morphogenetic response (SIMR) phenotypes; tolerance to multiple biotic and abiotic stress has been shown mediated by OsSRO1c and its interaction with various TFs (You et al., 2014).

The PARP domains of the SRO proteins from the sequenced genomes of *A. thaliana*, *A. lyrata*, *Vitis vinifera*, *Ricinus communis*, *Populus trichocarpa* and *Oryza sativa* ssp. *japonica* have been phylogenetically aligned and shown to cluster into two subgroups (Jaspers et al., 2010).

The first subgroup is found in all examined groups of land plants and consists of relatively long proteins with a WWE protein–protein interaction domain (Aravind, 2001) in the N terminus, the PARP catalytic domain and a C-terminal extension that contains an RST (RCD-SRO-TAF4) domain (Jaspers et al., 2010). These I-type SROs are present in all plant species and could be further divided into three subgroups according to the C-terminal RST domain.

The second subgroup is confined to the eudicot group of flowering plants. These proteins appear to be truncated relative to the first subgroup having lost the

N-terminal region with the WWE domain and retain only the catalytic domain and the RST domain.

SRO family members act as scaffolds bringing together TFs bound to their RST domains with other proteins. Members that contain WWE domains may recruit chromatin remodelling complexes through their WWE domains, and through the RST domain recruit several different TFs, especially those involved in abiotic stress response. The type of TFs bound by RCD1 and SRO family members are diverse, including members of the bZIP, WRKY, bHLH, HSF and AP2/ERF families (Broché et al., 2014).

Plants exposed to chronic, sublethal abiotic stress are characterized by reduced cell elongation, blockage of cell division in primary meristems and activation of secondary meristems (Potters et al., 2009). Plants displaying SIMR show an accumulation of antioxidants and other compounds modulating the stress responses. These changes allow the redistribution of resources to stress response pathways, permitting the acclimation to the environment. SIMR response shows accelerated flowering, a response associated with abiotic stresses, such as nutrient deficiency (Wada et al., 2010; Wada and Takeno, 2010) and salinity (Ryu et al., 2011) and is thought to guarantee reproduction before a lethality effect such as that produced by drought stress. SIMR has been hypothesized to be mediated by accumulation of ROS caused by the stressful conditions and subsequent alterations in auxin accumulation and signalling (Potters et al., 2007). Loss of RCD1 and SRO1 leads to a SIMR-like phenotype in normal conditions and to a cell death phenotype after apoplastic ROS production (Brochè et al., 2014).

Arabidopsis rcd 1 mutants were discovered based on their hypersensitivity to ozone and resistance to methyl viologen (Fujibe et al., 2004). rcd1 mutants are hypersensitive to other sources of apoplastic ROS, such as H_2O_2 (Teotia and Lamb, 2009) as well as salt (Teotia and Lamb, 2009). Conversely, rcd mutants are resistant to UV-B and the herbicide paraquat, which generate ROS in the plastid (Teotia and Lamb, 2009). In contrast, sro1-1 plants are not resistant to the chloroplastic ROS induced by paraquat but are resistant to apoplastic ROS and high salt levels (Teotia and Lamb, 2009). rcd1 single mutants have been shown to accumulate ROS in non-stress conditions. Loss of either RCD1 or SRO1 confers resistance to osmotic stress (Teotia and Lamb, 2009). RCD1 and SRO1 contribution to abiotic stress is complex and that the two genes may have some independent functions. In addition, loss of RCD1 or SRO1 alters responses to a number of different abiotic stresses, suggesting that these genes have broad functions (Zhu et al., 2013).

rcd1; sro1 double-mutant plants show low viability. Double-mutant seedlings do display some photobleaching under normal light conditions, suggesting they are under photooxidative stress (Teotia and Lamb, 2009). Most rcd1-3; sro1-1 plants die as embryos (Teotia and Lamb, 2009) and for the few that germinate only 10%–15% produce more than 2–3 true leaves (Jaspers et al., 2010; Teotia and Lamb, 2009) in stress conditions.

SRO5 expression is relatively low under normal conditions but its expression has been shown to be induced by salt treatment (Borsani et al., 2005) and repressed by high light (Khandelwal et al., 2008). sro5 plants were more sensitive to H_2O_2-mediated oxidative stress and to salt stress (Borsani et al., 2005). SRO5 has also been implicated in regulation of proline metabolism under salt stress both at the small RNA level and by counteracting ROS accumulation caused by proline accumulation (Borsani et al., 2005). Inhibiting ROS accumulation may be a core function of the SRO family.

15.1.8 Involvement of PARP-domain proteins in control of the flavonoid biosynthesis pathway

The core flavonoid structure consists of 15 carbon atoms, represented as C6–C3–C6, such as in diphenylpropane: two aromatic rings with six carbon atoms (ring A, B) are linked by a heterocyclic ring containing three carbon atoms (ring C). Depending on the substitution group in the central carbon ring, flavonoids are divided into subclasses such as chalcones, flavones, flavonols, flavanones, anthocyanins, pronathocyanidins/tannins and isoflavonoids. The condensation of a single 4-coumaroyl-CoA molecule with three malonyl-CoA units by chalcone synthase (CHS) yields the naringenin-chalcone. Chalcone isomerase (CHI) then produces naringenin, which serves as common precursor for a large number of flavonoids. From this step the pathway diverges into several branches, each resulting in a different class of flavonoids (flavones such as luteolin, flavonols such as quercetin, anthocyanins and proanthocyanidins/tannins). Consecutive enzymes of this biosynthesis pathway are organized into macromolecular complexes and associate with endomembranes. In their colocalization, enzymes may be organized in a linear array, loosely associated to the cytosolic face of the endoplasmic reticulum and/or anchored through cytochrome P450-dependent mono-oxygenases, cinnamate-4-hydroxylase and flavonoid-3′-hydroxylase (F3H). It has also been proposed that flavonol synthase (FLS) may play a key role in the stabilization of a large complex formed by CHS, CHI, F3H and FLS.

Many abiotic stress conditions, such as cold, excessive light or nutrient deprivation induce the production and accumulation of stress protective molecules and anthocyanin antioxidants. The enzymes of the anthocyanin biosynthesis pathway are tightly regulated at the transcriptional level.

The transcription of flavonoid biosynthesis genes is controlled by the activity and interaction of specific TFs. The plant species maize, *Antirrhinum* and *Petunia* have been exploited to elucidate the polyphenol biosynthesis pathway and its controlled by regulatory genes.

Anthocyanins are produced by a specific branch of the flavonoid pathway differently regulated in monocot and dicot species. In the monocot maize, the anthocyanin biosynthesis genes are activated as a single unit by a ternary complex of MYB-bHLH-WD40 TFs (MBW complex) (Xu et al., 2014; Baudry et al., 2006). In dicots, such as *Arabidopsis*, anthocyanin biosynthesis genes can be divided into two subgroups: early biosynthesis genes (EBGs) are activated by co-activator-independent R2R3MYB TFs, and late biosynthesis genes (LBGs) are activated by an MBW complex (Petroni and Tonelli, 2011). For instance, Transparent Testa 8 (TT8), a bHLH TF, interacts with the R2R3MYB TT2 and with the WD40 TF TTG1. As a subtle variation to this orchestrated activity, the R2R3MYB subgroup 4 contains the C2 repressor motif clade of transcriptional repressors (MYB3, MYB4, MYB7, MYB32) that sequester bHLH/WD40 complexes by competition with the activating R2R3MYB in the binding to the bHLH-WD40 partners, thus inhibiting proanthocyanidin and flavonoid biosynthesis genes. Proteins with an R3-like domain often act as inhibitors by competing with R2R3MYBs for co-regulators, particularly the bHLH type. Although some R-MYBs have DNA-binding activity, others lack conserved amino acids that are normally required for DNA binding. Moreover, some R-MYBs have hydrophobic residues in the predicted DNA-recognition helix (Feller et al., 2011), or lack positively charged surface residues, presumably interfering with DNA binding. Another subgroup includes proteins related to snapdragon (*Antirrhinum majus*) AmRAD with R-MYBs that participate in flower and fruit development (Feller et al., 2011). As a result, R-MYBs appear to be functionally divergent from MYB3R or R2R3MYB proteins, and may control gene expression directly or indirectly through histone modifications and chromatin remodelling, rather than direct DNA contact.

Variations in the methylation status of the MYB 10 promoter can lead to sectoring of anthocyanin pigmentation (Dixon et al., 2013). A further layer of complexity in the flavonoid synthesis is due to small RNAs regulating the TFs involved in the synthesis of flavonoids. It is known that microRNAs control the stability and activity of mRNAs. For instance certain *MYB* mRNA is targeted and degraded by miR-319 in many plant species, and by miR-159 and miR-160 in maize.

15.1.9 PARP control of anthocyanin synthesis in stress tolerance

A typical downstream response to abiotic stress conditions, such as high light or cold, is the induction of genes that encode enzymes involved in the production of secondary metabolites, flavonols and anthocyanins, often involved in stress responses (Vanderauwera et al., 2005; Cominelli et al., 2008). Anthocyanins could act as ROS protective agents. The anthocyanin synthesis pathway genes are induced not only by ROS but also by several other signals such as deprivation of nutrients or high concentrations of sugars, particularly sucrose. A series of signals are linked to light signalling pathway and circadian rhythm. Additionally, hormone signalling through ABA, jasmonate, cytokinin and gibberellins contribute to the regulation of anthocyanin biosynthesis (Das et al., 2012).

The relation between PARP activity and anthocyanin accumulation was studied using 3-methoxybenzamide inhibition of PARP activity in *Arabidopsis* both in response to long-term reactive oxygen stress in chloroplasts and to high levels of exogenous sucrose (Schulz et al., 2012). The analysis of growth, photosynthesis, cellular redox status and the expression of anthocyanin biosynthetic genes showed that chemical down-regulation of PARP activity reduces the accumulation of anthocyanin as well as ascorbate, and improves stress tolerance in whole plants. PARPs inhibition repressed anthocyanin accumulation under stress conditions. The reduction in defence mechanisms was paralleled by enhanced biomass production.

Noteworthy, the studied genes and molecules were repressed by PARPs inhibition even in unstressed conditions. The reduced anthocyanin production was shown to be based on the repression of transcription of key regulatory and biosynthesis genes.

The only reported PARP-mediated effect on the transcriptional level was the induction of the MYB TF Production of Anthocyanin Pigment 1 (PAP1), a strong positive regulator of the anthocyanin biosynthesis pathway (Tohge et al., 2005), in PARP2::RNAi lines following 6 h of light stress (Vanderauwera et al., 2007). However, neither study investigated the whole pathway in more detail or with combined transcriptional and biochemical analysis.

The transcript levels of key regulatory TFs such as the R3R3MYB PAP1 and the bHLH TT8 and the biosynthesis key enzyme CHS were studied to understand the mechanism at the basis of the reduced anthocyanin accumulation. These analyses demonstrated that PARP inhibition reduced the expression of biosynthetic and regulatory genes under stress and that this effect was strongest for oxidative stress. Thus, the down-regulation of these genes correlated with different reductions in anthocyanin accumulation in oxidative and sucrose stress. Also, although the effect was less pronounced in sucrose stress, it showed the same general effect of PARP inhibition. The PAP1 over-expression line (35S::PAP1) was used and the relative anthocyanin content was evaluated. In this background it has been possible to overcome the transcriptional repression of the pathway. The result showed that PARP inhibition was not able to reduce the relative anthocyanin content significantly, supporting the idea that the effect is based on transcriptional regulation.

To confirm that the effect of reduced PARP activity on the anthocyanin accumulation was not specific to chemical inhibition, Schulz used homozygous T-DNA knock-out lines for all three PARP genes. The experiments showed that all the parp knock-outs have a similarly reduced anthocyanin accumulation under stress but that this reduction is not as strong as with chemical inhibition. Thus might the inhibition of multiple isoforms enforce the effects on TF complexes. It is noteworthy that the chemical inhibition in the parp2 and parp3 background under stress leads to a further reduced anthocyanin accumulation, but not in the parp1 background. These results showed that PARP inhibition influences anthocyanin accumulation at the transcriptional level and that the effect was strongest for the LBGs regulatory genes due to the observed down-regulation of the later steps of the flavonoid biosynthetic pathway.

The microarray expression profiling confirmed the impact of PARP inhibition on transcription of the TFs PAP1 and TT8, indicating that these genes have a great role in the transcription of anthocyanin LBGs.

Chemical PARP inhibition reduced anthocyanin accumulation to a greater degree in oxidative than in sucrose stress. This might be due to a hormonal effect on anthocyanin accumulation (Devoto et al., 2005) especially as ABA was shown to induce the production of anthocyanin in plants grown on sucrose (Loreti et al., 2008). Plants with reduced PARP activity also have an elevated ABA content (Vanderauwera et al., 2007). A possible explanation can be the differential involvement or activation of the ABA pathway during oxidative and sucrose stress conditions, hence explaining the smaller reduction of anthocyanin accumulation by PARP inhibition under sucrose stress. There is a need to consider not only PARPs but also PARP-domain proteins such as RCD1 and SRO1 as the targets of 3-methoxybenzamide as targets and effectors of inactivation of TF complexes controlling anthocyanin synthesis. In this context, a contribution to the effects could be attributed to RCD1 and SRO1 proteins and their capacity to interact with TFs through their WWE and RTS protein–protein interaction domains. The type of TFs bound by the SRO family members are diverse, including members of the bZIP, WRKY, bHLH, HSF and AP2/ERF families. A number of the identified TFs have been shown to be involved in abiotic stress responses (Lamb, 2010). Thus, it could be possible that several TF complexes may be assembling-defective due to inhibition of SRO1/RCD1, required for orchestrating the formation of TF interactions.

15.2 Conclusion

From the studies in higher eukaryotes, PARP enzymes and PARP-domain containing proteins in plants are show to exert a control of chromatin, nucleosomes, gene silencing and epigenetic regulation by TF complexes. In addition, PARP activity regulates NAD pool. Synthesis and stability of pADR polymer seems to influence cell survival or death. Plant PARP-domain proteins have been shown able to regulate the formation and activity of TF complexes involved in the activation of flavonoid biosynthesis genes, supporting anthocyanin accumulation, in stressed and non-stressed plants. Further studies will bring us novel insights on gene activation and epigenetic regulation of plant cell responses to environment.

Acknowledgement

This review was made possible thanks to a Japanese Society for Promotion of Science 2013 Short-Term Fellowship assigned to Italian senior scientists.

References

Adams-Phillips, L., Briggs, A.G., Bent, A.F., 2010. Disruption of poly(ADP-ribosyl)ation mechanisms alters responses of Arabidopsis to biotic stress. Plant Physiol. 152, 267–280.

Ahel, I., Ahel, D., Matsusaka, T., Clark, A.J., Pines, J., Boulton, S.J., et al., 2008. Poly(ADP-ribose)-binding zinc finger motifs in DNA repair/checkpoint proteins. Nature. 451, 81–85.

Aravind, L., 2001. The WWE domain: a common interaction module in protein ubiquitination and ADP ribosylation. Trends Biochem. Sci. 26, 273–275.

Baudry, A., Caboche, M., Lepiniec, L., 2006. TT8 controls its own expression in a feedback regulation involving TTG1 and homologous MYB and bHLH factors, allowing a strong and cell-specific accumulation of flavonoids in Arabidopsis thaliana. Plant J. 46, 768–779.

Borsani, O., Zhu, J., Verslues, P.E., Sunkar, R., Zhu, J.K., 2005. Endogenous siRNAs derived from a pair of natural cis-antisense transcripts regulate salt tolerance in Arabidopsis. Cell. 123, 1279–1291.

Briggs, A.G., Bent, A.F., 2011. Poly(ADP-ribosyl)ation in plants. Trends. Plant Sci. 16, 372–380.

Brochè, M., Blomster, T., Salojarvi, J., Ciu, F., Sipari, N., Leppala, J., et al., 2014. Transcriptomics and functional genomics of ROS-induced cell death regulation by Radical-Induced Cell Death 1. PLoS Genet. 10, e1004112.

Citarelli, M., Teotia, S., Lamb, R.S., 2010. Evolutionary history of the poly(ADP-ribose) polymerase gene family in eukaryotes. BMC Evol. Biol. 10, 308.

Chen, G., Guo, X., Lv, F., Xu, Y., Gao, G., 2008. p72 DEAD box RNA helicase is required for optimal function of the zinc-finger antiviral protein. Proc. Natl. Acad. Sci. USA. 105, 4352–4357.

Cominelli, E., Gusmaroli, G., Allegra, D., Galbiati, M., Wade, H.K., Jenkins, G.I., et al., 2008. Expression analysis of anthocyanin regulatory genes in response to different light qualities in *Arabidopsis thaliana*. J. Plant Physiol. 165, 886–894.

Dani, N., Stilla, A., Marchegiani, A., Tamburro, A., Till, S., Ladurner, A.G., et al., 2009. Combining affinity purification by ADP-ribose-binding macro domains with mass spectrometry to define the mammalian ADP-ribosyl proteome. Proc. Natl. Acad. Sci. USA. 106, 4243–4248.

Dantzer, F., Santoro, R., 2013. The expanding role of PARPs in the establishment and maintenance of heterochromatin. FEBS J. 280, 3508–3518.

Das, P.K., Shin, D.H., Choi, S.-B., Park, Y.-I., 2012. Sugar-hormone cross-talk in anthocyanin biossyntthesis. Mol. Cells. 34, 501–507.

Devoto, A., Ellis, C., Magusin, A., Chang, H.S., Chilcott, C., Zhu, T., et al., 2005. Expression profiling reveals COI1 to be a key regulator of genes involved in wound- and methyl jasmonate-induced secondary metabolism, defence, and hormone interactions. Plant Mol. Biol. 58, 497–513.

Dixon, R.A., Liu, C., Jun, J.H., 2013. Metabolic engineering of anthocyanins and condensed tannins in plants. Curr. Opin. Biotechnol. 24, 329–335.

Doege, C.A., Inoue, K., Yamashita, T., Rhee, D.B., Travis, S., Fujita, R., et al., 2012. Early-stage epigenetic modification during somatic cell reprogramming by Parp1 and Tet2. Nature. 488, 652–655.

Feller, A., Machemer, K., Braun, E.L., Grotewold, E., 2011. Evolutionary and comparative analysis of MYB and bHLH plant transcription factors. Plant J. 66, 94–116.

Fujibe, T., Saji, H., Arakawa, K., Yabe, N., Takeuchi, Y., Yamamoto, K.T., 2004. A methyl viologen-resistant mutant of Arabidopsis, which is allelic to ozone-sensitive rcd1, is tolerant to supplemental ultraviolet-B irradiation. Plant Physiol. 134, 275–285.

Gagné, J.P., Isabelle, M., Lo, K.S., Bourassa, S., Hendzel, M.J., Dawson, V.L., et al., 2008. Proteome-wide identification of poly(ADP-ribose) binding proteins and poly (ADP-ribose)-associated protein complexes. Nucleic Acids. Res. 36, 6959–6976.

Gao, G., Guo, X., Goff, S.P., 2002. Inhibition of retroviral RNA production by ZAP, a CCCH-type zinc finger protein. Science. 297, 1703–1706.

Hassa, P.O., Hottiger, M.O., 2008. The diverse biological roles of mammalian PARPs, a small but powerful family of poly-ADP-ribose polymerases. Front. Biosci. 13, 3046–3082.

Hayakawa, S., Shiratori, S., Yamato, H., Kameyama, T., Kitatsuji, C., Kashigi, F., et al., 2011. ZAPS is a potent stimulator of signaling mediated by the RNA helicase RIG-I during antiviral responses. Nat. Immunol. 12, 37–44.

Ishikawa, K., Ogawa, T., Hirosue, E., Nakayama, Y., Harada, K., Fukusaki, E., et al., 2009. Modulation of the poly(ADP-ribosyl)ation reaction via the Arabidopsis ADP-ribose/NADH pyrophosphohydrolase, AtNUDX7, is involved in the response to oxidative stress. Plant Physiol. 151, 741–754.

Ishikawa, K., Yoshimura, K., Harada, K., Fukusaki, E., Ogawa, T., Tamoi, M., et al., 2010. AtNUDX6, an ADP-ribose/NADH pyrophosphohydrolase in Arabidopsis, positively regulates NPR1-dependent salicylic acid signaling. Plant Physiol. 152, 2000–2012.

Jaspers, P., Overmyer, K., Wrzaczek, M., Vainonen, J.P., Blomster, T., Salojärvi, J., et al., 2010. The RST and PARP-like domain containing SRO protein family: analysis of protein structure, function and conservation in land plants. BMC Genomics. 11, 170.

Juszczynski, P., Kutok, J.L., Li, C., Mitra, J., Aguiar, R.C., Shipp, M.A., 2006. BAL1 and BBAP are regulated by a gamma interferon-responsive bidirectional promoter and are overexpressed in diffuse large B-cell lymphomas with a prominent inflammatory infiltrate. Mol. Cell Biol. 26, 5348–5359.

Kanai, M., Tong, W.M., Sugihara, E., Wang, Z.Q., Fukasawa, K., Miwa, M., 2003. Involvement of poly(ADP-Ribose) polymerase 1 and poly(ADP-ribosyl)ation in regulation of centrosome function. Mol. Cell Biol. 23, 2451–2462.

Khandelwal, A., Elvitigala, T., Ghosh, B., Quatrano, R.S., 2008. Arabidopsis transcriptome reveals control circuits regulating redox homeostasis and the role of an AP2 transcription factor. Plant Physiol. 148, 2050–2058.

Kim, M.Y., Mauro, S., Gevry, N., Lis, J.T., Kraus, W.L., 2004. NAD+-dependent modulation of chromatin structure and transcription by nucleosome binding properties of PARP-1. Cell. 119, 803–814.

Lamb, R.S., 2010. Abiotic stress responses in plants: a focus on the SRO family. In: Montanaro, G. (Ed.), Selected Plant Physiology Aspects. InTech, Rijeka, Croatia, pp. 3–22, ISBN: 978-953-51-0557-2.

Lamb, R.S., Citarelli, M., Teotia, S., 2012. Functions of the poly(ADP-ribose) polymerase superfamily in plants. Cell. Mol. Life Sci. 69, 175–189.

Lin, Y., Tang, X., Zhu, Y., Shu, T., Han, X., 2011. Identification of PARP-1 s one of the transcription factors binding to the repressor element in the promoter region of COX-2. Arch. Biochem. Biophys. 505, 123–129.

Liu, S., Liu, S., Wang, M., Wei, T., Meng, C., Wang, M., et al., 2014. A wheat SIMILAR TO RCD-ONE gene enhances seedling growth and abiotic stress resistance by modulating redox homeostasis and maintaining genomic integrity. Plant Cell. 26 (1), 164–180.

Loreti, E., Povero, G., Novi, G., Solfanelli, C., Alpi, A., Perata, P., 2008. Gibberellins, jasmonate and abscisic acid modulate the sucrose-induced expression of anthocyanin biosynthetic genes in Arabidopsis. New Phytol. 179, 1004–1016.

Masson, M., Niedergang, C., Schreiber, V., Muller, S., Menissier-de Murcia, J., de Murcia, G., 1998. XRCC1 is specifically associated with Poly(ADP-Ribose) polymerase and negatively regulates its activity following DNA damage. Mol. Cell Biol. 18, 3563–3571.

Menegazzi, M., Grassi-Zucconi, G., De Prati, A.C., Ogura, T., Poltronieri, P., Nyunoya, H., et al., 1991. Differential expression of poly (ADP-ribose) polymerase and DNA polymerase β in rat tissues. Exp. Cell. Res. 197, 66–74.

Miwa, M., Hanai, S., Masuda, H., Koyama, Y., Hayashi, T., Yoshida, Y., et al., 1995. Analysis of biological function of poly (ADP-ribosyl)ation in *Drosophila melanogaster*. Biochimie. 77, 466–471.

Nirodi, C., Nagdas, S., Gygi, S.P., Olson, G., Aebersold, R., Richmond, A., 2001. A role for poly(ADP-ribose) polymerase in the transcriptional regulation of the melanoma growth stimulatory activity (CXCL1) gene expression. J. Biol. Chem. 275, 9366–9374.

Ohashi, S., Kanai, M., Hanai, S., Uchiumi, F., Maruta, H., Tanuma, S., et al., 2003. Subcellular localization of poly(ADP-ribose) glycohydrolase in mammalian cells. Biochem. Biophys. Res. Commun. 307, 915–921.

Pétriacq, P., de Bont, L., Tcherkez, G., Gakière, B., 2013. NAD. Not just a pawn on the board of plant-pathogen interactions. Plant Signal. Behav. 8 (1), e22477.

Petroni, K., Tonelli, C., 2011. Recent advances on the regulation of anthocyanin synthesis in reproductive organs. Plant Sci. 181, 219–229.

Potters, G., Pasternak, T.P., Guisez, Y., Jansen, M.A., 2009. Different stresses, similar morphogenic responses: integrating a plethora of pathways. Plant Cell Environ. 32, 158–169.

Potters, G., Pasternak, T.P., Guisez, Y., Palme, K.J., Jansen, M.A., 2007. Stress-induced morphogenic responses: growing out of trouble? Trends. Plant. Sci. 12, 98–105.

Rouleau, M., Patel, A., Hendzel, M.J., Kaufmann, S.H., Poirier, G.G., 2010. PARP inhibition: PARP1 and beyond. Nat. Rev. Cancer. 10, 293–301.

Ryu, J.Y., Park, C.M., Seo, P.J., 2011. The floral repressor BROTHER OF FT AND TFL1 (BFT) modulates flowering initiation under high salinity in Arabidopsis. Mol. Cells. 32, 295–303.

Schreiber, V., Dantzer, F., Ame, J.C., de Murcia, G., 2006. Poly(ADP-ribose): novel functions for an old molecule. Nat. Rev. Mol. Cell Biol. 7, 517–528.

Schulz, P., Neukermans, J., Van Der Kelen, K., Muhlenbock, P., Van Breusegem, F., Noctor, G., et al., 2012. Chemical PARP inhibition enhances growth of Arabidopsis and reduces anthocyanin accumulation and the activation of stress protective mechanisms. PLoS One. 7 (5), e37287.

Shimizu, M., Masuo, S., Fujita, T., Doi, Y., Kamimura, Y., Takaya, N., 2012. Hydrolase controls cellular NAD, sirtuin, and secondary metabolites. Mol Cell Biol. 32, 3743–3755.

Teotia, S., Lamb, R.S., 2009. The paralogous genes RADICAL-INDUCED CELL DEATH1 and SIMILAR TO RCD ONE1 have partially redundant functions during Arabidopsis development. Plant Physiol. 151, 180–198.

Tohge, T., Nishiyama, Y., Hirai, M.Y., Yano, M., Nakajima, J., Awazuhara, M., et al., 2005. Functional genomics by integrated analysis of metabolome and transcriptome of Arabidopsis plants overexpressing an MYB transcription factor. Plant J. 42, 218–235.

Vanderauwera, S., Zimmermann, P., Rombauts, S., Vandenabeele, S., Langebartels, C., Gruissem, W., et al., 2005. Genome-wide analysis of hydrogen peroxi-deregulated gene expression in Arabidopsis reveals a high light-induced transcriptional cluster involved in anthocyanin biosynthesis. Plant Physiol. 139, 806–821.

Vanderauwera, S., De Block, M., Van de Steene, N., van de Cotte, B., Metzlaff, M., Van Breusegem, F., 2007. Silencing of poly(ADP-ribose) polymerase in plants alters abiotic stress signal transduction. Proc. Natl. Acad. Sci. USA. 104 (38), 15150–15155.

Vyas, S., Chesarone-Cataldo, M., Todorova, T., Huang, Y.-H., Chang, P., 2013. A systematic analysis of the PARP protein family identifies new functions critical for cell physiology. Nat. Commun. 4, 2240.

Wacker, D.A., Ruhl, D.D., Balagamwala, E.H., Hope, K.M., Zhang, T., Kraus, W.L., 2007. The DNA binding and catalytic domains of poly(ADP-ribose) polymerase 1 cooperate in the regulation of chromatin structure and transcription. Mol. Cell Biol. 27, 7475–7485.

Wada, K.C., Takeno, K., 2010. Stress-induced flowering. Plant Signal. Behav. 5, 944–947.

Wada, K.C., Yamada, M., Shiraya, T., Takeno, K., 2010. Salicylic acid and the flowering gene FLOWERING LOCUS T homolog are involved in poor-nutrition stress induced flowering of *Pharbitis* nil. J. Plant Physiol. 167, 447–452.

Xu, W., Grain, D., Bobet, S., Le Gourrierec, J., Thavenin, J., Kelemen, Z., et al., 2014. Complexity and robustness of the flavonoid transcriptional regulatory network revealed by comprehensive analyses of MYB-bHLH-WDR complexes and their targets in Arabidopsis seed. New Phytol. 202, 132–144.

You, J., Zong, W., Du, H., Hu, H., Xiong, L., 2014. A special member of the rice SRO family, OsSRO1c, mediates responses to multiple abiotic stresses through interaction with various transcription factors. Plant Mol. Biol. 84, 693–705 (Epub ahead of print).

Zhang, Y., Wang, J., Ding, M., Yu, Y., 2013. Site-specific characterization of the Asp- and Glu-ADP-ribosylated proteome. Nat. Methods. 10, 981–986.

Zhu, Y., Gao, G., 2008. ZAP-mediated mRNA degradation. RNA Biol. 5, 65–67.

Zhu, Y., Du, B., Qian, J., Zou, B., Hua, J., 2013. Disease resistance gene-induced growth inhibition is enhanced by rcd1 independent of defense activation in Arabidopsis. Plant Physiol. 161, 2005–2013.

Applied oilseed rape marker technology and genomics

16

Christian Obermeier and Wolfgang Friedt
Department of Plant Breeding, Justus Liebig University, Giessen, Germany

16.1 Global importance of oilseed rape

Oilseed rape (OSR) (*Brassica napus* L.), a member of the Brassicaceae mustard family, is the world's second oilseed crop after soybean (Carré and Pouzet, 2014). The global acreage of *B. napus* OSR/canola together with related 'rapeseed' brassicas, i.e. turnip rape (*B. rapa*) and mustard species such as Ethiopian (*B. carinata*) and Indian mustard (*B. juncea*), is about 30 million ha, with a current production of around 50 million t. Major growing regions are China (6.6–7.4 million ha in 2008–2010), India (5.5–6.3 million ha, mainly Indian mustard), Canada (6.4–6.5 million ha), France (1.4–1.5 million ha), Germany (1.4–1.5 million ha) and Australia (1.7 million ha). Other countries with significant production are the Ukraine, Russia, the United States and Pakistan (FAOSTAT, 2013, http://faostat.fao.org). In Europe, OSR is mostly grown as a winter crop on a total acreage of 8.2 million ha. To date, most OSR is produced within the EU member states, i.e. 6.1–6.9 M ha in 2008–2010, with 18.9–21.4 million tons of seed harvested. The highest seed and oil yields are achieved in Central and Western Europe, where the annual yield of winter OSR (WOSR) crops varies between 3.3 and 4.3 t/ha. In Germany, average farm yields currently are around 3.5 t/ha, with a peak of 4 t/ha in 2004 (Statistisches Bundesamt, http://www.destatis.de). In central Europe WOSR generates the highest seed and oil yields in comparison to other oil crops; therefore it is the most important oilseed in this region. A great future potential is still seen in Eastern Europe, e.g. Russia, for increasing rapeseed production, depending on climatic and economic conditions. The yield potential in other major rapeseed growing regions of the world is much lower, e.g. below 2 t/ha in Canada and China (http://faostat.fao.org/site/339/default.aspx). This is due, e.g. to the harsh environment and short growing season of spring canola in Western Canada and the small-scale production of alternative rapeseed types as a second crop in Central China. The prospects of enhancing yield under such production conditions seem to be rather limited.

16.1.1 Broadening genetic diversity for OSR breeding

Genetic diversity of a species is generally valuable and particularly needed as a basis for breeding and crop enhancement. Minimal genetic distance is a prerequisite

Applied Plant Genomics and Biotechnology.

for achieving heterotic effects (yield advantage of F1 vs. parents, see below) and therefore necessary for breeding hybrid cultivars with high yield potential. Genetic variance can be created by sexual hybridization (crossing) of released cultivars, adapted breeding lines, landraces or even alien exotic germplasm. In addition, entirely new variation can be created by interspecific or intergeneric hybridization within the Brassicaceae family. Rapeseed (*B. napus*, $2n = 38$) is a natural hybrid between turnip rape (*B. rapa*, genome AA, $2n = 20$) and cabbage (*B. oleracea*, CC, $2n = 18$). Correspondingly, novel forms of rapeseed, including forage and oil types can be created by interspecific hybridization between selected genotypes of the two parental species. It has been demonstrated that the progeny of such interspecific hybrids ('resynthetic [RS] rapeseed') can give rise to completely new progenies differing from commercial OSR cultivars and elite material, as demonstrated by multivariate analysis (e.g. Seyis et al., 2003; Figure 16.1). At the same time, different interspecific families show obvious genetic distance and form separate subclusters within the interspecific progenies. The development of inbred lines derived from such wide crosses can lead to new test hybrids (F1) with elevated yield potential (Friedt et al., 2004; Gehringer et al., 2007; Seyis et al., 2003). Therefore, the use of

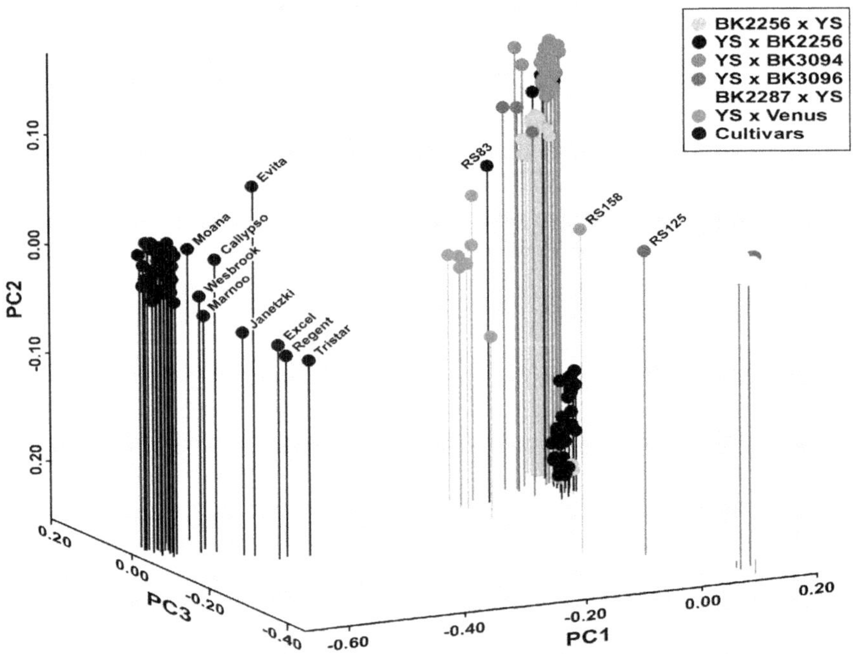

Figure 16.1 Three-dimensional principle component analysis based on morphological characteristics showing genetic relatedness of spring rapeseed (*B. napus*) cultivars and different 'resynthetic rapeseed' families. BK, Blumenkohl (cauliflower, *B. oleracea*); YS, Yellow Sarson (*B. rapa*) (Friedt et al., 2004; Seyis et al., 2003). (For interpretation of the references to colour in this figure, the reader is referred to the web version of this book.)

RS material for the development of new parental material for hybrid breeding can lead to new single-cross hybrids out-yielding former variety types regarding seed and oil yield.

Impressive cases of using exotic germplasm for developing improved breeding material are new cultivars expressing disease resistance derived from interspecific hybrids. For example, in 1987 the clubroot (*Plamodiophora brassicae*) resistant kale (*B. oleracea* ECD-15) was crossed to turnip rape (*B. rapa* ECD-04) to create new RS *B. napus* 15/04 ($2n = 38$). This progeny was crossed to WOSR cv. Falcon. F1 pollen were cultured and DH plants regenerated thereof. The selected DH line 47/19 was then two-times backcrossed to cv. Falcon to generate a BC2F1 population segregating for three dominant, race-specific clubroot resistance (CR) genes. After further crossing and dihaploidization steps, the resistant WOSR cv. Mendel could be released in 2001 (Diederichsen et al., 2009). Using this cultivar in further breeding, new CR cultivars such as Mendelson have been generated (Anonymus, 2013).

16.1.2 Current OSR breeding goals, methods and variety types

Major goals of OSR breeding are based on the needs in cultivation, processing and use of the products. Important agronomical traits include vigorous growth in the juvenile stage, winter hardiness (in case of WOSR), rapid regeneration in spring, adequate stress tolerance and flowering time, extensive branching and flowering, adequate silique formation, high fertility and seed number per pod and plant, sufficient resistance against fungal diseases and insect pests, lodging resistance, high seed yield per area, suitable seed size and oil content. For the seed processor, the size and physical and chemical composition of the seeds are relevant to adjust the processing technology and achieve high product yields, i.e. oil and by-products. For the end user, the composition of the seeds determines its quality, i.e. high oil content and suitable oil quality (fatty acids, secondary components such as tocopherols) combined with high digestibility and good protein quality (essential amino acids) determines the value of extraction meal or cake for animal nutrition.

Although brassicas typically have self-incompatibility systems to avoid self-fertilization and inbreeding, *B. napus* is a facultative outcrossing species. Therefore, it is able to cross-breed, but can also be self-pollinated or used for controlled hybridization to create hybrids (F1). Correspondingly, crop cultivars either represent open pollinated (OP) varieties, inbred lines or F1 hybrids. Nowadays, an increasing part of the registered cultivars represent F1 or single-cross hybrids versus OP line varieties. While inbred lines are classically produced via repeated self-pollination under isolation (e.g. bags), today breeders use *in vitro* culture of pollen (microspores) routinely to generate haploid and doubled haploid (DH) progenies to develop homozygous breeding lines for ultimate variety development.

The development of hybrid seed via cross-fertilization is best done using a genetic male sterility system such as 'Männliche Sterilität Lembke' (MSL) or the INRA ogu. In both cases, male-sterile female lines and fertile male lines are required. They are selected on the basis of extensive tests for their *per se* performance and 'combining ability'. Females and males are stripe-cultivated in production fields to harvest F1 seed produced by the mother after cross-fertilization by the male father. Since the introduction of the first restored WOSR hybrid variety in 1995 (Paulmann and Frauen, 1997) the proportion of hybrids has grown steadily, e.g. in Germany 43 current hybrids comprise almost 50% of the registered WOSR varieties (Anonymus, 2013). In terms of cultivation, there is a trend towards an increased use of hybrids in all major canola and OSR growing areas worldwide. For example, according to breeders' information, at least 70%–80% of the WOSR crop acreage in Germany today represents hybrids (Dr. M. Frauen, NPZ Lembke KG, pers. comm.). A total of 84 OSR cultivars are currently listed in Germany, 77 of which are winter types (Anonymus, 2013). Two of the WOSR cultivars are low glucosinolate but high erucic (0+) types and two are HOLLi lines (>75% oleic, <5% alpha-linolenic acid), while all others represent 00 types (zero erucic acid, low seed glucosinolates, equivalent to *Canola* (Canadian oil, low acid) quality.

Under farm conditions hybrid varieties tend to out-yield line varieties (open-pollinated, OP varieties). The yield advantage of hybrids has been estimated at 8%–16%, depending on the grain yield level (Christen and Friedt, 2012). The newest generation of WOSR hybrids display improved yield performance and stability while also achieving high and stable oil contents (Anonymus, 2013). Therefore, recent hybrid varieties such as 'Avatar', 'Raptor' and 'Mercedes' achieve enhanced farm yields as a prerequisite for profitable rapeseed production. Other major advantages of hybrids versus OP lines are their faster and more vigorous development in autumn leading to a better crop establishment before winter, their better nutrient (e.g. nitrogen) use ability resulting in reduced and more efficient application of fertilizers and their generally better winter survival and stronger growth in spring.

16.2 Rapeseed yield and quality and potential improvements

The seed oil of OSR/canola with 00 quality is widely used for both human nutrition and non-food purposes, either for generating energy (motor fuel) or renewable materials (e.g. hydraulic or lubrication oil). Today, almost all of the OSR production in Europe is of 00 quality. A relatively small amount of *high erucic acid rapeseed* (HEAR, also called 'traditional rapeseed') is still grown in Canada, Europe and probably in other regions for industrial end-uses. Due to its favourable fatty acid composition the oil extracted from current 00 rapeseed varieties is particularly suitable for both the production of edible oils and fats (e.g. salad oil, margarine) and of 'biodiesel' (*rapeseed methyl ester*, RME). Its relatively high mono-unsaturated fatty acid (oleic acid, 18:1) content together with relatively low contents of poly-unsaturated fatty acids (linoleic acid, 18:2; alpha-linoleic acid, 18:3) are determinants of high RME quality (European standard: DIN EN 14214). On the other hand, the combination of approximately 60% oleic and 10% linoleic acid determine the high nutritional and dietary value of rapeseed oil (Table 16.1).

Table 16.1 **Typical fatty acid compositions of selected genotypes of three major oilseed crops**

Crop	Variant	Origin	Fatty acid composition[a]									
			12:0	14:0	16:0	18:0	18:1	18:2	18:3	20:1	22:1	Rest
OSR	High erucic	Wild type	–	–	3	1	11	12	9	8	52	4
	00/canola	Natural mutation	–	–	4	2	60	21	10	1	1	1
	Low linolenic	Induced mutant	–	–	4	2	61	28	3	1	–	1
	Lauric	Genetic engineering	37	4	3	1	33	12	7	–	–	3
	High myristic	Genetic engineering	–	18	23	2	34	15	4	–	–	4
	HOLL/HOLLi	Induced mutant			4	1	84	5	3	1	–	2
Sunflower	Conventional	Wild type	–	–	7	5	19	68	–	–	–	1
	High oleic	Induced mutant	–	–	3	2	92	2	–	–	–	1
Linseed	Conventional	Wild type	–	–	6	4	18	14	58	–	–	–
	Low linolenic	Induced mutant	–	–	6	3	15	73	3	–	–	–

[a]12:0 Lauric, 14:0 Myristic, 16:0 Palmitic, 18:0 Stearic, 18:1 Oleic, 18:2 Linoleic, 18:2 Linolenic, 20:1 Eicosenic, 22:1 Erucic acid.
Source: From Wittkop et al. (2009).

The achievable oil yield from a given amount of harvested seeds depends primarily on its oil content. Further progress regarding oil content and yield can be expected from continued breeding activities. However, a large number of genes or quantitative trait loci (QTL) are involved in seed oil biosynthesis, and a strong environmental modification of seed oil content has been observed (Nesi et al., 2008). Delourme et al. (2006) comprehensively compared the seed oil content QTL among six available populations. Because of the limited marker information and map quality available at the time, the authors could only identify four linkage groups (A1, A3, A8 and C3) on which QTL were detected in multiple populations. Jiang et al. (2014) studied variation in the seed oil content in two related *B. napus* populations, the TN DH population and its derivative 'reconstructed' F2 population. Their results indicate that oil content in OSR is less affected by heterosis than other quantitative traits such as seed yield. By genetic mapping of these populations many QTL could be identified, where a large number showed epistatic interactions. By using common markers on different genetic maps and the TN map as a reference, QTL from three additional populations could be projected onto the TN genetic map. Altogether, 46 distinct QTL regions that control seed oil content could be identified on 16 of the 19 linkage groups of *B. napus* by using the TN map. Furthermore, 18 QTL were detected in multiple populations (Jiang et al., 2014). Their findings will be useful for breeding OSR with improved oil content and oil yield. But further investigations are essential to obtain deeper insight into the control of this complex trait in order to further improve oil content in high-yielding cultivars. In this context, the identification and utilization of genes contributing to oil content via genetic analysis in comparative QTL mapping in different genetic backgrounds will help to identify gene loci with a key function on this complex trait by knowledge-based breeding approaches. For example, in a comparative study homeologous genomic regions involved in oil content in different genetic backgrounds could be identified (Delourme et al., 2006), and novel alleles were found in individual genotypes. Recently, Zhu et al. (2012) identified 39 genes which were differentially expressed in rapeseed lines differing in a QTL influencing oil content. Among these genes, six were differentially expressed regardless of temperature, indicating their major relevance for oil content. Such research work is a basis and one recent example for the potential of the marker-assisted combination of favourable alleles at different genetic loci (see further detailed discussion below) to increase seed oil content in modern OSR by biotechnology-based breeding.

16.3 Future potentials of OSR variety and crop improvement

Studies on the future crop potential have to be necessarily based on the current knowledge and available material, i.e. existing varieties. The yield potential of OSR is generally expected to be much higher than the average farm yield obtained until now. Many own field trials (e.g. Gehringer et al., 2007) as well as breeders' observations and farmers' experience show that the seed yield potential of modern WOSR cultivars is definitely higher than 6 t/ha. While maximum yield requires optimum plant treatment and fertilization, reasonable yields can still be achieved at reduced N-fertilization levels and sub-optimal field conditions (i.e. soil quality, etc.). Whereas the variety type, i.e. OP or hybrid, is less important at marginal sites or low-input conditions, maximal seed and oil yield under high-fertility conditions tends to be more easily and safely achieved with hybrids than with OP varieties. While emissions due to N leaching under low input conditions will be naturally low, such emissions can be avoided under high-input conditions by highly productive varieties which combine a strong nutrient (N) uptake ability due to an efficient root system and a pronounced N-use or translocation efficiency.

Certainly it will be possible to develop new rapeseed varieties with improved nitrogen use efficiency (NUE) in the near future. In replicated field trials at widely different locations (Gehringer et al., 2007) F1 hybrids tended to have the highest yield in replicated multi-environment field experiments in comparison to other variety types under all nitrogen regimes studied, including no mineral N fertilization. Nevertheless, some new experimental hybrids (Test-Hy) were very close to the commercial hybrids under different fertilization levels. Therefore, it is deduced that not only the nutrient (N) uptake, but also the translocation and metabolization efficiency of rapeseed can be improved by breeding, making use of extremely broad genetic material. In a recent study, Ulas et al. (2012) have attributed genotypic differences of nitrogen efficiency to root-growth characteristics. By comparing the N-efficient cv. Apex with cv. Capitol they found a higher root-length density and more living fine roots in the former variety. The authors concluded that genotypes which particularly invest into root growth in the vegetative stage may be more N efficient than others. By using such genotypes for improving on-farm N efficiency, the sustainability of rapeseed cultivation and use would be greatly enhanced. Furthermore, OSR is known to have beneficial effects as a breaking crop in cereal-dominated rotations. For example, in an extensive evaluation of more than 700 Austrian field locations, Kirkegaard et al. (2008) has found grain yield advantages in winter wheat and winter barley grown in rotation with WOSR: when wheat and barley were planted after rapeseed, they yielded 615 kg/ha (wheat) and 430 kg/ha (barley) more than after a cereal pre-crop (http://plantbreeding.boku.ac.at/oilcrops/raps2001). Results of various German studies show even higher effects of rapeseed cultivation on the grain yield of subsequent wheat (or cereal) crops ranging from 0.7 to 2 t/ha (for more details refer http://www.ufop.de/agrar). These findings indicate that OSR is an indispensible field crop in agricultural farm systems otherwise dominated by cereals as is usually the case in many rapeseed producing regions of the world today.

In the last decades breeding procedures and commercial production of OSR cultivars have been strongly affected by biotechnology including cell and tissue culture, DH line production and hybrid breeding. Molecular marker technology has become an integral part of the implementation of these biotechnological approaches within commercial OSR breeding schemes. Molecular markers enable detailed genome analyses, genetic mapping and gene pyramiding towards knowledge-based breeding and marker-assisted selection (MAS) in plants and OSR. A variety of new commercial medium- to high-throughput genotyping-by-hybridization single nucleotide polymorphism (SNP) and genotyping-by-sequencing (GBS) platforms and approaches have been implemented in plant science within the last decade and have become available in *B. napus* research within the last five years. New emerging molecular marker-based approaches (e.g. genomic selection [GS] employing genome-wide molecular markers) have already been initially tested in *B. napus* and will further accelerate commercial exploitation of highly complex quantitative traits (e.g. yield and quality) through OSR biotechnology and molecular breeding in the near future.

16.3.1 Molecular marker systems

The types of molecular marker systems and the marker and sample throughput within genotyping assays increased strongly in the last three decades. Commonly applied marker types changed from hybridization-based markers analysed by gel electrophoresis in the 1980s to PCR-based markers analysed initially by gel electrophoresis and later by fluorescence detection in the 1990s up to more sophisticated highly parallel multi-enzyme- and multi-fluorescence detection-based genotyping-by-hybridization and GBS approaches in the last decade (Table 16.2, for details on marker technologies applied to *B. napus* also see reviews in Gail and Sharpe, 2012; Studer and Kölliker, 2013; Mir et al., 2014).

Most of these molecular marker systems have been established in human science before their application has been reported in the plant science including *B. napus* (Figure 16.2; Table 16.2). Based on the publication record an average delay of the first documented report of 15 commonly used marker types in plant science (3 years) and in *B. napus* (6 years) is apparent. It should be noted that this impression might be biased because the commercial use of new marker technologies for breeding in *B. napus* is not always documented immediately in scientific publications. However, the delay in the documented use of new marker systems in *B. napus* is also partly due to the highly complex and allopolyploid *B. napus* genome (Ganal et al., 2009, Trick et al., 2009, Huang et al., 2013). This results in a more time-consuming process for marker discovery and evaluation for the identification of useful locus-specific markers and is in strong contrast to these processes in other crop species with smaller and less complex diploid genomes (e.g. rice).

16.3.1.1 The forefathers of marker technology: hybridization- and PCR-based markers

The first type of molecular DNA marker used for *B. napus* genotyping in 1988 was the hybridization- and gel-based restriction fragment length polymorphism (RFLP) marker type (Song et al., 1988) (Table 16.2; Figure 16.2). This was replaced with easier to use PCR-based marker types beginning with random amplified dna polymorphism (RADP) markers in 1991 (Quiros et al., 1991; Halldén et al., 1994). Subsequently, amplified fragment length polymorphism (AFLP) markers were

Table 16.2 First description of the use of different common marker types in scientific publications (also see Figure 16.2)

Marker type (company)	Detection based on	Assay type	References for first description in science, plant science, *B. napus* genotyping
RFLP	Hybridization, gel	Restriction, single or multiple locus probes	Botstein et al. (1980), Rivin et al. (1983), Song et al. (1988)
SSR	PCR, gel	PCR, primers flanking 2–6 bp repeats	Litt and Luty (1989), Morgante and Olivieri (1993), Poulsen et al. (1993)
RAPD	PCR, gel	PCR, arbitrary primers	Williams et al. (1990), Williams et al. (1990), Quiros et al. (1991)
AFLP (Keygene)	PCR, gel	Restriction, ligation, selective PCR	Vos et al. (1995), Vos et al. (1995), Vos et al. (1995)
TaqMan SNP genotyping assay (Life Technologies)	PCR, fluorescence	5′ exonuclease, allele-specific probes, PCR	Holland et al. (1991), Salvi et al. (2001), Giancola et al. (2006)
SNaPshot (Life Technologies)	PCR, fluorescence, gel	Single base primer extension, PCR	Chen et al. (1999), Batley et al. (2003), Mikolajczyk et al. (2010)
GeneChip (Affymetrix)	Genotyping-by-hybridization	Allele-specific probes, solid surface array	Lipshutz et al. (1999), Cho et al. (1999)
DArT (DArT P/L)	Genotyping-by-hybridization	Random genomic clone probes, solid surface array	Jaccoud et al. (2001), Jaccoud et al. (2001), Raman et al. (2012)
GoldenGate assay (Illumina)	Genotyping-by-hybridization	Primer extension, ligation, bead array	Oliphant et al. (2002), Rostoks et al. (2006), Durstewitz et al. (2010)
High resolution melting (HRM) curve analysis	PCR, fluorescence	PCR, melting	2002: In: Reed et al. (2007), Lehmensiek et al. (2008), Lochlainn et al. (2011)
SNPlex genotyping system (Life Technologies)	PCR, fluorescence, gel	Oligonucleotide ligation, PCR	Tobler et al. (2005), Pindo et al. (2008), Chaouachi et al. (2008)
KASP assay (LGC Genomics)	PCR, fluorescence	5′ exonuclease, allele-specific PCR	Cuppen (2007), Clarke et al. (2013), Clarke et al. (2013)
Infinium assay (Illumina)	Genotyping-by-hybridization	Single base extension, bead array	2005: In: Gunderson et al. (2006, 2010): In: Ganal et al. (2014), Delourme et al. (2013)
RAD, GBS	Genotyping-by-sequencing	Restriction, ligation, NGS	Baird et al. (2008), Huang et al. (2010), Bus et al. (2012), Chen et al. (2013)

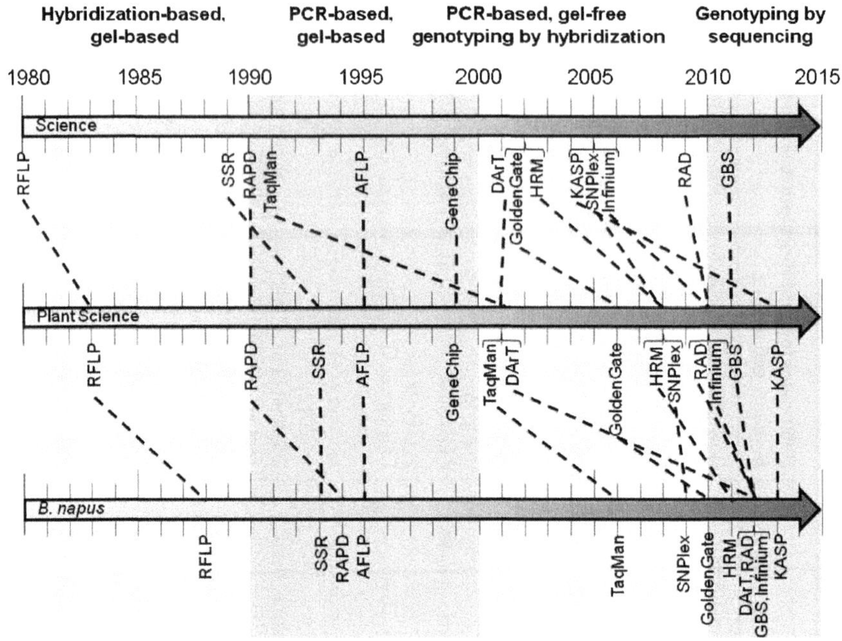

Figure 16.2 Timeline for the first scientific documentation for the application of different commonly used maker types in science, plant science and for *B. napus* genotyping. For details see Table 16.2.

developed for plant genotyping including *B. napus* (Vos et al., 1995). These PCR-based markers do not require pre-existing sequence information. They were complemented in *B. napus* genotyping by simple sequence repeat (SSR) or microsatellite markers in 1993 (Poulsen et al., 1993). Although discovery of SSR markers requires initial investment in sequencing, SSR markers were found to provide more accurate, reliable and rapid semi-automated analyses in general and in *B. napus* genotyping is specific (Kresovich et al., 1995; Mitchell et al., 1997). In addition, as SSR markers are highly polymorphic and co-dominant they widely replaced other PCR-based marker types for genotyping in plants and in *B. napus* beginning in the mid 1990s (Gali and Sharpe, 2012).

Due to increasing genome information of the A, B and C genomes of *Brassica* species by 2005 several hundred of SSR loci from *Brassica* species including *B. napus* (exhibiting an amphidiploid AC genome structure) were identified (Gali and Sharpe, 2012). The increasing availability of *B. napus* transcriptome data (EST collections), genome survey sequences (GSS) and BAC clones has made the development of GSS-SSRs, EST-SSRs and other marker types for *B. napus* possible (Cheng et al., 2009; Trick et al., 2009; Wang F et al., 2012; Li et al., 2013; Ganal et al., 2014).

By today, 799 public domain SSR loci have been published on a webpage for ‘microsatellite information exchange for Brassica’ at http://www.brassica.info

(maintained for the Multinational *Brassica* Genome Project [MBGP]) provided by 11 different sources including commercial consortia. Many more SSR primer sequences for *B. napus* and other *Brassica* species are available from other sources (e.g. Suwabe et al., 2004; Qiu et al., 2006; Radoev et al., 2008; Iniguez-Luy et al., 2009; Cheng et al., 2009; Nagaoka et al., 2010; Wang F et al., 2012; Li et al., 2013). In commercial breeding, SSR markers are up to this day still commonly used in many genotyping applications including MAS in *B. napus* (Table 16.6).

High-resolution melting (HRM) curve analysis is a simple high-throughput genotyping method based on the melting behaviour of amplified PCR products in the presence of a fluorescent dye. HRM can be used for genotyping of SNPs and other polymorphisms (Studer and Kölliker, 2013). The application of HRM for genotyping was first documented in the scientific literature using *B. rapa* by Lochlainn et al. (2011). However, HRM genotyping has been applied since at least 2010 for MAS in commercial rapeseed breeding in Germany (Table 16.5).

16.3.1.2 Marker technology coming of age: enzyme- and array-based genotyping-by-hybridization

Following the success of SSR markers in the early 1990s, since the early 2000s SNPs have become the polymorphism of choice for molecular marker development (Snowdon and Iniguez-Luy, 2012). A SNP is the smallest unit of inheritance representing a nucleotide difference between alleles at a specific locus (Studer and Kölliker, 2013). They are the most abundant form of polymorphism with frequencies in *B. napus* estimated at one SNP every 250–600 bp in the genome (Westermeier et al., 2009, Hayward et al., 2012) and one SNP every 50 bp in the transcriptome (Trick et al., 2009). Initially, enzyme-based genotyping-by-hybridization assays allowing medium throughput were developed, e.g. the TaqMan assay in the early 1990s by the Cetus Corporation company (now provided by Life Technologies; see Holland et al., 1991). The TaqMan assay is based on the 5′ exonuclease activity of the *Taq* polymerase and the use of allele-specific probes coupled to a quencher molecule and a fluorescent dye. The KASP (KASPar) assays provided by the company KBioScience/LGC Genomics since 2005 (Cuppen, 2007, Semagn et al., 2014) is a slightly modified method to the TaqMan method using a unique form of allele-specific PCR (Studer and Kölliker, 2013). TaqMan and KASP assays are well suited for genotyping a low to medium number of SNPs for a high number of samples (Syvanen, 2001). Scientific reports on the application of the TaqMan assay in *B. napus* originate from 2006 and on the KASP assay from 2013 (Giancola et al., 2006; Clarke et al., 2013).

Other enzyme-based low- to medium-throughput genotyping assays which are commonly used in plant science include SNPlex and SNaPShot assays (Applied Biosystems/Life Technologies). These SNP assays are based on single base primer extension or oligonucleotide ligation with fluorescently labelled nucleotides and capillary gel electrophoresis (Studer and Kölliker, 2013). The application of SNPlex and SNaPShot assays for genotyping was first documented in *B. napus* by Chaouachi et al. (2008) and Mikolajczyk et al. (2010), respectively. SNPlex genotyping for *B. napus* has been applied for a number of years by companies offering genotyping services (e.g. TraitGenetics Germany, DNA Landmarks Canada). Similarly, several of the SNP genotyping assay types listed above and in Table 16.2 have been used in *B. napus* genotyping in breeding companies several years before their application first has been reported in scientific publications.

Beginning in the late 1990s and up to the mid-2000s solid phase array-based marker systems (e.g. Affymetrix GeneChips, Diversity Arrays Technology DArT) and bead-based marker systems (e.g. Illumina GoldenGate and Infinium bead arrays) were established (Table 16.2). For *B. napus* the application of a growing number of these technologies were reported by Raman H et al. (2012) for DArT markers and by Durstewitz et al. (2010), Huang et al. (2013) and by Cao and Schmidt (2013) for the Illumina GoldenGate SNP bead array. Illumina Infinium SNP bead custom arrays with medium coverage (2,000–7,000 SNPs) have been developed for *B. napus* (e.g. Delourme et al., 2013; Clarke et al., 2013). An Infinium high-coverage SNP bead array for *B. napus* has been developed by the company Illumina in collaboration with an International *Brassica* SNP consortium. Sequence data was contributed by 16 academic and commercial partners from Australia, China, Europe and North and South America. A 6 K SNP array has been delivered to consortium partners in early 2012, and a 60 K SNP array has been delivered to consortium partners since the mid 2012 (Snowdon and Inguez-Luy, 2012; Edwards and Wang, 2012). The 60 K Infinium SNP bead array version has been used for high-density genetic mapping and genome-wide association analysis (GWAS) for *B. napus* seed quality traits (Liu et al., 2013; Li et al., 2014). Compared to their application in human genetics and animal breeding, the use of SNP assays is currently still relatively limited in crops including *B. napus* breeding, but is expected to grow rapidly as a variety of new SNP assays which can be tailored in cost efficiency to medium to high throughput for different breeding applications and crops recently became available (Snowdon and Inguez-Luy, 2012).

16.3.1.3 Emerging marker technology: GBS

One of these new emerging highly adaptable SNP technologies is based on GBS. It exploits the steep decrease in costs for next-generation sequencing (NGS) technologies in the last years (Snowdon and Iniguez-Luy, 2012; Studer and Kölliker, 2013) and will clearly become the marker genotyping platform of choice in the coming years (Mir and Varshney, 2013). Baird et al. (2008) described a marker technique called restriction-site-associated DNA marker (RAD, RADSeq), which is based on the complexity reduction of genomes by digestion with a restriction enzyme, ligation of size-selected DNA fragments to adapters, PCR amplification and high-throughput NGS. The sequenced fragments from a breeding population, a diversity set of germplasm or a natural population can be assembled *de novo* or aligned to a reference for SNP calling and diversity analyses. In addition to SNP calling RAD also allows the assembly of sequenced DNA fragments into mini contigs and is thus suited for application in species for which no reference genome is available. Modifications of the RAD technology are the GBS (Elshire et al., 2011) or double digest RAD (ddRAD, Peterson et al., 2012) methods which use two restriction enzymes leading to two defined ends of genomic fragments used in paired-end sequencing approaches for alignment to a reference. For SNP discovery in complex genomes – as for many important crop species including *B. napus* – GBS may increase cost efficiency compared to RAD as it allows an improvement of alignment accuracy with a lower sequence coverage for local paired-end alignment using a reference genome sequence. RAD/GBS and related technologies have been successfully used in plants (recent review in Hirsch et al., 2014) for marker development (e.g. Barchi et al., 2011 in eggplant), genetic diversity studies (e.g. Xu et al., 2014 in cucurbits), population genetics studies (e.g. Morris et al., 2013 in sorghum) and in genetic mapping and GS studies, e.g. in barley and wheat (Poland et al., 2012). In *B. napus* RAD has been first used for diversity detection by Bus et al. (2012). A ddRAD/GBS-like approach using complexity reduction with two restriction enzymes also has been used for the first time in *B. napus* by X. Chen et al. (2013) for the construction of a genetic map by developing a software pipeline suited for SNP calling in the complex allopolyploid *B. napus* genome.

Although GBS has the potential to produce SNP data from a much larger number of loci than array-based SNP assays, it is not clear yet if GBS will replace array-based SNP genotyping approaches completely in the near future (Ganal et al., 2014). This is partly due to the current lack of standardized bioinformatics analysis pipelines for GBS data and the error inherent in raw GBS data, which requires a significant investment in bioinformatics (Spindel et al., 2013), especially in crop species with a complex genome like *B. napus* where no or only low quality reference genome sequences are available. Cost efficiency for SNP arrays and for GBS approaches will be judged by the plant breeding industry based on their ability to easily compare genotyping experiments across different projects including time and cost requirements for automated reproducible data processing on a per SNP data point basis. However, with further decreasing sequencing costs and the availability of high-quality genomic reference sequences and bioinformatic analysis tools, sequencing-by-genotyping approaches will partly replace array-based SNP genotyping approaches, although this might require a longer time scale than sometimes anticipated due to the complex inheritance patterns of polyploid plants (Dufresne et al., 2013).

16.3.2 SNP marker discovery

The advent of high-throughput NGS technology in the mid-2000s lead to a reinforcement of high-throughput SNP discovery and analyses pipelines in many crop species including *B. napus* (Hayward et al., 2012; Clarke et al., 2013). SNP discovery in plants has been addressed by transcriptome sequencing (Gilles, 2013), amplicon sequencing (Sexton and Shapter, 2013), reduced genome representation sequencing (Davey et al., 2011) and whole-genome sequencing (Edwards, 2013). For a review on recently finished and ongoing sequencing projects in *B. napus* for SNP discovery and genotyping also see Table 16.1 in Snowdon and Inguez-Luy

(2012). SNP discovery has rapidly progressed in many plant species since the *A. thaliana* genome was completed in 2000.

In *B. napus* SNP discovery by transcriptome sequencing has first been performed by Trick et al. (2009) and later by a number of others (Parkin et al., 2010; Bancroft et al., 2011; Hu et al., 2012). SNPs detected within the *B. napus* transcriptome have contributed most to the currently used SNP markers for genotyping in *B. napus* (Durstewitz et al., 2010; Hayward et al., 2012). Amplicon sequencing for SNP detection in *B. napus* has been performed by Durstewitz et al. (2010) and Gholami et al. (2012). Delourme et al. (2013) applied a sequence capture enrichment approach using EST contigs for sequence capture, 454 sequencing and SNP detection to construct a custom array for *B. napus* genotyping. Clarke et al. (2013) applied a sequence capture enrichment approach using DNA from 47 genomic regions for sequence capture, 454 sequencing and array development. Reduced genome representation sequencing was applied to SNP discovery in *B. napus* using GBS approaches by Bus et al. (2012) and by Chen et al. (2013). Whole-genome sequencing for SNP discovery was done within the International *Brassica* SNP consortium of academic and commercial partners for the development of a 60 K Infinium SNP bead array (see above). SNP discovery *in B. napus* within the discussed approaches faced some difficulties due to the allopolyploid nature of the rapeseed genome. Homeologous pairs of genes within the *B. napus* genome derived from the A and C genome progenitor of the amphidiploid *B. napus* genome will result in transcripts which differ in sequence, on average, by only approximately 3.5% (Trick et al., 2009). This will make alignment to a genomic reference and automated SNP detection in *B. napus* difficult, resulting in the calling of a large number of 'hemi' SNPs. Using transcript-derived cDNA or genomic DNA fragments and commonly applied short-read NGS technology (e.g. 2×150 bp paired-end sequencing) the alignment to a reference will result for the reads obtained from any homozygous genotype at a given alignment position in the detection of 'hemi'-SNPs. A 'hemi'-SNP (also called an interhomeologue polymorphism or intergenomic SNP) represents a polymorphism where one base in the aligned sequence region corresponds to the A genome homeologue and the other to the C genome homeologue (Cao and Schmidt, 2013). These 'hemi'-SNPs are much more frequent in *B. napus* (about 80% of total SNPs) compared to 'simple' SNPs (intrahomeologue polymorphism) where the A and C genome can be easily distinguished using automated SNP detection (Trick et al., 2009). In addition to these 'simple' SNPs (homologous sequence variants) and 'hemi'-SNPs (homeologous sequence variants, HSVs) paralogous sequence variants (PSVs) are present, which are derived from duplicated genes that have diverged from a common ancestral sequence within one of the A and C subgenomes (Kaur et al., 2012). The discrimination of these different SNP types will be a challenge to adapt SNP calling software for *B. napus* for efficient discovery of locus-specific SNPs and reduction of SNP validation cost to identify SNP markers useful for genetic mapping and other breeding applications. Resequencing of homozygous parents of mapping populations for SNP discrimination and discovery using optimized alignment parameters and yet to be established standardized data analysis pipelines for short sequence reads obtained from genotypes of the mapping populations might be necessary to improve SNP calling efficiency in *B. napus* (Kaur et al., 2012). Inbred lines used for genetic mapping approaches might also exhibit residual heterozygosity, thus complicating the SNP calling procedures.

16.3.3 Molecular marker applications

The increasing availability of biochemical and molecular DNA-based marker systems during the last three decades has greatly facilitated their application for the characterization of genetic plant resources, gene functions and in marker-assisted breeding schemes in many crop species including *B. napus*.

The current standard and potential evolving applications of molecular markers in *B. napus* can be ranked by relevance based on the occurrence of the corresponding key words in scientific publications within the publicly accessible Google Scholar database (Figure 16.3) and in other scientific databases. Applying nine categories of marker use in plants modified after Henry (2013), the following ranking sorted by

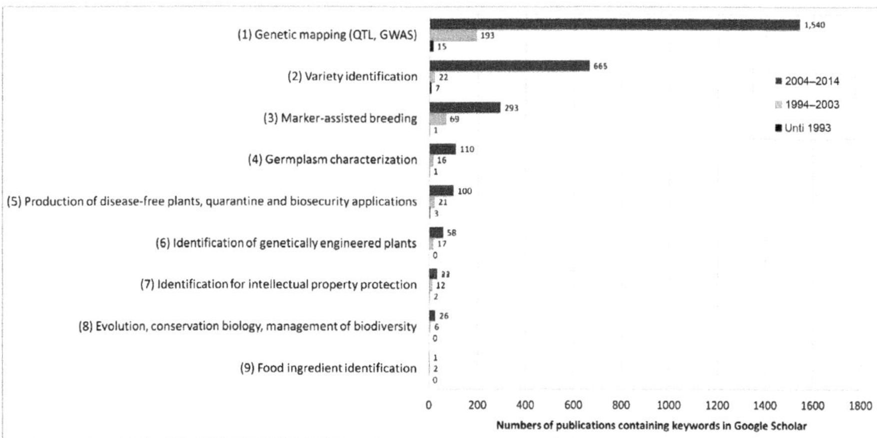

Figure 16.3 Standard and potential evolving applications of molecular markers in *B. napus* estimated based on search hits found in the Google Scholar database and sorted by relevance. Google Scholar accessed on 23 February 2014 with search terms: '*Brassica napus*' and any combination of the following terms: (1) 'molecular marker', 'genetic mapping'; (2) 'molecular marker', 'variety identification'; (3) 'molecular marker', 'marker-assisted breeding'; (4) 'molecular marker', 'germplasm characterization'; (5) 'molecular marker', 'quarantine'; (6) 'molecular marker', 'GMO'; (7) 'molecular marker', 'intellectual property rights'; (8) 'molecular marker', 'evolution', 'conservation biology', 'biodiversity'; (9) 'molecular marker', 'food testing'. Note that reported numbers only represent a rough estimate of the application of markers because key words might only be present once within the publication for cross-linking or might refer to the unsubstantiated predictions on future use of markers.

total numbers of publications on these subjects for *B. napus* in the Google Scholar database was observed:

1. Genetic mapping
2. Variety identification
3. Marker-assisted breeding
4. Germplasm characterization
5. Production of disease-free plants, quarantine and biosecurity applications
6. Identification of genetically engineered plants
7. Identification for intellectual property protection
8. Evolution, conservation biology, management of biodiversity
9. Food ingredient identification

The absolute numbers of publications on certain subjects might be biased based on the search algorithms used by the Google Scholar database (e.g. Beel and Gipp, 2010) and may be partly overlapping similar rankings found using other scientific databases (e.g. PubMed, AGRICOLA). Also it should be noted that some marker applications with a high commercial relevance might be published with a delay or not at all in scientific publications. However, in all of the application categories listed above, molecular markers have been used or have been suggested to be useful in *B. napus* according to the scientific literature with a strong increase in the last decade (Figure 16.3). The most common and economically relevant applications of molecular markers in *B. napus* research and breeding are summarized and discussed below.

16.3.3.1 Germplasm characterization

The characterization of *B. napus* and related *Brassica* species germplasm and the understanding of their genetic diversity can help breeders to specifically select parental lines for the production of new improved OSR varieties (Bonnema, 2012). Initially morphological and biochemical markers, e.g. seed oil fatty acid profiles, have been used to evaluate the genetic diversity. However, these markers are sensitive to environmental influences and with the advent of more robust molecular DNA-based markers they have been rapidly replaced. Molecular DNA markers have been used intensively by rapeseed breeders for analysis of genetic diversity of OSR cultivars, breeding lines and germplasm from related *Brassica* species for selection of parents for heterotic hybrid combinations and for introduction of new agronomically valuable alleles into the gene pool of elite breeding material. The gene pool of recent *B. napus* cultivars is narrow due to intensive selection during the last decades of OSR breeding (see above, Snowdon and Friedt, 2004). Different types of molecular markers have been applied, but especially SSR markers have been proven to be extremely useful for identification of genetic distances in germplasm collections due to their highly polymorphic, co-dominant and reproducible nature. SSR markers have also been found to provide better resolution than SNP markers for analysis of population structure and diversity in maize germplasm with about 10 times less markers necessary (Hamblin et al., 2007; van Inghelandt et al., 2010). Recent examples for germplasm characterization and diversity studies using

SSR markers for fingerprinting of *B. napus* germplasm and applications in practical breeding are summarized by Bonnema (2012) in general and by Rahman (2013) for spring-type OSR.

16.3.3.2 Variety identification

New crop varieties must undergo testing for distinctness, uniformity and stability (DUS) in countries that are members of the International Union for Protection of Plant Varieties (UPOV). DUS testing is the basis of intellectual property protection in the form of Plant Breeders' Rights (PBR) and Plant Variety Protection. For distinctness and uniformity tests the UPOV has accepted morphological characteristics that are unrelated to quality or commercial value (Jones et al., 2014).

However, morphological characteristics used by the UPOV scheme for *B. napus* do have low discriminative power as OSR cultivars are often difficult to distinguish based on their morphology (Bonnema, 2012). Molecular markers are required by the UPOV to be directly predicting traditional morphological characteristics, e.g. by the application of gene-based diagnostic molecular markers. Molecular markers producing anonymous DNA fragments (e.g. RAPD, AFLP) are not considered applicable in DUS testing for intellectual property rights protection as they may not provide legally sufficient information. SSR markers have been suggested to be useful for distinctness and uniformity testing for *B. napus* varieties (Tommasini et al., 2003). However, SSR markers produce length polymorphisms due to DNA replication errors resulting in multiple alleles per SSR locus, but mostly not in functional gene- or locus-specific information. The most suitable marker system for DUS testing fulfilling legal requirements might be SNP markers.

The usefulness of molecular markers for diagnosis of differences between varieties and their implementation in DUS testing has been evaluated by the UPOV in several crops including *B. napus* in the last five years and is still under discussion (Jones et al., 2014). However, when genotyping a collection of 335 OP OSR varieties with 18 selected SSR markers a disappointingly low correlation between 21 morphological characteristics and molecular marker-based distances was detected (UPOV 2008, http://www.upov.int/edocs/mdocs/upov/en/bmt/11/bmt_11_11_add.pdf).

The key to implementation of molecular markers for *B. napus* variety identification in the UPOV scheme in the future will be the identification of genomic *B. napus* regions and the use of functional SSR and SNP markers associated with morphological useful and discriminative DUS traits.

16.3.3.3 Genetic linkage mapping

Genetic mapping or linkage mapping in plants is commonly based on the measurement of co-segregation patterns of hundreds to millions of polymorphic molecular markers evaluated for hundreds of individuals from a biparental cross (a mapping population). The process of constructing a genetic map is based on statistical algorithms involving a stepwise approach: first grouping of markers on linkage groups, then ordered of markers within these groups and finally estimating the distance between the markers (Cheema and Dicks, 2009). Genetic map distances are usually expressed as recombination frequency in centimorgans (cM). One application of genetic maps is the identification of genomic regions linked to quantitative traits termed quantitative trait locus/loci (QTL) mapping. Genetic maps can also provide the basis for map-based cloning of major genes involved in important agronomic traits and the development of markers for MAS. Other applications include the dissection of genome and trait organization, the understanding of evolutionary relationships between and within species and the assistance in genome sequence assembly.

The first genetic map was constructed in 1913 using phenotyping markers applied to a segregating population of the fruit fly *Drosophila melanogaster*. The rapid evolution of DNA manipulation technology since the early 1980s led to a steady increase of the application of DNA-based molecular markers in diversity and genetic mapping studies in many crop species (Schlötterer, 2004). In the beginning of genetic mapping studies since 1988, RFLP markers have been intensively used to analyse the genome structure and evolutionary relationship between *Brassica* species (Figdore et al., 1988). In *B. napus* RFLP markers were applied for the first time by Landry et al. (1991) to construct a genetic linkage map from an F2 segregating mapping population. A few years later, a landmark paper by Parkin et al. (1995) described the comparative mapping of RFLP markers between different *Brassica* species including *A. thaliana*.

Genetic linkage mapping requires the production of segregating mapping populations by crossing two parents with phenotypic difference(s) in at least one trait of interest. Common types of mapping populations in *B. napus* genetic linkage mapping consist of F2, DH, recombinant inbred line (RIL) and backcross (BC) populations (Gali and Sharpe, 2012). A list of *B. napus* mapping populations used for genetic linkage mapping in the past has been made available by the MBGP at http://www.brassica.info. A summary on genetic linkage maps of *Brassica* species published from 2006 to 2012 is listed in Table 5–1 in Gali and Sharpe (2012). Genetic linkage maps published since 2012 for *B. napus* are listed in Table 16.3.

Applications of genetic linkage maps in *B. napus* include biparental QTL analyses, multiple population QTL analyses and integration, comparative mapping with related *Brassica* species and map-based cloning. In recent years a strong increase in the number of published biparental QTL and association studies can be observed for *B. napus*. Major traits which have been studied in *B. napus* in the last two years by QTL mapping and by association mapping are listed in Tables 16.3 and 16.5. *B. napus* traits studied within the last two decades using marker-based approaches include yield-related traits and heterosis, quality-related traits (seed oil, glucosinolates, fatty acid composition), flowering time and disease resistance traits (fungus and virus resistance). For a review also see Table 16.1 in Snowdon and Friedt (2004).

In recent years, the increasing availability of molecular markers of different types allowed map alignments for multiple biparental crosses within *B. napus* and other *Brassica* species using bridge or anchor markers for comparative genome and

Table 16.3 Summary of *B. napus* linkage maps generated from biparental populations and genetic mapping published since 2012

Biparental cross	Population type and size	Number of mapped marker loci	Map length and number of linkage groups (LGs)	Significance of linkage map/QTL identified	Reference
B104-2 × Eyou Changjia	124 F10 RIL	840 (62 AFLP, 257 SRAP, 472 SSR, 49 gene-based markers)	1914 cM, 19 LGs	QTL for seed yield and yield-related traits and their responses to reduced phosphorus supply	Ding et al. (2012)
SI1300 × Eagle	184 F2:F3 families	617 (142 SRAP, 163 functional markers, 160 SSR, 117 AFLP)	2055 cM, 20 LGs	QTL for seed yield per plant, siliques per plant, seeds per silique, 1,000 seed weight, heterosis	Li Y et al. (2012)
SW Hickory × JA177	190 DH lines	403 (397 SSR, 6 gene-specific)	2126 cM, 19 LGs	Seed weight	Cai et al. (2012)
SG-DH (Sollux × Gaoyou)	282 DH lines	481 (193 SSR, 175 STS, 84 SSCP, 10 CAPS, 17 SRAP, 2 SCAR)	1949 cM, 19 LGs	QTL for oil content with 14 lipid related candidate gene loci	Zhao J et al. (2012)
TNDH, RC-F2, BC4-F2 BC3-F2	202 DH lines, 463 RC-F2 lines, 63 BC4-F2 lines, 265 BC3-F2 lines	790 (SSR, RFLP, SRAP, STS, gene-based)	2190 cM, 19 LGs	QTL for tocopherol content in seeds	Wang X et al. (2012)
BQDH, TNDH	200 DH lines, 202 DH lines	486 (468 SSR, 9 gene-based, 7 SRAP)	1874 cM, 19 LGs	QTL for seed yield under low and normal boron conditions	Zhao Z et al. (2012)

(*Continued*)

Table 16.3 (Continued)

Biparental cross	Population type and size	Number of mapped marker loci	Map length and number of linkage groups (LGs)	Significance of linkage map/QTL identified	Reference
SASDH (Skipton × Ag-Spectrum)	186 DH lines	256 (229 SSR, 22 SRAP, 4 candidate gene, 1 SCAR)	2672 cM, 24 LGs	QTL for qualitative and quantitative *L. maculans* resistance. A major qualitative locus *Rlm4*	Raman R et al. (2012a)
No. 2127 × ZY821	88 DH lines	560 (233 EST-SSR, 327 SSR)	1875 cM, 19 LGs	EST-SSR marker discovery from transcriptome data for genetic mapping	Wang F et al. (2012)
BnaLMDH (Lynx-037DH × Monty-028DH)	131 DH lines	548 (437 DArT, 135 SSR, 6 IP, 6 gene-based)	2288 cM, 19 LGs	First genetic linkage map construced with DArT markers	Raman H et al. (2012)
Maxol*1 × Westar	101 DH lines	328 (DArT, SSR)	1250 cM, 19 LGs	Locus for qualitative *L. maculans* resistance *Rlm1*	Raman R et al. (2012b)
H155 × Qva	276 DH lines	376 (271 SSR, 105 SRAP)	1383 cM, 19 LGs	QTL for pod-shatter resistance	Wen et al. (2012)
TNDH (Tapidor × Ningyou7)	202 DH lines	786 (SSR, SNP, STS, SSCP, RFLP, CAPS, AFLP, MS-AFLP)	2117 cM, 19 LGs	QTL for glucosinolates in seeds and leaves	Feng et al. (2012)
TNDH (Tapidor × Ningyou7)	186 DH lines	621 (RFLP, SSR, SNP, AFLP)	2060 cM, 19 LGs	QTL shoot and root growth	Shi et al. (2012)

B104-2 × Eyou Changjia	124 F10 RIL lines	840 (62 AFLP, 257 SRAP, 472 SSR, 49 gene-based markers)	1914 cM, 19 LGs	QTL for seed mineral accumulation	Ding et al. (2013)
KNDH (KenC-8 × N53-2)	384 DH lines	403 (275 SSR, 117 SRAP, 10 STS, 1 IFLP)	1784 cM, 19 LGs	QTL for oil content	Wang et al. (2013)
ExR53-DH (Express617 × R53)	214 DH lines	205 (150 SSR, 53 AFLP, 2 SCAR)	2026 cM, 21 LGs	QTL for *V. longisporum* resistance, phenylpropanoids in hypocotyl	Obermeier et al. (2013)
ExR53-DH (Express617 × R53)	229 DH lines	229 (80 SSR, 149 AFLP)	2283 cM, 21 LGs	QTL for seed dormancy, thousand kernel weight, ABA content	Schatzki et al. (2013)
DYDH (Darmor-bzh × Yudal)	280 DH lines	3497 (2664 SNP, 833 PCR)	2049 cM, 19 LGs	Genotyping of SNP from integrated map in *B. napus* collections to estimate polymorphism level and linkage disequilibrium (LD)	Delourme et al. (2013)
TNDH (Tapidor × Ningyou7)	94 DH lines	3594 (2763 Infinium SNP, 831 PCR)	1947 cM, 19 LGs		Delourme et al. (2013)
AADH (Aviso × Aburamasari)	87 DH lines	3385 Infinium SNP	3495 cM, 19 LGs		Delourme et al. (2013)
AMDH (Aviso × Montego)	96 DH lines	2301 Infinium SNP	1947 cM, 19 LGs		Delourme et al. (2013)

(*Continued*)

Table 16.3 **(Continued)**

Biparental cross	Population type and size	Number of mapped marker loci	Map length and number of linkage groups (LGs)	Significance of linkage map/QTL identified	Reference
Integrated map DYDH, TNDH, AADH, AMDH	557 DH lines	7367 (5764 Infinium SNP, 1603 PCR), 7650 (5986 Infinium SNP, 1664 SSR or sequence derived PCR)	2250 cM, 19 LGs		Delourme et al. (2013)
TNDH (Tapidor × Ningyou7)	176 DH lines	798 (RFLP, SSR, STS, 53 gene-based)	2075 cM, 19 LGs	QTL for root architectural and biomass traits	Shi et al. (2013)
SASDH (Skipton × Ag-Spectrum)	186 DH lines	674 (SSR, SRAP, candidate genes, SCAR, 495 DArT)	4515 cM, 24 LGs	QTL for flowering time	Raman et al. (2013)
8008 × 4942C-5	181 DH lines	385 (224 SSR, 44 IP, 117 AFLP)	1979 cM, 19 LGs	QTL for silique length, number of seeds per silique, thousand seed weight	Qi et al. (2014)
TNDH (Tapidor × Ningyou7), TNRC-F2	202 DH lines, 404 F2 lines	TNDH: 786 (SSR, SNP, STS, SSCP, RFLP, CAPS, AFLP, MS-AFLP)	TNDH: 2117 cM, 19 LGs	QTL for seed oil content on TNDH reference and QTL projected from populations DY, RNSL, SG	Jiang et al. (2014)

AFLP, amplified fragment length polymorphism; CAPS, cleavage amplification polymorphism; DArT, diversity array technology; InDel, insertion–deletion polymorphism; Infinium SNP, Illumina Infinium bead array; IP, intron polymorphism; MS-AFLP, metyhlation-sensitive AFLP; RAD, restriction-enzyme associated DNA polymorphism; RFLP, restriction fragment length polymorphism; SCAR, sequenced characterized amplified region polymorphism; SNP, single nucleotide polymorphism; SRAP, sequence-related amplified polymorphism; SSCP, single strand conformation polymorphism; SSR, simple sequence repeat polymorphism; STS, sequence tagged site polymorphism.
Source: For linkage maps published from 2006 to 2012 see Gali and Sharpe (2012).

trait analyses. QTL analyses in multiple populations and integrative QTL or MetaQTL analysis for the production of consensus maps across multiple populations and comparison of QTL regions recently has become more and more common in *B. napus* (e.g. Raman et al., 2013; Wang et al., 2013; Zhou et al., 2014). QTL mapping in multiple populations directly targeted to breeding approaches has also been undertaken (Würschum et al., 2012). Comparative genetic linkage mapping using genetic and genomic resources from related species such as *B. rapa* and *A. thaliana* has shown to be very useful in the genetic and physical fine mapping of a number of agronomically important gene loci, e.g. for disease resistance (Mayerhofer et al., 2005), male sterility (Xie et al., 2012; Lu et al., 2013) and seed oil content (Zhao et al., 2012). For the study of yield-related genes in *B. napus* a comparative genomics approach using information from the rice genome has been applied (Li et al., 2012). Genetic linkage mapping combined with comparative genomics has been suggested to be very useful for map-based cloning and marker development from functional genes involved in the trait(s) of interest.

16.3.3.4 Fine mapping and map-based cloning

Map-based cloning is based on the fine-scale linkage mapping of a trait of interest with tightly flanked markers within a large segregating mapping population and identification of the gene(s) by detailed analysis of large contig clones and chromosome walking (Navabi and Parkin, 2012). Map-based cloning approaches in *B. napus* have been accelerated in recent years by the increasing availability of genomic sequences for *Brassica* species, e.g. the release of the *B. rapa* genome sequence (X. Wang et al., 2011). However, examples for successful implementation of map-based cloning for gene identification in *Brassica* species (Navabi and Parkin, 2012) and *B. napus* are recent and rare (Table 16.4), e.g. in *B. napus* the identification of candidate genes and development of diagnostic markers for a genic male sterility locus (Li et al., 2012), and for the *LepR3* race-specific resistance gene against Phoma (*Leptosphaeria maculans)* (Larkan et al., 2013). Although the *B. napus* genome exhibits a high level of genome duplications, chromosomal rearrangements are creating challenges in developing markers and identifying candidate genes based on the microsynteny with *Arabidopsis* (Snowdon and Friedt, 2004); all recent successful map-based cloning studies for major genes in *B. napus* have been using *Arabidopsis* data (Yi et al., 2010; Dun et al., 2011; Larkan et al., 2013). These studies first identified gene-linked markers on *B. napus* linkage groups by genetic mapping, then identified collinear regions within the genome of *A. thaliana* and *Brassica* species by alignment with marker sequences and finally employed fine-scale mapping of the target gene in *B. napus* using *A. thaliana*-derived markers in large segregating *B. napus* populations. The upcoming availability of genome-wide data from a variety of *Brassica* species (see below) will result in a more efficient and rapid success of map-based cloning approaches as well as of QTL for more complex traits controlled by multiple genes in the near future.

Table 16.4 **Summary on fine mapping and map-based cloning in *B. napus***

Population	Population type and size	Number of cosegregated or closely linked marker loci	Interval length	Chromosome	Map-based cloning	Significance	Reference
DH12075 × PSA12, Shiralee × PSA12	340 and 2548 BC1	1 RAPD, 3 RFLP	3.1 cM, 3.2 cM	N7	No	Fine mapping of *L. maculans* resistance *LmR1*	Mayerhofer et al. (2005)
S45AB, mutant from cv. Oro	1947 NILs	7 AFLP, 4 SCAR	0.4 cM	N7	No	Fine mapping of genic male-sterile gene *BnMs1*	Yi et al. (2006)
S45AB, mutant from cv. Oro	2650 NILs	12 AFLP, 5 SCAR	0.075 cM	N16	No	Fine mapping of genic male-sterile gene *BnMs2*	Lei et al. (2007)
7365AB	2000 NILs	17 AFLP, 5 SCAR	0.4 cM	N19	No	Fine mapping of genic male-sterile gene *BnMs3*	Huang et al. (2007)
9012AB	4136 NILs	5 AFLP, 4 SCAR, 2 PCR	0.14 cM	N19	No	Fine mapping of genic male-sterile gene *BnMs3*	He et al. (2008)
S45AB, S4516AB	4132 NILs, 2650 NILs	SCAR	0.4 cM, 0.038 cM	N7, N16	Yes	Fine mapping of genic male-sterile genes *BnMS1*, *BnMS2*	Yi et al. (2010)
7365AB, 750AB	2134 NILs, 5022 NILs	4 IP	0.04 cM	N19	Yes	Fine mapping of genic male-sterile gene *Bnms3*	Dun et al. (2011)

HZ396 × Y106	151 NILs (BC3F1), 807 NILs (BC4F1)	3 AFLP, 2 SCAR, 3 SSR	5.72 cM	N8, N19	No	Fine mapping of QTL for seeds per silique	Zhang et al. (2012)
9012AB, DH195AB, GosAB	16,703 NILs, 4,054 NILs, 3,316 NILs	3 SCAR	0.14 cM	N19	Yes	Fine mapping of genic male-sterile gene *Bnms3*	Li J et al. (2012)
GH06 × P174, F7 RIL138 × P174	232 F9 872 BC1F2	SCAR	0.75 cM	N9	No	Fine mapping for acid detergent lignin content in seeds	Liu L et al. (2012)
(1141A × Huayehui) × Wu108	3662 BC14F1	4 SSR, 9 SCAR	1.39 cM	N9	No	Fine mapping of nuclear restorer gene for *pol* CMS	Liu Z et al. (2012)
SG-DH (Sollux × Gaoyou)	NILs, 428 BC5F4	2 STS, 7 SSCP	4.3 cM	N1	No	Fine mapping of an oil content QTL	Chen Y et al. (2013)
Topas DH16516 × Surpass500	97 BC1, 349 BC3	6 SSR, 7 InDel	0.8 cM	N10	Yes	Fine mapping of *L. maculans LepR3* reistance gene	Larkan et al. (2013)
EMS mutant T6 × B409	3472 BC8, 5288 BC4	18 (SCAR, AFLP)	0.35 cM	N17	No	Fine mapping of a chlorophyll-deficient mutant (*BnaC.ygl*)	Zhu et al. (2014)

AFLP, amplified fragment length polymorphism; BC, backcross; DH, doubled haploid lines; InDel, insertion–deletion polymorphism; IP, intron polymorphism; NILs, near isogenic lines; RFLP, restriction fragment length polymorphism; SCAR, sequenced characterized amplified region polymorphism; SSCP, single strand conformation polymorphism; SSR, simple sequence repeat polymorphism; STS, sequence tagged site polymorphism.

16.3.3.5 Association mapping

With the advent of large numbers of SNP markers, Linkage Disequilibrium (LD) mapping or association mapping approaches based on the genetic analysis of populations have been initiated as a possible alternative to the discovery of candidate regions or genes based on biparental QTL mapping in plants (reviewed in Huang and Han, 2014) and in *B. napus* (Gali and Sharpe, 2012). Also for *B. napus*, high-density genetic SNP maps have been produced and applied for comparative biparental genetic linkage mapping, e.g. for disease resistance to, and the identified SNP markers applied in genetically diverse *B. napus* collections for association mapping.

GWAS is based on the use of unrelated populations or lines to investigate genome-wide associations between marker polymorphisms (mostly SNPs) and phenotypes (Zhu et al., 2008). Association mapping with candidate gene markers and genome-wide markers have been applied in maize since 2001 (Zhu et al., 2008) and in *B. napus* since 2008 (Hasan et al., 2008; Table 16.5). In *B. napus* analysed trait-marker associations include seed oil content (Zou et al., 2010), *L. maculans* resistance (Jestin et al., 2011), seed tocopherol content (X. Wang et al., 2012) and seed weight and quality (Li et al., 2014). GWAS in *B. napus* also has been performed for seed glucosinolate content using SNPs derived from sequenced transcripts termed associative transcriptomics (Harper et al., 2012). So far, most association genetic studies in *B. napus* have focused on the mapping of traits rather than to apply these methods in commercial breeding programmes.

An approach which combines the advantages of biparental QTL mapping and population-based association mapping has been first applied in maize and is called Nested Association Mapping (NAM) (Zhu et al., 2008, Yu et al., 2008). NAM can increase the statistical power of QTL detection compared to GWAS for many important and complex agronomical traits which exhibit many small-effect QTL, e.g. seed and oil yield (Edwards et al., 2013). Different public–private consortia in Germany, France, Canada and China are currently developing mapping populations for NAM approaches; see Snowdon and Inguez-Luy 2008 and the MBGP Steering Committee Meeting at the Plant & Animal Genome Conference PAG, 12 January 2014, available at http://www.brassica.info/info/reference/minutes/MBGP-Minutes_PAG-2014.pdf).

16.3.3.6 MAS and breeding

To accelerate breeding the selection for a desired trait or phenotype of interest might be pursued indirectly by selecting for molecular marker polymorphisms closely linked to the wanted phenotype. This MAS approach requires cost-efficient, highly reproducible high-throughput genotyping platforms allowing selection of plants at an early growth stage. In the past, the identification of suitable markers that are highly polymorphic, are tightly linked to the trait of interest (e.g. within 1 cM distance) and are broadly applicable has been time-consuming (Hayward et al., 2012). Since marker systems which exhibit low polymorphism and/or poor

Table 16.5 **Summary on association mapping studies in *B. napus***

Diversity set	Population type and size	Number of mapped marker loci	Population structure, number of groups	Trait-marker associations	Reference
European winter and spring oilseed, fodder and vegetable rapeseed varieties	94 core set varieties, 46 winter-type varieties	559 SSR	1: spring types, 2: winter types, 3: fodder types	Seed glucosinolates	Hasan et al. (2008)
Canola quality winter rapeseed varieties	84 varieties	684 AFLP		14 traits	Honsdorf et al. (2010)
French and European recent and historical commercial varieties	128 varieties	208 SSR	8–9	*L. maculans* resistance	Jestin et al. (2011)
Western European winter-tye varieties	49 varieties	559 SSR	1: winter-type fodder, 2-4: winter-type oilseed	Seed phenolic acids	Rezaeizad et al. (2011)
European and Chinese varieties	248 varieties	52 SSR, 6 candidate gene SNPs	3,6	Flowering time	Wang N et al. (2011)
Diverse European and Chinese varieties and derived lines of 7 crop types	84 (53) accessions	62,980 SNPs called from RNA-Seq data based on unigenes	1: winter types, 2: spring, swede (rutabaga), fodder and Chinese ecotypes	Seed glucosinolate and erucic acid content	Harper et al. (2012)

(*Continued*)

Table 16.5 **(Continued)**

Diversity set	Population type and size	Number of mapped marker loci	Population structure, number of groups	Trait-marker associations	Reference
European and worldwide collection	96 core set varieties plus 133 varieties	52 SSR, 51 candidate gene SNPs, CAPS	4	Tocopherol content	Fritsche et al. (2012)
Worldwide inbred line collection	192 inbred lines	451 single-locus SSR, 740 AFLP	1: semi-winter types from China, 2: winter types from Europe and Canada	Plant height, first branch height, inflorescence length, silique length, seeds per silique, seed weight	Cai et al. (2014)
Inbred lines from the National Mid-term Gene Bank for Oil Crops of China	472 inbred lines	60K Illumina Infinium SNP array	1: Asian semi-winter types, 2: European winter-type lines, 3: European spring type lines, 4: mixed	Seed weight, seed quality, seed erucic acid content, seed glucosinolate content	Li et al. (2014)

genomic distribution (e.g. RFLP) or poor reproducibility (e.g. RAPD) have been replaced in *B. napus* genotyping by highly polymorphic and more reproducible SSR markers, the identification of tightly linked markers for marker-assisted breeding improved significantly. With the establishment of high-throughput SNP discovery pipelines and SNP genotyping platforms in recent years (see above) the identification of markers suitable for marker-assisted breeding in *B. napus* should improve considerably in the near future.

The implementation of MAS in breeding from biparental crosses and/or breeding populations involves a number of steps including the following (modified from Collard and Mackill, 2008):

1. Population development: identification of germplasm representing the genomic diversity in breeding pools, parental selection and hybridization
2. QTL mapping: linkage map construction or association mapping, phenotypic evaluation of traits, QTL analysis
3. QTL validation: confirmation of positions and effects, verification in independent populations, different backgrounds, fine mapping
4. Marker validation: testing of markers in relevant breeding material. Identification of a 'toolbox' of polymorphic markers.

Obstacles to the successful implementation of molecular markers in marker-assisted breeding applications can be (i) a lack of reliability and accuracy in biparental and in association QTL mapping studies; (ii) insufficient linkage between marker and QTL/gene; (iii) limited number of markers and limited polymorphism of markers in breeding material; (iv) effects of genetic background; (v) QTL × environment interaction effects; (vi) high cost of MAS and (vii) the 'Application gap' between research laboratories, plant breeding institutes and breeders' labs (Collard and Mackill, 2008). As the costs for genotyping using multiple platforms adapted to medium to high marker and sample throughput (see above) have been constantly decreasing, the limited availability of polymorphic markers and high costs are no longer factors that represent strong obstacles for application of marker-assisted breeding approaches in *B. napus*. Based on the scientific literature, the impact of MAS in plant breeding might be underestimated as MAS results may not be published. A gap between QTL and MAS studies in the scientific literature evaluated by a search in the Google Scholar database (Figure 16.4) suggests that only a subset of QTL studies leads to the successful transfer of derived molecular markers for use in MAS. This is partly due to the relatively low resolution of QTL regions in linkage maps using biparental populations. Thus, it is not surprising that only a low number of QTL studies have resulted in molecular markers that led to the commercial exploitation in the plant breeding industry and in commercial *B. napus* breeding. However, this gap might also be partly due to the lack of scientific publications on successful implementation of MAS in a commercial setup.

Thus, data on the use of molecular markers in their breeding programmes have been collected from German OSR breeders in a survey in 2011 and have been summarized in Table 16.6. Molecular markers are used in a commercial setup for qualitative as well as quantitative traits in MAS in OSR breeding including markers for

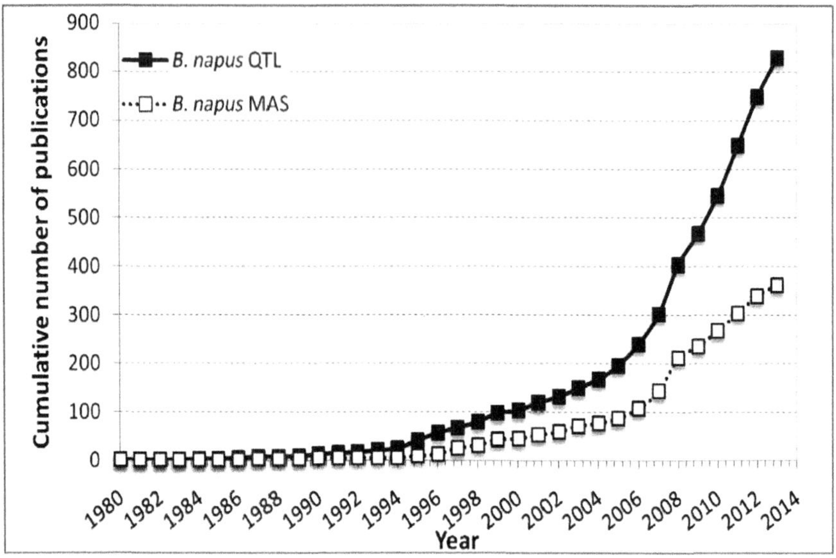

Figure 16.4 From quantitative trait loci (QTL) linkage mapping to MAS in OSR breeding: a gap between published QTL and MAS studies. Cumulative numbers of publications with the terms "*Brassica napus*" in title and "quantitative trait locus" or "quantitative trait loci" (QTL) and "marker-assisted selection" (MAS) are similar somewhere within the document (excluding citations and patents) obtained from Google Scholar accessed on 20 March 2014.

Table 16.6 Major OSR traits for which molecular markers are currently used in a commercial setting by (German) breeders for marker-assisted breeding[a]

Type of trait	Trait of interest	Types of molecular markers
Qualitative	Hybrid system (MSL)	SSR, InDel, SNP
	Hybrid system (Ogu-Rf CMS)	Isozyme, SSR, InDel, HRM, SNP
	Clearfield HT	SNP
	Dwarf growth type	SNP
Quantitative	Genetic distance	SSR, DArT, SNP
	High oleic/low linolenic acid in seeds	SNP
	Phoma (*Leptospheria maculans*) resistance	SSR, InDel
	Clubroot (*Plasmodiophora brassicae*) resistance	SSR
	V. longisporum resistance	SSR

[a]Anonymous survey amongst German oilseed rape breeders in 2011. DArT, diversity arrays technology marker; HRM, high-resolution melting curve analysis marker; InDel, insertion or deletion marker; Isozyme, isozyme marker; SNP, single nucleotide polymorphism marker (e.g. Illumina Infinium, KASP); SSR, simple sequence repeat marker.

hybrid breeding, oil yield and quality and disease resistance. In addition, molecular markers for a number of qualitative and quantitative traits are being tested by OSR breeders in pre-commercial schemes and in collaborative research projects with academic institutions. The types of molecular markers used in commercial setups include almost the whole range of marker types, ranging from traditional isozyme marker technology up to state of the art modern SNP marker technology. However, currently none of these approaches for MAS in OSR on a commercial scale uses genome-wide markers, but mostly gene- or QTL-linked markers.

16.3.3.7 Genomic selection

Current MAS methods are better suited for dealing with a few major-effect genes than with many small-effect genes (Heffner et al., 2009). For the analysis of many small-effect genes involved in highly complex traits the application of genome-wide markers in GS approaches as proposed by Meuwissen et al. (2001) is favoured.

GS or genome-wide selection (GWS) approaches employing genome-wide markers to predict genomic breeding values (*genomic estimated breeding values*, GEBVs) for complex traits have been pioneered in animal breeding. In MAS trait-associated markers are first identified and verified in a limited number of populations and then a subset of polymorphic markers are applied for selection in breeding populations. In contrast, GS and GWS is fundamentally different from MAS in that information from all available polymorphic markers and their simultaneously estimated effects are used for calculation of GEBVs and selection in breeding populations. GS has been evaluated for its usefulness in forest tree breeding (Grattapaglia, 2014) and also crop breeding (Nakaya and Isobe, 2012) and is an evolving field in *B. napus* breeding (Cowling et al., 2012, Würschum et al., 2014).

The efficiency of GS for a crop is depending on the complexity and heritability of the target trait(s), i.e. the number of controlling QTL, GxE interaction, the accuracy of phenotyping in the field, the allelic diversity and linkage disequilibrium, the breeding scheme applied, the level of inbreeding, the population size and the number and type of markers applied. For typical maize and wheat breeding programmes it has been estimated that genetic gain using GS even with low to moderate GEBV accuracies will be 2 to 3 times higher than with MAS approaches (Heffner et al., 2010). In contrast to animal (e.g. cattle) breeding where the allelic diversity and phenotype is documented across many generations with known pedigree for each individual, the analysis of genotypes and phenotypes in plant and *B. napus* breeding is usually only documented in the current breeding cycle and no identity-by-descent information is recorded (Cowling, 2014). In addition, in predominately self-pollinated crops like *B. napus* breeding populations are usually small and allelic diversity is low. This is in contrast to livestock breeding where effective population size is large, allelic diversity is high and the genotype and phenotype information for all individuals is available (Cowling, 2014). In this setting the value of GS is much higher than in the *B. napus* breeding setting. Thus, the efficient application of

GS in self-pollinating crops might also require the development of new statistical methods for including pedigree history of individuals.

Recently, the potential of GS in *B. napus* has first been evaluated using 253 SNP markers and nine families of a total number of 391 DH lines for morphological, quality- and yield-related traits controlled by multiple small-effect QTL (Würschum et al., 2014). Medium to high prediction accuracies for these traits were observed, which suggest that GS can be a powerful genomic tool for commercial rapeseed breeding in the future.

16.3.3.8 Genome analyses and comparative genomics

The sequencing of the model plant *A. thaliana* (Brassicaceae) and completion of the genome sequence in 2000 has provided first comparative insights into the genome of *Brassica* species and the opportunity for comparative genomics with *B. napus* (Edwards and Wang, 2012). Since then, due to the rapid advancement and cost-decrease in next-generation DNA sequencing technology within the last decade, a number of international genome sequencing initiatives have been initiated and sequencing of the genomes of a variety of *Brassica* species is nearing completion (Snowdon and Iniguez-Luy, 2012). A whole-genome *B. rapa* sequence of a Chinese cabbage (accession Chiifu-401-42) was published in 2011 (X. Wang et al., 2011). And very recently, the whole-genome sequence for a *B. oleracea* DH rapid-cycling line (TO1000) has been.

Also, the whole-genome sequence for other crop species from the Brassicaceae family, e.g. *Camelina sativa*, a low-input oil crop, has been completed recently and data applied for comparative genomics with *B. napus* and other *Brassica* species (Parkin, 2014).

A non-public whole-genome sequence for *B. napus* has been reported to be completed in a press release by Bayer Crop Science in 2009 (Snowdon and Inguez-Luy, 2012). Whole-genome sequences for a semi-winter OSR type cultivar ZS11 (Liu, 2014) and for a winter OSR type cultivar Darmor-Bzh have been recently (Chalhoub, 2014), and a whole-genome sequence for a spring type OSR type has been prepared and will soon be released (http://www.brassica.info/info/reference/minutes/MBGP-Minutes_PAG-2014.pdf).

Furthermore, resequencing of more than 200 diverse *B. napus* accessions is ongoing in a number of countries including Canada, China and Germany (MBGP Meeting at the Plant & Animal Genome Conference PAG, 12.1.2014, http://www.brassica.info/info/reference/minutes/MBGP-Minutes_PAG-2014.pdf). In addition, within the ongoing international BMAP Alignment Project the production of high-quality whole-genome sequences for 20 species from the order Brassicales has now started (http://www.brassica.info/resource/sequencing/bmap.php).

These whole-genome sequencing activities, comparative genomics resources and high-throughput genotyping technologies for screening large germplasm collections will accelerate gene identification and genomics-based breeding approaches in OSR within the next few years.

16.4 Conclusion

Today, OSR (*B. napus*) is a favourite crop plant for basic science and applied breeding research, making advantage from its close relationship to major model plants such as *Arabidopsis thaliana* and *B. rapa*. After the genome of the latter two species has been fully sequenced earlier, the genome of *B. napus* has been published recently. This event is going to boost gene discovery, the elucidation of genetic networks and the control of major characteristics of rapeseed including complex agronomic traits such as oil content and seed yield. In addition to its outstanding importance as a vegetable oil source, OSR is an important crop for biofuel production today and is expected to be so in the future because of the following reasons: (i) Rapeseed is an important dicotyledonous crop plant with nutrient requirement as well as diseases and pests differing from monocots (grasses); therefore, it is grown as a major alternative to cereals in many parts of the world, particularly Canada, China and Europe. (ii) OSR is characterized by a high oil content and valuable oil quality, i.e. favourable fatty acid pattern, which allows its alternative use as a food, feed or fuel, giving the farmer the freedom to produce for different purposes and supply separate markets and industrial production chains. (iii) Besides seed oil as valuable major compound, the rapeseeds are also rich in protein, which determines the high nutritional value of rapeseed meals and cakes as a livestock feed (cf. Wittkop et al., 2009). (iv) The species *B. napus* comprises a broad morpho-physiological diversity including spring and winter growth types, seed oil as well as forage types, OP and hybrid cultivars, food or non-food use. (v) The rapeseed plant is very amenable to biotechnology applications in breeding. Therefore, OSR has a high potential for further improvements regarding yield and quality; novel germplasm can be developed by interspecific and intergeneric hybridization (or cell fusion) to create diverse elite material as a basis for the creation of better parental lines and hybrid cultivars. Marker technology and genomics will be the basis for making use of the great agronomical and industrial potential of the rapeseed plant.

References

Anonymus, 2013. Beschreibende Sortenliste 2013. Bundessortenamt, Hannover, Germany.

Baird, N.A., Etter, P.D., Atwood, T.S., Currey, M.C., Shiver, A.L., Lewis, Z.A., et al., 2008. Rapid SNP discovery and genetic mapping using sequenced RAD markers. PLoS One. 3, e3376.

Bancroft, I., Morgan, C., Fraser, F., Higgins, J., Wells, R., Clissold, L., et al., 2011. Dissecting the genome of the polyploid crop oilseed rape by transcriptome sequencing. Nat. Biotechnol. 29, 762–766.

Barchi, L., Lanteri, S., Portis, E., Acquadro, A., Valè, G., Toppino, L., et al., 2011. Identification of SNP and SSR markers in eggplant using RAD tag sequencing. BMC Genomics. 12, 304.

Batley, J., Mogg, R., Edwards, D., O'Sullivan, H., Edwards, K.J., 2003. A high-throughput SNuPE assay for genotyping SNPs in the flanking regions of *Zea mays* sequence tagged simple sequence repeats. Mol. Breed. 11, 111–120.

Beel, J., Gipp, B., 2010. Academic search engine spam and google scholar's resilience against it. J. Electron. Publishing. 13, 3.

Bonnema, G., 2012. Diversity and taxonomy of brassica oil crops. In: Edwards, D., Batley, J., Parkin, I., Kole, C. (Eds.), Genetics, Genomics and Breeding of Oilseed Brassicas. CRC Press, USA, pp. 47–72.

Botstein, D., White, R.L., Skolnick, M., Davis, R.W., 1980. Construction of a genetic linkage map in man using restriction fragment length polymorphisms. Am. J. Hum. Genet. 32, 314–333.

Bus, A., Hecht, J., Huettel, B., Reinhardt, R., Stich, B., 2012. High-throughput polymorphism detection and genotyping in *Brassica napus* using next-generation RAD sequencing. BMC Genomics. 13, 281.

Cai, D., Xiao, Y., Yang, W., Ye, W., Wang, B., Younas, M., et al., 2014. Association mapping of six yield-related traits in rapeseed (*Brassica napus* L.). Theor. Appl. Genet. 127, 85–96.

Cao, H.X., Schmidt, R., 2013. Screening of a *Brassica napus* bacterial artificial chromosome library using highly parallel single nucleotide polymorphism assays. BMC Genomics. 14, 603.

Carré, P., Pouzet, A., 2014. Rapeseed market, worldwide and in Europe. Oilseed Fats Crops Lipids. 21, D102.

Chaouachi, M., Chupeau, G., Berard, A., McKhann, H., Romaniuk, M., Giancola, S., et al., 2008. A High-Throughput multiplex method adapted for GMO detection. J. Agric. Food. Chem. 56, 11596–11606.

Cheema, J., Dicks, J., 2009. Computational approaches and software tools for genetic linkage map estimation in plants. Briefings in Bioinformatics. 10, 595–608.

Chen, X., Levine, L., Kwok, P.Y., 1999. Fluorescence polarization in homogenous nucleic acid analysis. Genome Res. 9, 492–498.

Chen, X., Li, X., Zhang, B., Xu, J., Wu, Z., Wang, B., et al., 2013. Detection and genotyping of restriction fragment associated polymorphisms in polyploid crops with a pseudo-reference sequence: a case study in allotetraploid *Brassica napus*. BMC Genomics. 14, 346.

Chen, Y., Qi, L., Zhang, X., Huang, J., Wang, J., Chen, H., et al., 2013. Characterization of the quantitative trait locus OilA1 for oil content in *Brassica napus*. Theor. Appl. Genet. 126, 2499–2509.

Cheng, X., Xu, J., Xia, S., Gu, J., Yang, Y., Fu, J., et al., 2009. Development and genetic mapping of microsatellite markers from genome survey sequences in *Brassica napus*. Theor. Appl. Genet. 118, 1121–1131.

Cho, R.J., Mindrinos, M., Richards, D.R., Sapolsky, R.J., Anderson, M., Drenkard, E., et al., 1999. Genome-wide mapping with biallelic markers in *Arabidopsis thaliana*. Nat. Genet. 23, 203–207.

Christen, O., Friedt, W. (Eds.), 2012. Winterraps – Das Handbuch für Profis, second ed. DLG-Verlags-GmbH.

Christen, O., Sieling, K., Hanus, H., 1992. The effect of different preceding crops on the development, growth and yield of winter wheat. Eur. J. Agron. 1, 21–28.

Clarke, W.E., Parkin, I.A., Gajardo, H.A., Gerhardt, D.J., Higgins, E., Sidebottom, C., et al., 2013. Genomic DNA enrichment using sequence capture microarrays: a novel approach to discover sequence nucleotide polymorphisms (SNP) in *Brassica napus* L. PLoS One. 8, e81992.

Collard, B.C.Y., Mackill, D.J., 2008. Marker-assisted selection: an approach for precision plant breeding in the twenty-first century. Phil. Trans. R. Soc. B. 363, 557.

Cowling, W.A., 2014. Accelerating plant breeding in the genomics area. The 19th Crucifer Genetics Workshop: Genetic Improvement of Brassicacea Crops in the Era of

Genomics. Brassica 2014. March 30–April 2, Wuhan, People's Republic of China. Program and Abstract Book, p. 202.

Cowling, W.A., Cullis, B.R., Beeck, C.P., Nelson, M.N.G., 2012. Towards genomic selection in oilseed rape. In: Edwards, D., Batley, J., Parkin, I., Kole, C. (Eds.), Genetics, Genomics and Breeding of Oilseed Brassicas. CRC Press, USA, pp. 219–229.

Cuppen, E., 2007. Genotyping by allele-specific amplification (KASPar). CSH Protoc. Available from: http://dx.doi.org/doi:10.1101/pdb.prot4841.

Davey, J.W., Hohenlohe, P.A., Paul, D., Etter, P.D., Boone, J.Q., Catchen, J.M., et al., 2011. Genome-wide genetic marker discovery and genotyping using next-generation sequencing. Nat. Rev. Genet. 12, 499–510.

Delourme, R., Falentin, C., Huteau, V., Clouet, V., Horvais, R., Gandon, B., et al., 2006. Genetic control of oil content in oilseed rape (*Brassica napus* L.). Theor. Appl. Genet. 113, 1331–1345.

Delourme, R., Falentin, C., Fomeju, B.F., Boillot, M., Lassalle, G., André, I., et al., 2013. High-density SNP-based genetic map development and linkage disequilibrium assessment in *Brassica napus* L. BMC Genomics. 14, 120.

Diederichsen, E., Frauen, M., Linders, E.G.A., Hatakeyama, K., Hirai, M., 2009. Status and perspectives of clubroot resistance breeding in crucifer crops. J. Plant Growth. Regul. 28, 265–281.

Ding, G., Zhao, Z., Liao, Y., Hu, Y., Shi, L., Long, Y., et al., 2012. Quantitative trait loci for seed yield and yield-related traits, and their responses to reduced phosphorus supply in *Brassica napus*. Ann. Bot. 109, 747–759.

Ding, G., Shi, L., Zhao, H., Cai, H., Liu, K., Xu, F., 2013. Genetic analysis of seed mineral accumulation affected by phosphorus deprivation in *Brassica napus*. Euphytica. 193, 251–264.

Dufresne, F., Stift, M., Vergilino, R., Mable, B.K., 2013. Recent progress and challenges in population genetics of polyploidy organisms: an overview of current state-of-the-art molecular and statistic tools. Mol. Ecol. 23, 40–69.

Dun, X., Zhou, Z., Xia, S., Wen, J., Yi, B., Shen, J., et al., 2011. BnaC.Tic40, a plastid inner membrane translocon originating from *Brassica oleracea*, is essential for tapetal function and microspore development in *Brassica napus*. Plant J. 68, 532–545.

Durstewitz, G., Polley, A., Plieske, J., Luerssen, H., Graner, E.M., Wieseke, R., et al., 2010. SNP discovery by amplicon sequencing and multiplex SNP genotyping in the allopolyploid species *Brassica napus*. Genome. 53, 948–956.

Edwards, D., Wang, X., 2012. Genome sequencing initiatives. In: Edwards, D., Batley, J., Parkin, I., Kole, C. (Eds.), Genetics, Genomics and Breeding of Oilseed Brassicas. CRC Press, USA, pp. 152–157.

Edwards, D., Batley, J., Snodwon, R.J., 2013. Accessing complex crop genomes with next-generation sequencing. Theor. Appl. Genet. 126, 1–11.

Edwards, M., 2013. Whole-genome sequencing for marker discovery. In: Henry, R.J. (Ed.), Molecular Markers in Plants. Wiley-Blackwell, pp. 21–34.

Elshire, R.J., Glaubitz, J.C., Sun, Q., Poland, J.A., Kawamoto, K., Buckler, E.S., et al., 2011. A robust, simple genotyping-by-sequencing (GBS) approach for high diversity species. PLoS One. 6, e19379.

Feng, J., Long, Y., Shi, L., Shi, J., Barker, G., Meng, J., 2012. Characterization of metabolite quantitative trait loci and metabolic networks that control glucosinolate concentration in the seeds and leaves of *Brassica napus*. New Phytol. 193, 96–108.

Figdore, S.S., Kenard, W.C., Song, K.M., Slocum, M.K., Osborn, T.C., 1988. Assessment of the degree of restriction fragment polymorphism in *Brassica*. Theor. Appl. Genet. 75, 833–840.

Friedt, W., Seyis, F., Snowdon, R.J., Lühs, W., 2004. Resynthetic rapeseed germplasm as a source of novel genetic variation for breeding of *Brassica napus*. GCIRC Bull. 21, 38–46.

Fritsche, S., Wang, X., Li, J., Stich, B., Kopisch-Obuch, F.J., Endrigkeit, J., et al., 2012. A candidate gene-based association study of tocopherol content and composition in rapeseed (*Brassica napus*). Front Plant Sci. 3, 129.

Gali, K.K., Sharpe, A.G., 2012. Molecular linkage maps: strategies, resources and achievements. In: Edwards, D., Batley, J., Parkin, I., Kole, C. (Eds.), Genetics, Genomics and Breeding of Oilseed Brassicas. CRC Press, USA, pp. 85–129.

Ganal, M.W., Altmann, T., Röder, M.S., 2009. SNP identification in crop plants. Curr. Opin. Plant Biol. 12, 211–217.

Ganal, M.W., Wieseke, R., Luerssen, H., Durstewitz, G., Graner, E.-M., Plieske, J., et al., 2014. High-throughput SNP profiling of genetic resources in crop plants using genotyping arrays. In: Tuberosa, R., Graner, A., Frison, E. (Eds.), Genomics of Plant Genetic Resources, vol. 1. Managing, Sequencing and Mining Genetic Resources. Springer, New York/Heidelberg/Dordrecht/London, pp. 113–130.

Gehringer, A., Snowdon, R.J., Spiller, T.H., Basunanda, P., Friedt, W., 2007. New oilseed rape (*Brassica napus*) hybrids with high levels of heterosis for seed yield under oligotrophic conditions. Breed. Sci. 57, 315–320.

Giancola, S., McKhann, H.I., Bérard, A., Camilleri, C., Durand, S., Libeau, P., et al., 2006. Utilization of the three high-throughput SNP genotyping methods, the GOOD assay, Amplifluor and TaqMan, in diploid and polyploid plants. Theor. Appl. Genet. 112, 1115–1124.

Gilles, S., 2013. Transcriptome sequencing for marker discovery. In: Henry, R.J. (Ed.), Molecular Markers in Plants. Wiley-Blackwell, pp. 57–66.

Grattapaglia, D., 2014. Breeding forest trees by genomic selection: current progress and the way forward. In: Tuberosa, R., Graner, A., Frison, E. (Eds.), Genomics of Plant Genetic Resources, vol. 1. Managing, Sequencing and Mining Genetic Resources. Springer, New York/Heidelberg, Germany/Dordrecht, the Netherlands/London, pp. 1651–1682.

Gunderson, K.L., Kuhn, K.M., Steemers, F.J., Ng, P., Murray, S.S., Shen, R., 2006. Whole-genome genotyping of haplotype tag single nucleotide polymorphisms. Pharmacogenomics. 7, 641–648.

Halldén, C., Nilsson, N.O., Rading, I.M., Säll, T., 1994. Evaluation of RFLP and RAPD markers in a comparison of *Brassica napus* breeding lines. Theor. Appl. Genet. 88, 123–128.

Hamblin, M.T., Warburton, M.L., Buckler, E.S., 2007. Empirical comparison of simple sequence repeats and single nucleotide polymorphisms in assessment of maize diversity and relatedness. PLoS One. 2, e1367.

Harper, A.L., Trick, M., Higgins, J., Fraser, F., Clissold, L., Wells, R., et al., 2012. Associative transcriptomics of traits in the polyploid crop species *Brassica napus*. Nat. Biotechnol. 30, 798–802.

Hasan, M., Friedt, W., Pons-Kühnemann, J., Freitag, N.M., Link, K., Snowdon, R.J., 2008. Association of gene-linked SSR markers to seed glucosinolate content in oilseed rape (*Brassica napus* ssp. *napus*). Theor. Appl. Genet. 116, 1035–1049.

Hayward, A., Mason, A., Dalton-Morgan, J., Zander, M., Edwards, D., Batley, J., 2012. SNP discovery and applications in *Brassica napus*. J. Plant Biotechnol. 39, 1–12.

He, J., Ke, L., Hong, D., Xie, Y., Wang, G., Liu, P., et al., 2008. Fine mapping of a recessive genic male sterility gene (*Bnms3*) in rapeseed (*Brassica napus*) with AFLP- and *Arabidopsis*-derived PCR markers. Theor. Appl. Genet. 117, 11–18.

Heffner, E.L., Sorrells, M.E., Jannink, J.-L., 2009. Genomic selection for crop improvement. Crop Sci. 49, 1–12.

Heffner, E.L., Lorenz, A.J., Jannink, J.-L., Sorrells, M.E., 2010. Plant breeding with genomic selection: gain per unit time and cost. Crop Sci. 50, 1681–1690.

Henry, R.J., 2013. Evolution of DNA marker technology in plants. In: Henry, R.J. (Ed.), Molecular Markers in Plants. Wiley-Blackwell, pp. 3–20.

Hirsch, C.D., Evans, J., Buell, C.R., Hirsch, C.N., 2014. Reduced representation approaches to interrogate genome diversity in large repetitive plant genomes. Brief. Funct. Genomics. Available from: http://dx.doi.org/doi:10.1093/bfgp/elt051. (First published online: 6 January 2014).

Holland, P.M., Abramson, R.D., Watson, R., Gelfand, D.H., 1991. Detection of specific polymerase chain reaction product by utilizing the 5′-3′ exonuclease activity of *Thermus aquaticus* DNA polymerase. PNAS. 88, 7276–7280.

Honsdorf, N., Becker, H.C., Ecke, W., 2010. Association mapping for phenological, morphological, and quality traits in canola quality winter rapeseed (*Brassica napus* L.). Genome. 53, 899–907.

Hu, Z., Huang, S., Sun, M., Wang, H., Hua, W., 2012. Development and application of single nucleotide polymorphism markers in the polyploid *Brassica napus* by 454 sequencing of expressed sequence tags. Plant Breed. 131, 293–299.

Huang, S., Deng, L., Guan, M., Li, J., Lu, K., Wang, H., et al., 2013. Identification of genome-wide single nucleotide polymorphisms in allopolyploid crop *Brassica napus*. BMC Genomics. 14, 717.

Huang, X., Han, B., 2014. Natural variations and genome-wide association studies in crop plants. Annu. Rev. Plant Biol. 65, 4.1–4.21.

Huang, X., Wei, X., Sang, T., Zhao, Q., Feng, Q., Zhao, Y., et al., 2010. Genome-wide association studies of 14 agronomic traits in rice landraces. Nat. Genet. 42, 961–967.

Huang, Z., Chen, Y.F., Yi, B., Xiao, L., Ma, C.Z., Tu, J.X., et al., 2007. Fine mapping of the recessive genic male sterility gene (*BnMs3*) in *Brassica napus* L. Theor. Appl. Genet. 115, 113–118.

van Inghelandt, D., Melchinger, A., Lebreton, C., Stich, B., 2010. Population structure and genetic diversity in a commercial maize breeding program assessed with SSR and SNP markers. Theor. Appl. Genet. 120, 1289–1299.

Jaccoud, D., Peng, K., Feinstein, D., Kilian, A., 2001. Diversity arrays: a solid state technology for sequence information independent genotyping. Nucleic Acids Res. 29, e25.

Jestin, C., Lodé, M., Vallée, P., Domin, C., Falentin, C., Horvais, R., et al., 2011. Association mapping of quantitative resistance for *Leptosphaeria maculans* in oilseed rape (*Brassica napus* L.). Mol. Breed. 27, 271–287.

Jiang, C., Shi, J., Li, R., Long, Y., Wang, H., Li, D., et al., 2014. Quantitative trait loci that control the oil content variation of rapeseed (*Brassica napus* L.). Theor. Appl. Genet. Available from: http://dx.doi.org/doi:10.1007/s00122-014-2271-5.

Jones, H., Norris, C., Cockram, J., Lee, D., 2014. Variety protection and plant breeders' rights in the 'DNA' era'. In: Lübberstedt, T., Varshney, R.K. (Eds.), Diagnostics in Plant Breeding. Springer, pp. 369–402.

Kaur, S., Francki, M.G., Forster, J.W., 2012. Identification, characterization and interpretation of single-nucleotide sequence variation in allopolyploid crop species. Plant Biotechnol. J. 10, 125–138.

Kirkegaard, J., Christen, O., Krupinsky, J., Layzell, D., 2008. Break crop benefits in temperate wheat production. Field Crops Res. 107, 185–195.

Kresovich, S., Szewc-McFadden, A.K., Blieck, S.M., McFerson, J.R., 1995. Abundance and characterization of simple-sequence repeats (SSRs) isolated from a size-fractionated genomic library of *Brassica napus* L. (rapeseed). Theor. Appl. Genet. 91, 206–211.

Landry, B.S., Hubert, N., Etoh, T., Harada, J.J., Lincoln, S.E., 1991. A genetic map for *Brassica napus* based on restriction fragment length polymorphisms detected with expressed DNA sequences. Genome. 34, 543–552.

Larkan, N.J., Lydiate, D.J., Parkin, I.A.P., Nelson, M.N., Epp, D.J., Cowling, W.A., et al., 2013. The *Brassica napus* blackleg resistance gene *LepR3* encodes a receptor-like protein triggered by the *Leptosphaeria maculans* effector AVRLM1. New Phytol. 197, 595–605.

Lehmensiek, A., Sutherland, M., McNamara, R., 2008. The use of high resolution melting (HRM) to map single nucleotide polymorphism markers linked to a covered smut resistance gene in barley. Theor. Appl. Genet. 117, 721–728.

Lei, S., Yao, X., Yi, B., Chen, W., Ma, C., Tu, J., et al., 2007. Towards-map-based cloning: fine mapping of a recessive genic male-sterile gene (*BnMs2*) in *Brassica napus* L. and syntenic region identification based on the *Arabidopsis thaliana* gene sequences. Theor. Appl. Genet. 115, 643–651.

Li, F., Ma, C., Chen, Q., Liu, T., Shen, J., Tu, J., et al., 2012. Comparative mapping reveals similar linkage of functional genes to QTL of yield-related traits between *Brassica napus* and *Oryza sativa*. J. Genet. 91, 163–170.

Li, F., Chen, B., Xu, K., Wu, J., Song, W., Bancroft, I., et al., 2014. Genome-wide association study dissects the genetic architecture of seed weight and seed quality in rapeseed (*Brassica napus* L.). DNA Res. Available from: http://dx.doi.org/doi:10.1093/dnares/dsu002 (Epub ahead of print).

Li, H., Younas, M., Wang, X., Li, X., Chen, L., Zhao, B., et al., 2013. Development of a core set of single-locus SSR markers for allotetraploid rapeseed (*Brassica napus* L.). Theor. Appl. Genet. 126, 937–947.

Li, J., Hong, D., He, J., Ma, L., Wan, L., Liu, P., et al., 2012. Map-based cloning of a recessive genic male sterility locus in *Brassica napus* L. and development of its functional marker. Theor. Appl. Genet. 125, 223–234.

Li, Y., Shen, J., Wang, T., Chen, Q., Zhang, X., Fu, T., et al., 2012. QTL analysis of yield-related traits and their association with functional markers in *Brassica napus* L. Aust. J. Agric. Res. 58, 759–766.

Lipshutz, R.J., Fodor, S.P.A., Gingeras, T.R., Lockhart, D.J., 1999. High density synthetic oligonucleotide arrays. Nat. Genet. 21, 20–24.

Litt, M., Luty, J.A., 1989. A hypervariable microsatellite revealed by *in vitro* amplification of a dinucleotide repeat within the cardiac muscle actin gene. Am. J. Hum. Genet. 44, 397–401.

Liu, L., Stein, A., Wittkop, B., Sarvari, P., Li, J., Yan, X., et al., 2012. A knockout mutation in the lignin biosynthesis gene *CCR1* explains a major QTL for acid detergent lignin content in *Brassica napus* seeds. Theor. Appl. Genet. 124, 1573–1586.

Liu, L., Qu, C., Wittkop, B., Yi, B., Xiao, Y., He, Y., et al., 2013. A high-density SNP map for accurate mapping of seed fibre QTL in *Brassica napus* L. PLoS One. 8, e83052.

Lochlainn, S.O., Amoah, S., Graham, N.S., Alamer, K., Rios, J.J., Kurup, S., et al., 2011. High resolution melt (HRM) analysis is an efficient tool to genotype EMS mutants in complex crop genomes. Plant Methods. 7, 43.

Lu, W., Liu, J., Xin, Q., Wan, L., Hong, D., Yang, G., 2013. A triallelic genetic male sterility locus in *Brassica napus*: an integrative strategy for its physical mapping and possible local chromosome evolution around it. Ann. Bot. (Lond). 111, 305–315.

Mayerhofer, R., Wilde, K., Mayerhofer, M., Lydiate, D., Bansal, V.K., Good, A.G., et al., 2005. Complexities of chromosome landing in a highly duplicated genome: toward map-based cloning of a gene controlling blackleg resistance in *Brassica napus*. Genetics. 171, 1977–1988.

Meuwissen, T.H.E., Hayes, B.J., Goddard, M.E., 2001. Prediction of total genetic value using genome-wide dense marker maps. Genetics. 157, 1819–1829.

Mikolajczyk, K., Dabert, M., Karlowski, W.M., Spasibionek, S., Nowakowska, J., Cegielska-Taras, T., et al., 2010. Allele-specific SNP markers for the new low linolenic mutant genotype of winter oilseed rape. Plant Breed. 129, 502–507.

Mir, R.R., Varshney, R.K., 2013. Future prospects of molecular markers in plants. In: Henry, R.J. (Ed.), Molecular Markers in Plants. Wiley-Blackwell, pp. 169–190.

Mir, R.R., Hiremath, P.J., Rierra-Lizarazu, O., Varshney, R.K., 2014. Evolving molecular marker technologies in plants: from RFLPs to GBS. In: Lübberstedt, T., Varshney, R.K. (Eds.), Diagnostics in Plant Breeding. Springer, pp. 229–247.

Mitchell, S.E., Kresovich, S., Jester, C.A., Hernandez, C.J., Szewc-McFadden, A.K., 1997. Application of multiplex PCR and fluorescence-based, semi-automated allele sizing technology for genotyping plant genetic resources. Crop Sci. 37, 617–624.

Morgante, M., Olivieri, A.M., 1993. PCR-amplified microsatellites as markers in plant genetics. Plant J. 3, 175–182.

Morris, G.P., Ramu, P., Deshpande, S.P., Hash, C.T., Shah, T., Upadhyaya, H.D., et al., 2013. Population genomic and genome-wide association studies of agroclimatic traits in sorghum. PNAS. 110, 453–458.

Nakaya, A., Isobe, S.N., 2012. Will genomic selection be a practical method for plant breeding? Ann. Bot. 110, 1303–1316.

Navabi, Z.K., Parkin, I.A.P., 2012. Physical mapping in *Brassica* species. In: Edwards, D., Batley, J., Parkin, I., Kole, C. (Eds.), Genetics, Genomics and Breeding of Oilseed Brassicas. CRC Press, USA, pp. 130–151.

Nesi, N., Delourme, R., Bregeon, M., Falentin, C., Renard, M., 2008. Genetic and molecular approaches to improve nutritional value of *Brassica napus* L. seed. C. R. Biol. 331, 763–771.

Obermeier, C., Hossain, M.A., Snowdon, R., Knüfer, J., von Tiedemann, A., Friedt, W., 2013. Genetic analysis of phenylpropanoid metabolites associated with resistance against *Verticillium longisporum* in *Brassica napus*. Mol. Breed. 31, 347–361.

Oliphant, A., Barker, D.L., Stuelpnagel, J.R., Chee, M.S., 2002. BeadArray technology: enabling an accurate, cost-effective approach to high-throughput genotyping. Biotechniques. (Suppl. 5), 56–58.

Parkin, I., 2014. Insights from sequencing *Camelina sativa* and exploring beyond the *Brassica* crops. The 19th Crucifer Genetics Workshop: Genetic Improvement of Brassicacea Crops in the Era of Genomics. Brassica 2014. March 30–April 2, Wuhan, People's Republic of China. Program and Abstract Book, p. 67.

Parkin, I.A., Clarke, W.E., Sidebottom, C., Zhang, W., Robinson, S.J., Links, M.G., et al., 2010. Towards unambiguous transcript mapping in the allotetraploid *Brassica napus*. Genome. 53, 929–938.

Parkin, I.A.P., Sharpe, A.G., Keith, D.J., Lydiate, D.J., 1995. Identification of the A and C genomes of amphidiploid *Brassica napus* (oilseed rape). Genome. 38, 1122–1131.

Paulmann, W., Frauen, M., 1997. The first fully restored F1-hybrids of winter oilseed rape. GCIRC Bull. 13, 24–25.

Peterson, B.K., Weber, J.N., Kay, E.H., Fisher, H.S., Hoekstra, H.E., 2012. Double digest RADseq: an inexpensive method for *de novo* SNP discovery and genotyping in model and non-model species. PLoS One. 7, e37135.

Pindo, M., Vezzulli, S., Coppola, G., Cartwright, D.A., Zharkikh, A., Velasco, R., et al., 2008. SNP high-throughput screening in grapevine using the SNPlex™ genotyping system. BMC. Plant Biol. 8, 12.

Poland, J.A., Brown, P.J., Sorrells, M.E., Jannink, J.-J., 2012. Development of high-density genetic maps for barley and wheat using a novel two-enzyme genotyping-by-sequencing approach. PLoS One. 7, e32253.

Poulsen, G.B., Kahl, G., Weising, K., 1993. Abundance and polymorphism of simple repetitive DNA sequences in *B. napus* L. Theor. Appl. Genet. 85, 994–1000.

Qi, L., Mao, L., Sun, C., Pu, Y., Fu, T., Ma, C., et al., 2014. Interpreting the genetic basis of silique traits in *Brassica napus* using a joint QTL network. Plant Breed. 133, 52–60.

Quiros, C.F., Hu, J., This, P., Chevre, A.M., Delseny, M., 1991. Development and chromosomal localization of genome-specific markers by polymerase chain reaction in *Brassica*. Theor. Appl. Genet. 82, 627–632.

Rahman, H., 2013. Review: breeding spring canola (*Brassica napus* L.) by the use of exotic germplasm. Can. J. Plant Sci. 93, 363–373.

Raman, H., Raman, R., Nelson, M.N., Aslam, M.N., Rajasekaran, R., Wratten, N., et al., 2012. Diversity array technology markers: genetic diversity analyses and linkage map construction in rapeseed (*Brassica napus* L.). DNA Res. 19, 51–65.

Raman, H., Raman, R., Kilian, A., Detering, F., Long, Y., Edwards, D., et al., 2013. A consensus map of rapeseed (*Brassica napus* L.) based on diversity array technology markers: applications in genetic dissection of qualitative and quantitative traits. BMC Genomics. 14, 277.

Raman, R., Taylor, B., Marcroft, S., Stiller, J., Eckermann, P., Coombes, N., et al., 2012a. Molecular mapping of qualitative and quantitative loci for resistance to *Leptosphaeria maculans* causing blackleg disease in canola (*Brassica napus* L.). Theor. Appl. Genet. 125, 405–418.

Raman, R., Taylor, B., Lindbeck, K., Coombes, N., Barbulescu, D., Salisbury, P., et al., 2012b. Molecular mapping and validation of *Rlm1* gene for resistance to *Leptosphaeria maculans* in canola (*Brassica napus* L.). Crop Pasture Sci. 63, 1007–1017.

Reed, G.H., Kent, J.O., Wittwer, C.T., 2007. High-resolution DNA melting analysis for simple and efficient molecular diagnostics. Pharmacogenomics. 8, 597–608.

Rezaeizad, A., Wittkop, B., Snowdon, R., Hasan, M., Mohammadi, V., Zali, A., et al., 2011. Identification of QTLs for phenolic compounds in oilseed rape (*Brassica napus* L.) by association mapping using SSR markers. Euphytica. 177, 335–342.

Rivin, C.J., Zimmer, E.A., Cullis, C.A., Walbot, V., 1983. Evaluation of genomic variability at the nucleic acid level. Plant Mol. Biol. Rep. 1, 9–16.

Rostoks, N., Ramsay, L., Mackenzie, K., Cardle, L., Bhat, P.R., Roose, M.L., et al., 2006. Recent history of artiffcial outcrossing facilitates whole-genome association mapping in elite inbred crop varieties. PNAS. 103, 18656–18661.

Salvi, S., Tuberosa, R., Phillips, R.L., 2001. Development of PCR-based assays for allelic discrimination in maize by using the 5′-nuclease procedure. Mol. Breed. 8, 169–176.

Schatzki, J., Schoo, B., Ecke, W., Herrfurth, C., Feussner, I., Becker, H.C., et al., 2013. Mapping of QTL for seed dormancy in a winter oilseed rape doubled haploid population. Theor. Appl. Genet. 126, 2405–2415.

Schlötterer, C., 2004. The evolution of molecular markers–just a matter of fashion? Nat. Rev. Genet. 5, 63–69.

Semagn, K., Babu, R., Hearne, S., Olsen, M., 2014. Single nucleotide polymorphism genotyping using Kompetitive Allele Specific PCR (KASP): overview of the technology and its application in crop improvement. Mol. Breed. 33, 1–14.

Sexton, T.R., Shapter, F.M., 2013. Amplicon sequencing for marker discovery. In: Henry, R.J. (Ed.), Molecular Markers in Plants. Wiley-Blackwell, pp. 35–56.

Seyis, F., Snowdon, R.J., Lühs, W., Friedt, W., 2003. Molecular characterization of novel resynthesised rapeseed (*Brassica napus* L.) lines and analysis of their genetic diversity in comparison to spring rapeseed cultivars. Plant Breed. 122, 473–478.

Seyis, F., Friedt, W., Lühs, W., 2006. Yield of *Brassica napus* L. hybrids developed using resynthesized rapeseed material. Field Crops Res. 96, 176–180.

Shi, L., Yang, J., Liu, J., Li, R., Long, Y., Xu, F., et al., 2012. Identification of quantitative trait loci associated with low boron stress that regulate root and shoot growth in *Brassica napus* seedlings. Mol. Breed. 30, 393–406.

Shi, L., Shi, T., Broadley, M.R., White, P.J., Long, Y., Meng, J., et al., 2013. High-throughput root phenotyping screens identify genetic loci associated with root architectural traits in *Brassica napus* under contrasting phosphate availabilities. Ann. Bot. 112, 381–389.

Snowdon, R., Friedt, W., 2004. Molecular markers in Brassica oilseed breeding: current status and future possibilities. Plant Breed. 123, 1–8.

Snowdon, R.J., Luy, F.L.I., 2012. Potential to improve oilseed rape and canola breeding in the genomics era. Plant Breed. 131, 351–360.

Song, K.M., Osborn, T.C., Williams, P.H., 1988. *Brassica* taxonomy based on nuclear restriction fragment length polymorphisms (RFLPs). Theor. Appl. Genet. 75, 784–794.

Spindel, J., Wright, M., Chen, C., Cobb, J., Gage, J., Harrington, S., et al., 2013. Bridging the genotyping gap: using genotyping by sequencing (GBS) to add high-density SNP markers and new value to traditional bi-parental mapping and breeding populations. Theor. Appl. Genet. 126, 2699–2716.

Studer, B., Kölliker, R., 2013. SNP genotyping technologies. In: Lübberstedt, T., Varshney, R.K. (Eds.), Diagnostics in Plant Breeding. Springer, pp. 187–228.

Syvanen, A.-C., 2001. Accessing genetic variation: genotyping single nucleotide polymorphism. Nat. Rev. Genet. 2, 930–942.

Tobler, A.R., Short, S., Andersen, M.R., Paner, T.M., Briggs, J.C., Lambert, S.M., et al., 2005. The SNPlex genotyping system: a flexible and scalable platform for SNP genotyping. J. Biomol. Tech. 16, 396–404.

Tommasini, L., Batley, J., Arnold, G.M., Cooke, R.J., Donini, P., Lee, D., et al., 2003. The development of multiplex simple sequence repeat (SSR) markers to complement distinctness, uniformity and stability testing of rape (*Brassica napus* L.) varieties. Theor. Appl. Genet. 106, 1091–1101.

Trick, M., Long, Y., Meng, J., Bancroft, I., 2009. Single nucleotide polymorphism (SNP) discovery in the polyploidy *Brassica napus* using Solexa transcriptome sequencing. Plant Biotechnol. J. 7, 334–346.

Ulas, A., Schulte, A., Erley, G., Kamh, M., Wiesler, F., Horst, W.J., 2012. Root-growth characteristics contributing to genotypic variation in nitrogen efficiency of oilseed rape. J. Plant Nutr. Soil. Sci. 175, 489–498.

Vos, P., Hogers, R., Bleeker, M., Reijans, M., van de Lee, T., Hornes, M., et al., 1995. AFLP: a new technique for DNA fingerprinting. Nucleic Acids Res. 23, 4407–4414.

Wang, F., Wang, X., Chen, X., Xiao, Y., Li, H., Zhang, S., et al., 2012. Abundance, marker development and genetic mapping of microsatellites from unigenes in *Brassica napus*. Mol. Breed. 30, 731–744.

Wang, J., Lydiate, D.J., Parkin, I.A.P., Falentin, C., Delourme, R., Carion, P.W.C., et al., 2011. Integration of linkage maps for the Amphidiploid *Brassica napus* and comparative mapping with *Arabidopsis* and *Brassica rapa*. BMC Genomics. 12, 101.

Wang, X., Wang, H., Wang, J., Sun, R., Wu, J., Liu, S., et al., 2011. The genome of the mesopolyploid crop species *Brassica rapa*, *Brassica rapa* Genome Sequencing Project Consortium. Nat Genet. 43, 1035–1039.

Wang, X., Zhang, C., Li, L., Fritsche, S., Endrigkeit, J., Zhang, W., et al., 2012. Unraveling the genetic basis of seed tocopherol content and composition in rapeseed (*Brassica napus* L.). PLoS One. 7, e50038.

Wang, X., Wang, H., Long, Y., Li, D., Yin, Y., Tian, J., et al., 2013. Identification of QTLs associated with oil content in a high-oil brassica napus cultivar and construction of a high-density consensus map for QTLs comparison in *B. napus*. PLoS One. 8, e80569.

Wen, Y.C., Zhang, S.F., Yi, B., Wen, J., Wang, J.P., Zhu, J.C., et al., 2012. Identification of QTLs involved in pod-shatter resistance in *Brassica napus* L. Crop Pasture Sci. 63, 1082–1089.

Westermeier, P., Wenzel, G., Mohler, V., 2009. Development and evaluation of single-nucleotide polymorphism markers in allotetraploid rapeseed (*Brassica napus* L.). Theor. Appl. Genet. 119, 1301–1311.

Williams, J.G., Kubelik, A.R., Livak, K.J., Rafalski, J.A., Tingey, S.V., 1990. DNA polymorphisms amplified by arbitrary primers are useful as genetic markers. Nucleic Acids Res. 18, 6531–6535.

Wittkop, B., Snowdon, J., Friedt, W., 2009. Status and perspectives of breeding for enhanced yield and quality of oilseed crops for Europe. Euphytica. 170, 131–140.

Würschum, T., Liu, W., Maurer, H.P., Abel, S., Reif, J.C., 2012. Dissecting the genetic architecture of agronomic traits in multiple segregating populations in rapeseed (*Brassica napus* L.). Theor. Appl. Genet. 124, 153–161.

Würschum, T., Abel, S., Zhao, Y., 2014. Potential of genomic selection in rapeseed (*Brassica napus* L.) breeding. Plant Breed. 133, 45–51.

Xie, Y., Dong, F., Hong, D., Wan, L., Liu, P., Yang, G., 2012. Exploiting comparative mapping among *Brassica* species to accelerate the physical delimitation of a genic male-sterile locus (*BnRf*) in *Brassica napus*. Theor. Appl. Genet. 125, 211–222.

Xu, P., Xu, S., Wu, X., Tao, Y., Wang, B., Wang, S., et al., 2014. Population genomic analyses from low-coverage RAD-Seq data: a case study on the non-model cucurbit bottle gourd. Plant J. 77430–77442.

Yi, B., Chen, Y., Lei, L., Tu, J., Fu, T., 2006. Fine mapping of the recessive genic male-sterile gene (*Bnms1*) in *Brassica napus* L. Theor. Appl. Genet. 113, 643–650.

Yi, B., Zheng, F.Q., Lei, S.L., Chen, Y.N., Yao, X.Q., Wen, J., et al., 2010. Two duplicate CYP704B1 homologous genes *BnMs1* and *BnMs2* are required for pollen exine formation and tapetal development in *Brassica napus*. Plant J. 63, 925–938.

Yu, J., Holland, J.B., McMullen, M.D., Buckler, E.S., 2008. Genetic design and statistical power of nested association mapping in maize. Genetics. 178, 539–551.

Zhang, L., Li, S., Chen, L., Yang, G., 2012. Identification and mapping of a major dominant quantitative trait locus controlling seeds per silique as a single Mendelian factor in *Brassica napus* L. Theor. Appl. Genet. 125, 695–705.

Zhao, J., Huang, J., Chen, F., Xu, F., Ni, X., Xu, H., et al., 2012. Molecular mapping of *Arabidopsis thaliana* lipid-related orthologous genes in *Brassica napus*. Theor. Appl. Genet. 124, 407–421.

Zhao, Z., Wu, L., Nian, F., Ding, G., Shi, T., Zhang, D., et al., 2012. Dissecting quantitative trait loci for boron efficiency across multiple environments in *Brassica napus*. PLoS One. 7, e45215.

Zhou, Q.-H., Fu, D.-H., Mason, A.S., Zeng, Y.-J., Zhao, C.-X., Huang, Y.-J., 2014. In silico integration of quantitative trait loci for seed yield and yield-related traits in *Brassica napus*. Mol. Breed. 33, 881–894.

Zhu, C., Gore, M., Buckler, E.S., Yu, J., 2008. Status and prospects of association mapping in plants. Plant Genome. 1, 5–20.

Zhu, Y., Cao, Z., Xu, F., Huang, Y., Chen, M., Guo, W., et al., 2012. Analysis of gene expression profiles of two near-isogenic lines differing at a QTL region affecting oil content at high temperatures during seed maturation in oilseed rape (*Brassica napus* L.). Theor. Appl. Genet. 124, 515–531.

Zou, J., Jiang, C., Cao, Z., Li, R., Long, Y., Chen, S., et al., 2010. Association mapping of seed oil content in *Brassica napus* and comparison with quantitative trait loci identified from linkage mapping. Genome. 53, 908–916.

Index

Note: Page numbers followed by "*f*", and "*t*" refers to figures and tables respectively.

Edwards Brothers Malloy
Ann Arbor MI. USA
June 26, 2015